Structures

From theory to practice

Cover Photograph

The Gateshead Millennium Bridge, 2002 (foreground) extends the great tradition of engineering in the Tyneside region. Its slender arch is seen here matching the parabolic profile of the 1928 High Level Tyne Bridge. In receiving a Structural Steel Design Award in 2002 it was said to have 'visual daring and elegance in its closed position, giving way to theatre and power in operation'. Photo: Graeme Peacock (www.graeme-peacock.com).

Structures

From theory to practice

Alan Jennings

Spon Press
Taylor & Francis Group

LONDON AND NEW YORK

First published 2004
by Spon Press
2 Park Square, Milton Park, Abingdon, Oxfordshire OX14 4RN

Simultaneously published in the USA and Canada
by Spon Press
29 West 35th Street, New York, NY 10001

Spon is an imprint of the Taylor & Francis Group

Typeset in Sabon by
Newgen Imaging Systems (P) Ltd, Chennai, India
Printed and bound in Great Britain by
MPG Books Ltd, Bodmin

British Library Cataloguing in Publication Data
A catalogue record for this book is available from the British Library

Library of Congress Cataloging in Publication Data
        Structures: from theory to practice / Alan Jennings.
            p. cm.
        Includes bibliographical references and index.
        1. Structural design. I. Title.

    TA658.2.J46 2004
    624.1'7–dc22   2003027278

ISBN 0–415–26842–7 (alk. paper)
ISBN 0–415–26843–5 (pb: alk. paper)

Contents

WITHDRAWN

Foreword

This is an entry-level text-book on structural analysis with a difference – as the title indicates, the focus is very much on gaining a thorough understanding of structures and structural behaviour rather than on the intricacies of structural analysis alone.

I have always believed that structural analysis – whilst a rigorous and academically demanding subject in its own right – is a means to an end and should not be studied in isolation. What is needed is a constant reminder that analysis leads to design, and marrying these two every step of the way brings structures teaching and study to life.

Alan Jennings, a distinguished academic and teacher, has a clear vision of this duality of purpose, and his new text introduces real-life examples from almost the first pages of the book. These include, I am pleased to say, numerous historical and recent important structural milestones, which should be the required knowledge for every structural engineer.

His easy-going, almost conversational style makes the subject accessible and enjoyable but the intellectual rigour is there in full. As a life-long practising designer there is much to interest me here, and the broad coverage given to issues such as fatigue and dynamics, as well as the large number of worked examples, leave me in no doubt that his text is a worthy and worthwhile addition to the armoury of young engineers as they embark on gaining a real understanding of structures. Put away your computer programmes and learn how to analyse structures using simple mathematics, without recourse to codes of practice. The information in this book will stand you in good stead for the rest of your career.

Dr John Roberts, FREng, BEng, PhD, FICE, FIStructE,
Director, Babtie Group
President, Institution of Structural Engineers – 1999–2000
Fellow, Royal Academy of Engineering

Preface

This text has been developed to accompany first- and second-year structures courses at universities and to indicate the relevance of theory to practice. Linking practical items to the development of theory throughout the text has provided a fascinating challenge which has had an influence on the way the theoretical concepts have been sequenced and presented. Lecturers may wish to consult this text when planning courses or setting assignments and may find it a useful reference source if students have problems understanding the theory, its scope or its relevance to practice. However, the book is primarily intended to assist students with self-learning including, where relevant, project and problem-based exercises by providing a more complete picture of fundamental aspects than most texts have space for. Practising engineers, particularly recent graduates, might also find this approach stimulating. Most of the important subjects within the theory of structures are introduced in such a way that the reader may appreciate how specialist topics link into the general scene. The main emphasis is towards civil engineering structures, but some of the items relating to practice indicate the breadth of the subject.

The twentieth century has seen lecture courses and design procedures being increasingly based on theory and scientific experiment. The twenty-first century may see, through IT, more emphasis being placed on practical experience. In the words of Handy (1994), 'Focused intelligence, the ability to acquire and apply knowledge and know-how is the new source of wealth'. Good design will involve not only sound scientific methods, but also utilising an increasing amount of available information about the performance of comparable structures. When introducing practical applications, problems and failures as well as successful structures are discussed. Generally, the failures have been better investigated and documented and contain the most important lessons. Petroski (1985) reported Barry LePatnier as saying 'Good judgement is usually the result of experience. And experience is frequently the result of bad judgement'.

The various aspects of structural theory have been developed in such a way that their basis in mathematics and mechanics is clear. This should provide the firmest foundation on which students may later study topics in greater depth via the vast wealth of structures-related literature. Other objectives have been

- to develop theoretical concepts in a rigorous way;
- to relate theory to structural behaviour and good design practice as well as analysis;
- to consolidate understanding of equilibrium before dealing with elastic behaviour;
- to ensure that three-dimensional aspects are not overlooked;
- to stimulate readers to take an interest in the structures they see around them or hear about;
- to promote an appreciation of the responsibilities of the structural engineer;
- to provide some landmarks which could give a framework for developing a deeper interest in the history of structures;
- to encourage those who are keen to study the subject further.

However, the overriding objective in writing the book has been to develop understanding. Sparkes (1991) said that 'Once acquired, understanding – like skills, but unlike knowledge – tends not to be forgotten'. It is only by understanding the basis of theories and why structures are designed in specific ways that engineers become proficient and in control of the tools, such as computer programs and codes of practice, that they use. For all the scientific content, it is important to recognise structural engineering as an art requiring entrepreneurial skills coupled with sound judgement. The following sobering thought emphasises the importance of treating the theory of structures and its application with the highest respect: 'Structural Design is the art of moulding materials we do not wholly understand, into shapes we cannot precisely analyse, so as to withstand forces we cannot really assess; in such a way that the public at large has no reason to suspect the extent of our ignorance' (A.R. Dykes, Chairman's Address, IStructE Scottish Branch 1979).

Alan Jennings

Acknowledgements

The author gratefully acknowledges help in obtaining material and also useful suggestions from many people regarding the script and illustrations. In particular thanks are due to Dr John Hill, formerly of Doran Consulting and President of the Institution of Structural Engineers – 2000–2001, Dr Stephen Gilbert and Dr Raymond Gilfillan, both of Queen's University, for comments on the whole of the first draft. Also for comments regarding or assistance with parts of the script, thanks are due to Brian Campbell (Consulting Engineer); Alan Cooper of Ferguson & McIlveen; Professor Sir Bernard Crossland, formerly President of the Institution of Mechanical Engineers; Peter Curran of Gifford and Partners; Pat Dallard, Michele Janner and Zygmunt Lubkowski of Arup; Dr Phil Donald, formerly of John Graham (Dromore) Ltd; Marcus Kennedy (Architectural Engineering student, Queen's University); Professor Chengi Kuo and Professor Colin McFarlane, Department of Naval Architecture and Marine Engineering, University of Strathclyde; Desmond Mairs of Whitby Bird and Partners; John Marshall (architect) of RMJM; Carol Morgan (Archivist, The Institution of Civil Engineers); Dr Roger Pope of British Constructional Steel Association; Peter Rhodes (retired structural engineer); David Riordan of Bombardier Aerospace and Dr John Roberts of Babtie Group Ltd.

Thanks also go to many at Queen's University, Belfast and elsewhere who have helped with the collection of illustrations or who have given assistance or encouragement; Kay Millar, Angela Brady and Cherith McFarland for word-processing skills; Newgen Imaging Systems and also Ernest Patterson for skills with computer images; Spon Press for help in many ways and Margery Jennings for all her support and patience.

Credits for illustrations

The author and publishers acknowledge the following, where appropriate with permission to reproduce their copyright illustrations.

Figures 2.17 and 2.18: From Mendelssohn (1974) by permission of Eurospan Group, London.

Figures 2.23, 2.54 and 4.74: Institution of Civil Engineers.

Figure 2.29: From *Ocean and Coastal Management*, Vol. 30. Leopoldo Franco, Ancient Mediterranian Harbours, pp. 132 and 133, 1996, by permission from Excerpta Medica Inc.

Figure 2.30: Reprinted From *Manual on the Use of Rock in Coastal and Shoreline Engineering*, CIRIA Special Publication 83, 1991 by permission of CIRIA.

Figure 2.40: From Page (1993) by permission of Transport Research Laboratory, Crowthorne, UK.

Figure 2.47: From Cresy's *Treatise on Bridges, Vaults and Arches*, 1839.

Figures 2.50 and 2.51: From Home (1931) by permission of Gospatrick Home.

Fig 2.53: From Shirley-Smith (1964).

Figures 2.55, 9.31 and 9.32: From the New Civil Engineer, London with art work by Anthea Carter.

Figure 2.56: From Sandström (1970).

Figures 3.65 and 3.66: From Cohen (1955) by permission of HMSO, London.

Figure 4.70: From Beckett (1980).

Figure 4.73: By permission of Lloyd's Register of Shipping.

Figures 4.78 and 4.79: From Barber *et al.* (1971) by permission of the State of Victoria, Australia.

Figure 5.3: R.J. Welch by permission of National Museums and Galleries of Northern Ireland.

Figure 5.6: From Harriss (1975).

Figure 5.32(b): Jill Jennings.

Figures 5.37 and 5.38: From CargoLifter A.G.

Figure 5.42: From a print owned by Richard Parker, copyright of the Civil Engineering Dept, Imperial College, London.

Figure 5.44: From Koerte (1992).

Figures 5.66 and 5.67: By permission of Anthony Hunt Associates.

Figures 5.68–5.73: From Naesheim (1981).

Figures 6.68 and 6.69: From Levy and Salvadori (1992).

Figures 6.73–6.75, 6.77 and 6.78: By permission of Gifford Consulting.

Figure 7.21: By permission of the National Museum of American History.

Figures 7.22 and 7.23: From Schneider (1908).

Figure 7.28: From Short (1962).

Figure 7.47: By permission of MSCUA, University of Washington Libraries (image UW21413).

Figure 7.50: From Shirley-Smith (1964).

Figure 9.5: By permission of Dewhurst, MacFarlane and Partners; Photography by Kenji Kobayshi.

Figures 9.28–9.30: By permission of Earthquake Engineering Field Investigation Team, Institution of Structural Engineers.

Figures 9.33–9.40: From McAllister (2002).

Chapter 1

Rigid body mechanics

Rigid body mechanics is concerned with movements of bodies as a whole (i.e. ignoring any distortions) and the forces to which they are subjected. The principles considered in this chapter are so basic to the study of structures that readers with prior experience are advised to first check that they can answer the questions before proceeding further. Those who find this chapter too difficult should consult books primarily devoted to mechanics (statics in particular) or the relevant mathematics. See for instance Roberts, A.P. (2003) or Das *et al.* (1994).

On completion of this chapter you should be able to do the following:

- Identify forces acting on structures and determine the resultants of pressures and distributed loads.
- Appreciate how structures may be restrained to prevent body movements.
- Investigate both two- and three-dimensional equilibrium situations and also quasi-equilibrium situations where bodies are accelerating.
- Take account of frictional forces.
- Cope with compound systems of rigid bodies.
- Deduce direct methods which may not only give simpler solutions but also provide better insight into behaviour.

1.1 Resultants of parallel forces and pressures

Forces arise in distributed form either throughout bodies (e.g. gravitational, inertia or magnetic forces) or over surfaces (e.g. fluid pressures or pressures between bodies). In equilibrium investigations, it is convenient to work with concentrated forces which are idealised 'resultants' of the distributed forces. There are three parameters which are important for force resultants namely,

a   magnitude (typical units being Newtons);
b   line of action;
c   point of action.

The point along its line of action at which a force acts is not of concern until internal forces are discussed in Chapter 3. Consideration of the lines of action of forces will involve the concept of the 'moment' of a force, typical units of which are Newtons × metres. The important criterion to be satisfied is that the 'moment' of the resultant force about any point should equal the aggregate of the moments of the distributed forces which it is representing. Lines of action of resultants can usually be evaluated by considering centres of gravity or centroids of areas. However, for more complicated cases it may be necessary to resort to mathematical integration. The following examples are graded in difficulty with integration being required for the last one.

Ex. 1.1
A rectangular tank having a base $3\,m \times 4\,m$ and sides of height $2\,m$ is completely filled with water of weight $9.81\,kN/m^3$. Determine the magnitudes and lines of action of the resultant fluid pressures acting on the base of the tank and also each of the $4\,m \times 2\,m$ sides.

Answer
The water pressure at any point depends on the weight of water above it. Thus the pressure on the bottom is uniform at $9.81 \times 2 = 19.62\,kN/m^2$. Since this acts over an area of $3\,m \times 4\,m$, the resultant force is $19.62 \times 3 \times 4 = 235.44\,kN$ (i.e. the volume of the pressure cuboid shown in Fig. 1.1(a)). Because the pressure is uniform, the line of action must be vertically downwards through the centre of the base.

The pressure variation over the sides is linear, varying from zero at the top to $19.62\,kN/m^2$ at the bottom. Hence the pressure acting on a long side can be represented by the pressure wedge shown in Fig. 1.1(a), the volume of which is $1/2 \times 19.62 \times 4 \times 2 = 78.48\,kN$.

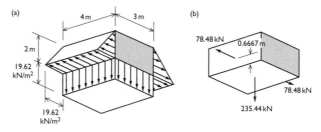

Figure 1.1 Water pressures and resultant forces acting on the bottom and long sides of a tank: (a) pressure distributions; (b) resultant forces.

The resultant force on the side of the tank is therefore 78.48 kN and it must act outwards with its line of action through the centre of gravity of the triangular wedge, that is, at height 0.6667 m above the base as shown in Fig. 1.1(b).

Ex. 1.2
A beam of length 8 m carries distributed loading as shown in Fig. 1.2(a). Determine the magnitude of the resultant force and the distance it acts from A.

Answer
The two loading blocks may each be resolved into their resultant forces as shown in Fig. 1.2(b). The magnitude of the overall resultant force is their sum equal to 1340 N. If this resultant is assumed to act at distance $\bar{x}$ from A, for its moment about A to equal the sum of the moments of the block resultants,

$$1340\bar{x} = 1100 \times 2.5 + 240 \times 6.5$$

$$\therefore \bar{x} = 3.216\,\text{m}$$

The position of the resultant is shown in Fig. 1.2(c).

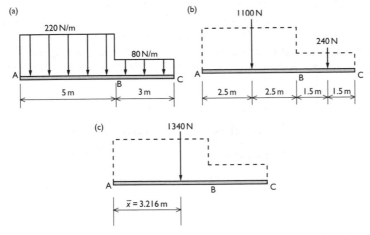

Figure 1.2 Resultant of stepped distributed loading: (a) loading; (b) resultants of block loads; (c) overall resultant.

NB   You are invited to check that the same line of action would have been obtained if moments had been equated about any other point (such as B or C).

Ex. 1.3

Figure 1.3(a) shows the side view of an elevated platform which has a central column support. If the maximum downwards loading that can occur anywhere on the platform is 8 kN/m span and no upwards loading is anticipated, determine

a   the maximum download which the column may need to support;
b   the maximum moment which the column may need to support.

(Self-weight of the platform may be neglected.)

Answer

The largest download on the column is obtained when the whole platform carries the maximum load of 8 kN/m giving a resultant force of 80 kN acting in the line of the column (see Fig. 1.3(b) and (c)). Although this has no moment about the column position, a moment can be generated by loading eccentrically. The maximum moment is obtained when one side only is fully loaded. In this case, the resultant

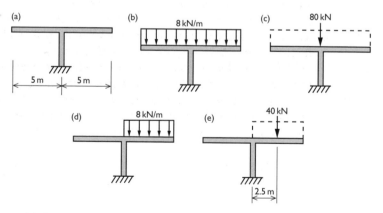

Figure 1.3 Different loading cases for an elevated platform: (a) platform; (b) maximum download case; (c) maximum download; (d) maximum column moment case; (e) resultant force for maximum column moment.

force is 40 kN at a distance 2.5 m from the column centreline giving a moment of 100 kNm. When the right-hand side is loaded as
shown in Fig. 1.3(d) and (e), the moment is clockwise, but a similar anticlockwise moment will occur if only the left-hand side is
loaded.

Comment
Designers of structures need to be careful to recognise how different loading
cases may differently affect various parts of the structure. For the column
support above, the maximum moment case (b) is likely to be more critical
than the maximum load case (a).
   The following example illustrates the use of integration for a more difficult
problem. (A table of integrals is required for all who are not mathematical
wizards.)

Ex. 1.4
A horizontal pipe of radius r is blocked by means of a flange plate. The
pipe contains a fluid of density ρ with a free surface at height $2r$ above
its centreline. Expressions are required for the resultant force on the
flange plate and the position of its line of action.

Answer
The force acting on a horizontal strip of the flange plate is easily
obtained because the pressure will be constant across the strip (see
Fig. 1.4(a)). Consider a strip of thickness dx and length ℓ at a height
x above the centreline of the pipe. The force acting on the strip is
$dP = \rho g(2r - x)\ell\, dx$. In order to form a solvable integral, it is necessary to relate ℓ to x which may be done via the angular coordinate θ at
which the strip intersects the pipe (θ being measured from the vertical
as shown in Fig. 1.4(b)). With the strip intersecting an arc $r\, d\theta$ of the
pipe circumference, it can be shown that $x = r\cos\theta$, $dx = r\sin\theta\, d\theta$,
$\ell = 2r\sin\theta$.
   Hence, by using tables of integrals (e.g. Dwight, 1961), it is possible
to show that the resultant force and moment about the axis of the
pipe are

$$P = \int_{\theta=0}^{\pi} dP = 2\pi\rho g r^3$$

and

$$M = \int_{\theta=0}^{\pi} x\, dP = \frac{-\pi \rho g r^4}{4}$$

Hence, the distance of, $\bar{x}$, the line of action of the resultant force above the pipe centreline is $\bar{x} = M/P = -r/8$ (see Fig. 1.4(c)).

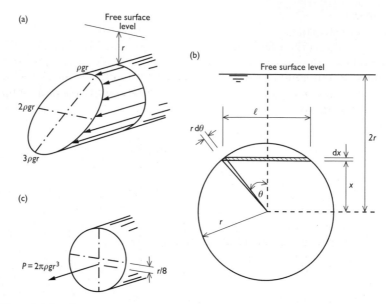

Figure 1.4 Loading on a pipe flange plate: (a) pressure distribution; (b) geometry of strip; (c) resultant force.

NB   The resultant force is equal to the area of the flange plate times the pressure $2\rho g r$ at the centroid of the plate. This result, but not its line of action, could have been deduced from inspection of the pressure diagram.

Q.1.1

The axle loads of a heavy goods vehicle are shown in Fig. 1.5. Determine the magnitude of the resultant force and the distance of its line of action behind the leading axle.

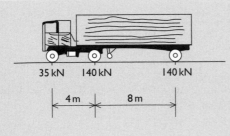

35 kN   140 kN   140 kN

|← 4 m →|← 8 m →|

Figure 1.5

Q.1.2
Figure 1.6 represents a possible loading on the floor beam of a silo.
Determine the magnitude of the resultant force and the distance of its
line of action from C.

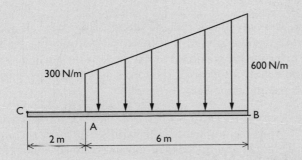

300 N/m   600 N/m

C   B

A

|← 2 m →|←   6 m   →|

Figure 1.6

Q.1.3
A barrier is to be placed across a diversion channel whose cross-
sectional dimensions are shown in Fig. 1.7. Consider the case where
water is up to the full height of 4 m on the upstream face of the barrier with the downstream channel empty. The water pressure can be
assumed to act horizontally at right angles to the barrier with a value
of $10y \, \text{kN/m}^2$ where y is the depth below the upstream water surface
in metres. Determine the magnitude of the resultant force acting on
the barrier due to water pressure and also its height above the bed of
the channel.

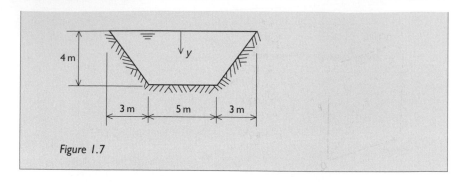

Figure 1.7

1.2 Newton's laws of motion

Newton's laws of motion may be stated as follows:

First law: A particle on which no nett forces are acting has zero acceleration.

Second law: A particle on which a force is acting experiences an acceleration in the direction of that force, proportional to the force and inversely proportional to the mass of the particle.

Third law: Action equals reaction, or the forces acting between particles are equal and oppositely directed.

The idea that force is proportional to acceleration had previously been put forward by Galileo. However, in Sir Isaac Newton's *Principia*, the importance of mass was identified and his laws of motion enabled a unified and logical development of mechanics, both statics and dynamics, to emerge.

Most investigations of structures are conducted assuming that they are at rest. Hence the first and third laws, which directly pertain to statics, would seem to be of most relevance. However, the first law is only a special case of the crucial second law, and it is this which establishes the nature of forces. Accelerations like displacements and velocities are known to be vector quantities. Hence, from the second law, forces are also seen to be vector quantities. Thus, if two forces P_1 and P_2 act on a particle Q, their resultant R may be obtained as the diagonal of a parallelogram with the lengths of the sides and their directions corresponding to P_1 and P_2 as shown in Fig. 1.8(a) or, alternatively, the closure of the triangle of forces (which is just one half of the parallelogram of forces) as shown in Fig. 1.8(b).

The parallelogram and triangle of forces, and their extension to cater for more than two forces (e.g. Fig. 1.8(c) and (d)), were particularly useful in the nineteenth and the first half of the twentieth centuries when graphical methods were used extensively for structural analysis. However, with the facility of pocket calculators and computers now available, graphical methods are no longer normally used. Instead, problems are solved by 'resolving' forces

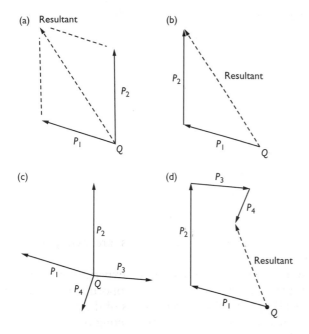

Figure 1.8 Graphical methods for finding resultants of forces acting at a point: (a) parallelogram of forces; (b) triangle of forces; (c) four forces acting at a point; (d) corresponding polygon of forces.

in different directions. The directions chosen are often orthogonal, but do not need to be. The two main problems that resolution of forces may be used for are

a   finding resultants of force systems;
b   finding the magnitudes of forces required to maintain equilibrium of particles, bodies or systems.

These methods will be developed and utilised in the ensuing sections of this chapter.

1.3   Forces acting at a point

If forces P_x and P_y act along orthogonal x and y axes, the parallelogram of forces gives the resultant as a force $P = (P_x^2 + P_y^2)^{1/2}$ acting at an angle θ to the x axis, where $\tan \theta = P_y/P_x$ (see Fig. 1.9). Conversely, if a force P acts in the xy plane at an angle θ to the x axis, this force may be replaced by its components $P_x = P \cos \theta$ and $P_y = P \sin \theta$ acting in the x and y directions.

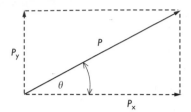

Figure 1.9 Resolution of a planar force into orthogonal components.

These two complementary techniques can be combined to determine the resultant of any set of forces acting at a point as follows:

a   Define directions for resolution of forces (two are required for a plane problem and three for a 3D problem).
b   Resolve each force into its components in these directions.
c   Add all components acting in the same direction.
d   Use the summed components to determine the resultant.

Although Newton restricts discussion to forces acting on a particle, the technique is also applicable to forces acting on any body. If the forces all act through the centre of gravity, there will be no tendency for the body to develop any rotational acceleration.

Ex. 1.5
At one instant of descent, a bungee jumper is subject to a gravity force of 0.6 kN and a bungee force of 0.4 kN acting at an angle $\tan^{-1} 0.25$ from the upwards vertical as shown in Fig. 1.10(a). Find the resultant force acting on the person.

Answer
Noting that the bungee force acts along the diagonal of a $1 : 4 : \sqrt{17}$ right-angled triangle, it must have components $P_x = -0.4 \times 1/\sqrt{17}$ kN and $P_y = 0.4 \times 4/\sqrt{17}$ kN. All the component forces are shown on the first two rows of Table 1.1. The last row shows the addition of these components and the evaluation of the resultant using the theorem of Pythagorus. The resultant force of 0.233 kN has an inclination α from the vertical such than $\tan \alpha = 0.097/0.212$, i.e. $\alpha = 24.6°$. Figure 1.10(b) shows the resultant force and its line of action.

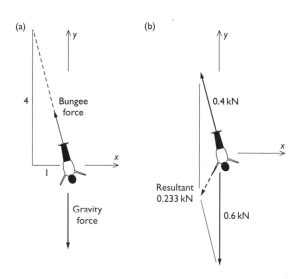

Figure 1.10 Forces acting on a bungee jumper: (a) forces acting; (b) parallelogram of forces.

Table 1.1 Forces acting on bungee jumper

Force	Magnitude	Components		Resultant
		x	y	
Bungee	0.4	−0.09701	0.38806	
Gravity	0.6	0	−0.6000	
Summation		−0.09701	−0.21194	0.23309

NB   A sketch of the parallelogram of forces gives a useful check on
the computation to ensure that any gross errors due, for instance, to a
wrong sign have not been made.

The more common problem in statics, however, is to determine the
magnitudes of forces acting on a body which are necessary to maintain
equilibrium. When all the forces are acting in one plane, there are two
independent directions (which could be orthogonal x and y directions) for
possible movement and hence two force variables will be necessary for equilibrium. The analysis procedure can follow that used for the bungee jumper,
except that the unknown forces need to be carried into the summation of
components, which should each sum to zero.

Ex. 1.6
A stone block weighing 60 kN is to be temporarily held at rest on
a plane of inclination 30°. If the reaction, R, between the plane and
the block can make a maximum deviation of 20° from the normal
without slippage occurring, find the least force P acting parallel to the
plane which will prevent the block from slipping downwards. Can the
magnitude of P be reduced by changing its angle?

First answer
Figure 1.11(a) shows all the forces acting on the block when it is only
just prevented from slipping. The direction of all the forces are known,
but the magnitudes of P and R are unknown. Table 1.2 lists the force
components with P and R as variables. The condition that the x and y
components both sum to zero give

$$P \cos 30° - R \cos 80° = 0$$

$$P \cos 60° + R \cos 10° = 60 \, \text{kN}$$

The answer, $P = 11.09 \, \text{kN}$ and $R = 55.30 \, \text{kN}$, is obtained by solving
these simultaneous equations.

Comment:
From this analysis, it would be difficult to answer the last part of the
question. More insight into the physical behaviour is obtained, and
often less computation is required, if more direct solution methods
can be devised as shown next.

Second answer
Resolve perpendicular to R, that is in the direction w in Fig. 1.11(b),
in order to ensure that the resulting equation does not involve R. Thus

$$P \cos 20° - 60 \cos 80° = 0$$

Hence $P = 11.09 \, \text{kN}$.
  Furthermore, if the inclination of P to the direction w is varied, this
equation becomes

$$P \cos \alpha - 60 \cos 80° = 0$$

  It is evident that the minimum value for P occurs when $\cos \alpha$ has
its maximum value, that is when $\alpha = 0°$ and the force P is acting
perpendicular to the direction of R. Thus

$$P_{min} = 60 \cos 80° = 10.41 \, \text{kN}$$

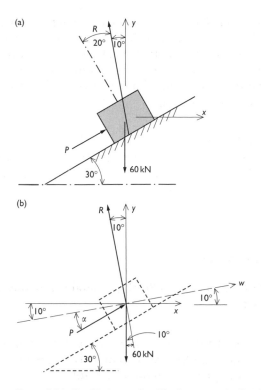

Figure 1.11 Equilibrium of a block on an inclined plane: (a) forces acting; (b) forces relative to direction *w*.

Table 1.2 Forces acting on block

Force	Magnitude	Components	
		x	*y*
Weight	60	0	−60
Reaction	R	−R cos 80°	R cos 10°
Applied force	P	P cos 30°	P cos 60°

The magnitude of R may not be required, but if it is, it may be obtained directly by resolving perpendicular to the line of action of P. Thus, in the original problem

$$R \cos 20° = 60 \cos 30°$$

Q.1.4
A bollard on a wharf carries anchor ropes from three ships which exert forces in the xz plane as shown in Fig. 1.12. Determine the magnitude and direction of the resultant force on the bollard.

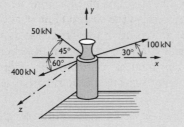

Figure 1.12

Q.1.5
An aircraft weighing 8000 kN is climbing at a uniform velocity at an angle of 13°. Apart from the self-weight, the forces acting on the aircraft are the engine thrust T, drag D and lift L, which act in the directions shown in Fig. 1.13. The drag D is known to be 350 kN. Use the condition that the aircraft is in equilibrium to determine the necessary engine thrust for this manoeuvre.

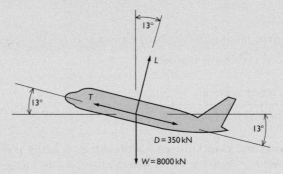

Figure 1.13

Q.1.6
A weight of 350 kN, hanging on a cord, is pulled sideways a distance 1.2 m by a horizontal force P, applied to the cord at a distance 4 m below the point of support. Determine the value of the force P. What is the least force that can be applied to this point in order to obtain the same lateral displacement?

1.4 Resultants of forces acting in a plane

When all the forces acting on a body do not pass through the same point, it is necessary to consider their turning effects (or 'moments'). A single force P acting at point A is equivalent to a force P, having the same magnitude, acting on another point O, together with a turning moment such that $M = Pd$, where d is the perpendicular distance of O from the original line of action of the force, as shown in Fig. 1.14. Thus, any number of planar forces acting on a body can be resolved into forces acting at a chosen point together with moments. The resultant of the forces can then be found by the techniques discussed in the previous two sections and the resultant moment can be obtained by summation.

If all the forces are transferred by this means to the centre of gravity, the resultant force will give rise to a translational acceleration in the direction of its line of action and the resultant moment will result in an angular acceleration.

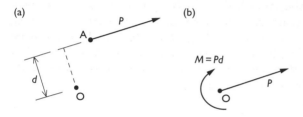

Figure 1.14 Changing the line of action of a force: (a) a single force; (b) an equivalent force plus moment.

Ex. 1.7
Figure 1.15(a) shows the plan view of a ship which is being pulled into a mooring. Consider the situation where the fore and aft mooring ropes exert forces of 42 kN and 40 kN, respectively, and the water pressure on the ship exerts a resultant force of 64.5 kN. With the geometry as shown in the figure, determine the overall resultant force through the centre of gravity of the ship together with the resultant moment.

Answer
In Table 1.3, the forces have been resolved into their x and y components and the clockwise moments about the centre of gravity of each of these components derived from their 'arms' (i.e. the distance

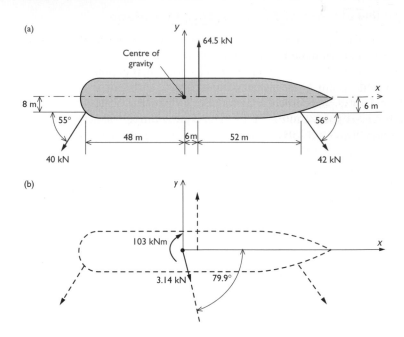

Figure 1.15 A ship in a mooring manoeuvre: (a) forces acting; (b) resultant force and moment.

Table 1.3 Forces acting on ship (clockwise moments positive)

Force (kN)	x component (kN)	Arm (m)	Moment contribution (kNm)	y component (kN)	Arm (m)	Moment contribution (kNm)
40	−22.94	−8	184	−32.77	−48	−1573
64.5	0	0	0	64.50	6	−387
42	23.49	−6	−141	−34.82	58	2020
Totals	0.55		43	−3.09		60

of the centre of gravity from their lines of action). From the summations in the table, it follows that the resultant force acting at the centre of gravity is $(0.55^2 + 3.09^2)^{1/2} = 3.14\,\text{kN}$ acting at an angle of $\tan^{-1}(-3.09/0.55) = -79.9°$ together with an anticlockwise resultant moment of $43 + 60 = 103\,\text{kNm}$ as shown in Fig. 1.15(b). The inertia properties of the ship would be required to compute its accelerations from these resultants.

A couple

A 'couple' is a pair of equal and opposite forces with parallel lines of action. If the forces are of magnitude F and their lines of action are distance h apart as shown in Fig. 1.16, resolving them to any point O gives zero as the resultant force and Fh as the resultant moment (clockwise for the configuration shown in the figure). It is noteworthy that the moment, Fh, is the same wherever point O is situated. Similarly, a moment on its own (which can be represented by a couple) is not position dependent.

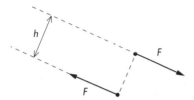

Figure 1.16 A couple with moment Fh.

Q.1.7
Specify the magnitude and direction of a single force acting through point A and also a moment which are together statically equivalent to the three forces shown in Fig. 1.17.

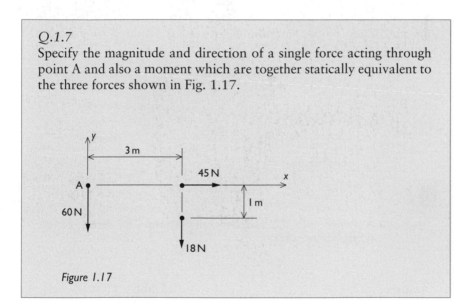

Figure 1.17

1.5   Planar equilibrium of rigid bodies

There are three independent planar equilibrium conditions required to prevent each of three possible independent planar movements of a rigid body. The force components should sum to zero in two directions and there should be no resultant moment. Normally, it will be necessary to have three variables whose magnitudes can be set to satisfy the equilibrium conditions. However, where there are no force components acting in one direction, it may be possible to dispense with one of the variables. A four stage process for solving equilibrium problems is as follows:

(i) Specify all the forces acting on the body together with their lines of action.
(ii) See whether the number of unknowns matches the number of equilibrium conditions to be satisfied.
(iii) If so, specify appropriate equilibrium equations.
(iv) Solve these equations to determine the variables.

All the following examples illustrate this technique for planar problems.

Ex. 1.8
For the loaded L-shaped unit shown in Fig. 1.18(a), determine the magnitude of the variables P, R and α required to maintain equilibrium.

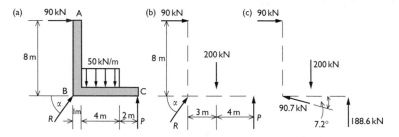

Figure 1.18 A loaded unit: (a) geometry and loading; (b) forces acting; (c) equilibrium solution.

Answer
The distributed force may be replaced by its resultant to give the forces and their lines of action as shown in Fig.1.18(b). Since there are three variables and three equilibrium conditions are required, a solution should be possible. Resolving horizontally and vertically gives

$$90 + R \cos \alpha = 0, \quad R \sin \alpha + P = 200\,\text{kN}$$

and taking moments about B (NB any point could have been chosen) gives

$$90 \times 8 + 200 \times 3 = P \times 7 \, \text{kNm}$$

Solving these equations gives $P = 188.6 \, \text{kN}$, $R \sin \alpha = 11.4 \, \text{kN}$ and $R \cos \alpha = -90 \, \text{kN}$.

Because $\sin^2 \alpha + \cos^2 \alpha = 1$, summing the squares of the last two equations gives $R = 90.7 \, \text{kN}$, and since $\tan \alpha = 11.4/(-90)$, $\alpha = -7.2°$ or $172.8°$. However, to satisfy the correct signs for $\sin \alpha$ and $\cos \alpha$, $\alpha = 172.8°$ must be the only correct solution as shown in Fig. 1.18(c).

Ex. 1.9
A child's swing weighing 50 N supports a child of weight 75 N sitting centrally. Figure 1.19 shows the seat deflected sideways by a force of 40 N acting in line with the bottom of the seat. The chain supports both make the same angle α with the vertical. With the seat dimensions shown in the figure, determine the forces P_1 and P_2 in the chains and their angle of inclination.

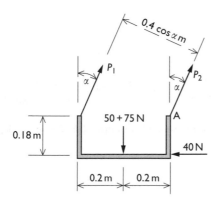

Figure 1.19 A child's swing.

Answer
Since the number of unknowns matches the number of equilibrium equations to be satisfied a solution should be possible. Resolving

horizontally and vertically gives

$$(P_1 + P_2)\sin\alpha = 40\,\text{N}, \quad (P_1 + P_2)\cos\alpha = 125\,\text{N}$$

and taking moments about point A gives

$$0.4P_1\cos\alpha + 40 \times 0.18 = 125 \times 0.2\,\text{Nm}$$

the solution of which gives $P_1 = 46.7\,\text{N}$, $P_2 = 84.6\,\text{N}$ and $\alpha = 17.7°$.

Ex. 1.10
A pontoon in the form of a cuboid has a plan area of 5 m × 3 m. Under its self-weight of 16 kN, the pontoon floats horizontally. Estimate the amount of tilt which will take place if a load of 6 kN is placed at the centre of one end of the pontoon. (The water pressure on the sides and ends of the pontoon may be ignored and the water pressure on the bottom may be assumed to act vertically. Also, it may be assumed that the deck of the pontoon remains above the water level and horizontal movements of the loads are negligible.)

Answer
A side view of the tilting pontoon is shown in Fig. 1.20(a). The hydrostatic pressure on the bottom of the pontoon will give an uplift which varies linearly from the shallow end, A, to the deep end, B, depending on the depths to which the pontoon sinks in the water. If the pressures at the shallow and deep end are p_1 and p_2 per unit span respectively, the hydrostatic loading for the pontoon will be as shown in Fig. 1.20(b). Since no horizontal forces are assumed to be present, there are only two equilibrium conditions to be satisfied and these should be sufficient to determine p_1 and p_2.

Splitting the hydrostatic pressure block into triangles, ABD and BCD gives resultant uplifts of $7.5p_1$ and $7.5p_2$ acting at the one-third span positions. Referring to Fig. 1.20(c), equilibrium equations for the pontoon are as in Fig. 1.20.

Resolving vertically, we have

$$7.5(p_1 + p_2) = 22\,\text{kN}$$

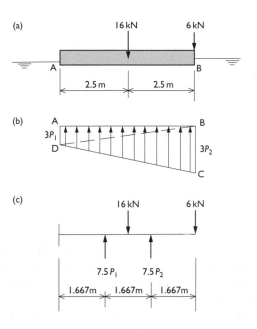

Figure 1.20 A floating pontoon: (a) geometry and imposed loading; (b) hydrostatic pressure block; (c) equilibrium of forces.

and taking moments about the centre of gravity, we obtain

$$7.5 \times 0.8333(p_2 - p_1) = 6 \times 2.5 \text{ kNm}$$

which yield $p_1 = 0.2667 \text{ kN/m}^2$ and $p_2 = 2.6667 \text{ kN/m}^2$. Hence the pontoon is estimated to sink to depths of 27 mm and 271 mm at the shallow and deep ends respectively with an angle of tilt of 2.8°.

Q.1.8
Figure 1.21 shows a plan view of the same ship investigated in Ex. 1.7, but this time the mooring ropes fore and aft hold it static against the water pressures. If the aft rope is known to have an orientation at 55° to the longitudinal axis of the ship, determine the forces in the ropes and the orientation of the fore rope.

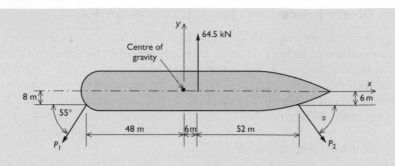

Figure 1.21

Q.1.9
Noting the fact that the unloaded end of the pontoon of Ex. 1.10 will lift out of the water when the eccentric load is increased from 6 kN to 12 kN, determine the new angle of tilt. The assumptions specified in brackets for Ex. 1.10 may still be assumed to apply.

Q.1.10
A girder bridge of span 8 m and weight 260 kN has end A located at its correct position, whilst end D is being lowered slowly onto its bearing pad by means of a cable EF. When the bridge is inclined at 15°, the geometry is as shown in Fig. 1.22 in which G is its centre of gravity. Determine the horizontal and vertical components of reaction, H_A and V_A, and also the pull, P, in the cable.

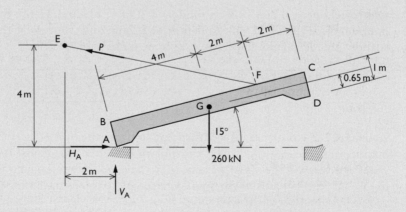

Figure 1.22

1.6   Direct procedures for planar equilibrium

To gain insight into equilibrium situations, it is generally much more effective to find direct techniques to determine required variables than to set up and solve the full set of equilibrium equations. Where one of three force variables is required, its magnitude may be obtained directly by taking moments about the intersection point of the lines of action of the other two variables. When the other two variables have parallel lines of action, it is simply necessary to resolve forces perpendicular to them (which is equivalent to taking moments about their theoretical intersection point at infinity).

Ex. 1.11
A ladder AB of length 6 m and weight 200 N rests at an inclination of 60° against a vertical wall. Assume that the ladder's self-weight is uniformly distributed, that the wall at B is smooth (i.e. only providing a horizontal reaction) and the ground at A provides both horizontal and vertical reaction.

a   Determine a direct method of finding the maximum horizontal ground reaction when a person of weight 600 N stands one quarter of the way up the ladder.
b   Identify whether the friction force will increase when the wall is leaning backward or forward at the point of contact.

Answer
a   Let the reaction at A be specified in terms of its horizontal and vertical components, H_A and V_A, respectively. With the wall reaction as H_B, the complete set of forces are shown in Fig. 1.23(a). To find H_A directly, moments need to be taken about the intersection of the lines of action of H_B and V_A which is C. Thus

$$200 \times 1.5 + 600 \times 0.75 = 5.196 H_A$$

giving $H_A = 144.3$ kN.
b   Where the wall leans backward, the above analysis is the same except that the wall reaction is no longer horizontal, resulting in C moving upwards (see Fig. 1.23(b)). The only difference this makes to the equation for H_A is that its coefficient (the distance AC)

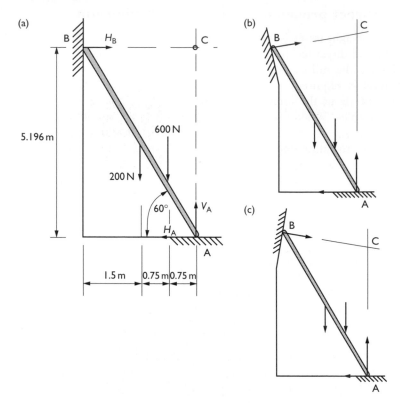

(a)

B H_B C

5.196 m   600 N

200 N

60° V_A (c)

H_A

A

1.5 m   0.75 m 0.75 m

(b)   B   C

A

B   C

A

Figure 1.23 A ladder problem: (a) applied forces; (b) with the wall leaning back; (c) with the wall leaning forward.

increases, resulting in a smaller value for H_A. Correspondingly, when the wall leans forward, the distance EC decreases resulting in an increased value for H_A (see Fig. 1.23(c)) so this is the more critical situation.

Ex. 1.12
A signboard, supported from a wall by three bars, has the geometry shown in Fig. 1.24(a). The sign is of uniform density and weighs 220 N. The bars have negligible weight.

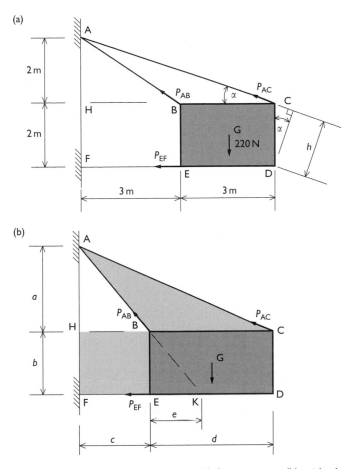

Figure 1.24 A signboard with bar supports: (a) forces acting; (b) with changed dimensions.

a   Find the magnitude of the force exerted on the sign by the bar AC.
b   If the geometry of a similar sign is as shown in Fig. 1.24(b), identify what geometrical condition needs to be satisfied for the force in bar AC to pull rather than push on the signboard.

Answer
a   The signboard is subject to its own self-weight and three bar forces P_{AC}, P_{AB} and P_{EF}. Since these are the only variables, the problem should be solvable. To determine P_{AC} directly, moments need to be taken about the intersection of AB and EF, namely D.

Thus

$$220 \times 1.5 + P_{AC}h = 0\,\text{Nm}$$

where $h = 2 \cos \alpha = 1.897\,\text{m}$
Hence $P_{AC} = -174\,\text{N}$.
At first sight it seems strange that this force acts in the opposite sense to that shown in Fig. 1.24(a). However, if AC were not present, the board would be free to rotate about D and would start to move in an anticlockwise direction. Bar AC needs to push on the board to prevent this happening, hence the negative sign.

b   Since P_{AC} can be obtained directly by taking moments about K in Fig. 1.24(b), it follows that P_{AC} changes from negative to positive if K lies to the left rather than the right of G. Thus bar AC will pull if

$$\frac{d}{2} > e$$

However, from the similarity of triangles ABH and BKE, it follows that

$$\frac{c}{a} = \frac{e}{b}$$

Hence by eliminating e between these two conditions: $ad/2 > bc$. Thus the area of triangle ABC needs to be greater than the area of the rectangle HBEF.

The concurrency theorem

This may be stated as: 'If a body is in equilibrium under the action of only three forces, they must be concurrent' (i.e. their lines of action must all pass through the same point).

By taking moments about the intersection of the lines of action of two of the three forces, it follows that, in order to avoid a resultant moment, the line of action of the other force must also pass through the same point.

Ex. 1.13
Use the concurrency theorem to determine the angle of the fore mooring rope in Q.1.8.

Answer
Project the line of action of P_1 to meet the line of action of the resultant water pressure at point C, which is found to be 54 tan 55° − 8 = 69.12 m from the axis of the ship. From the concurrency theorem, the line of action of force P_2 must also pass through C. Hence in metre units:

$$52 \tan \alpha = 69.12 + 6$$

giving $\alpha = 55.3°$ (see Fig. 1.25).

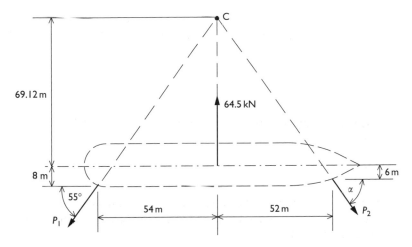

Figure 1.25 Concurrency of forces for the ship mooring problem.

Ex. 1.14
Find the angle of the reaction at B for the L-shaped unit of Ex. 1.8.

Answer
If the 90 kN and the 200 kN forces are combined into one resultant force where their lines of action intersect at D (Fig. 1.26), their resultant must be inclined at an angle β from the horizontal where tan β = 200/90. Hence it must intercept the line of action of force P at E, which is 4 tan β − 8 = 0.889 m below C. Because the number of forces have now been reduced to three, the concurrency theorem

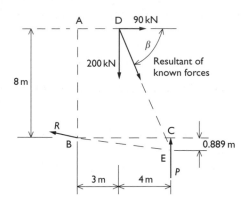

Figure 1.26 Concurrency of forces for the unit of Fig. 1.18.

indicates that the force R must pass through E and must be inclined at $\tan^{-1}(0.889/7) = 7.2°$ as shown in Fig. 1.26.

Q.1.11
The bed of a mobile crane weighs 138 kN and the jib weighs 24 kN. With the jib inclined at 40°, the horizontal distances of the hook and the centres of gravity from the wheels of the crane are as shown in Fig. 1.27.

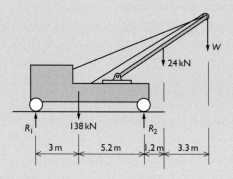

Figure 1.27

a   Determine the magnitude of the smaller of the two vertical wheel axle forces when the crane picks up a weight of $W = 56$ kN.

b   What magnitude of W would cause the crane to overturn if the
    driver tried to lift it?

Q.1.12
Figure 1.28 shows an alternative bar arrangement for supporting the
signboard of Ex. 1.12(a).

a   Determine the magnitude of the pull exerted on the signboard by
    the bar AB.
b   Identify which of the supporting forces may change sign if the
    geometry is altered.

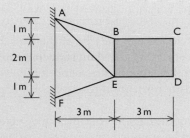

Figure 1.28

Q.1.13
For the ladder problem of Ex. 1.11;

a   Determine the angle of inclination from the vertical of the ground
    reaction at A when the person is one quarter of the way up the
    ladder.
b   Determine the maximum value of this angle for the case where the
    person slowly climbs the ladder to the three-quarter point.

Q.1.14
What does the concurrency theorem imply if two of the three forces
are parallel?

Q.1.15
When only two forces act on a body in equilibrium, what can be said
about these forces?

1.7 Restraints and reactions

The nature of restraints

A body which remains at rest when subject to applied forces must have external 'restraints' to prevent movement. Forces exerted by restraints are called 'reactions'. One form of restraint is by cable, rope or chain as illustrated by the child's swing (Ex. 1.9) or the moored ship (Exs 1.7 and 1.13). Another involves attachments such as bars as for the signboards (Ex. 1.12). In these cases, the line of action of the reaction is along the cable, rope, chain or bar. Other types of restraint, however, may provide more complex force systems. Contact between two surfaces provide a reaction whose direction can vary depending on the amount of friction present (Ex. 1.11). Other situations are more complex still.

Consider the support of a cupboard which is screwed to a wall. Downwards movement of the cupboard must be restrained by an upwards reaction of equal magnitude, V, whose line of action can only be at the wall. However, the gravity force and its reaction are misaligned so producing a turning effect. To stop both rotation and also horizontal movement of the cabinet, it is necessary to have equal and opposite horizontal reactions, H_A and H_B, as shown in Fig. 1.29(a).

The reaction H_A, necessary to prevent the cabinet rotating outwards about its bottom right-hand corner, can only be supplied by screws. It is therefore important that sufficient screws are provided near to the top of the cabinet to enable this reaction to occur. The reaction, H_B, preventing the cabinet moving into the wall will arise through bearing pressure. There is no need for screws to supply this force. The reaction, V, preventing vertical movement, will be supplied by a combination of friction and forces in the screws. If there are only sufficient screws to supply the reaction H_A, however, no friction forces will be able to develop near the top of the cabinet where it is trying to pull away from the wall. Although friction may develop near to the bottom

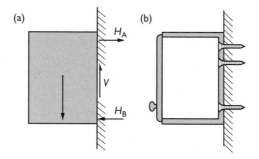

Figure 1.29 Support of a wall cabinet: (a) fully restrained; (b) cross-section showing possible screw fixings.

of the cabinet where there is bearing pressure, this alone is unlikely to be sufficiently large to sustain V. Extra screws are therefore needed which help to develop V through

a   'dowel action', that is by screws crossing the interface and therefore being a barrier to slippage;
b   developing extra friction forces through tightening the screws so that the cabinet and wall are locally squeezed together.

In considering where to put any extra screws required to support V, it is worth considering that a screw placed near the bottom of the cabinet would help to keep the cabinet in place if it receives a sharp upward force (see Fig. 1.29(b)).

In many situations, the support conditions are not so much complex as uncertain. A floor joist resting in a wall recess provides uncertainty about the reactions if the recess is not manufactured accurately to the shape of the end of the joist, some possible situations are being contrasted in Fig. 1.30.

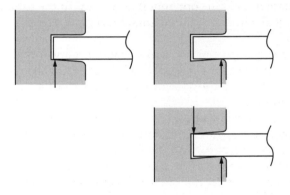

Figure 1.30 Uncertainties in a roughly fabricated support for a floor joist.

Bearings

In major structures, such uncertainty is unacceptable. Bearing supports in the form of hinges or rollers have generally been carefully designed and manufactured so that the restraint conditions and the reactions can be determined as accurately as possible. Single span girder bridges, for example, are normally 'simply-supported', that is supported at one end to allow for rotation only and at the other end to allow for rotation and horizontal movement. The main reason to allow horizontal movement at one end is so that thermal expansion and contraction of the bridge can take place. You should, if possible, have a look under girder bridges or look at photographs of such bridges to see what type of bearings have been used. Figure 1.31 shows a

Figure 1.31 A footbridge at Durham, England: (a) general view; (b) one component of the central joint fabricated in bronze (the ring diameters are 125 mm).

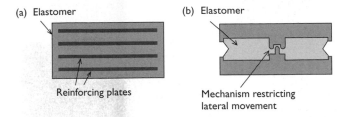

(a) Elastomer

(b) Elastomer

Reinforcing plates

Mechanism restricting
lateral movement

Figure 1.32 Forms of elastomeric bearing: (a) allowing horizontal movement and rotation; (b) allowing rotation only.

Table 1.4 Some different support conditions

Kind of support	Movement allowed			Possible reactions		
	u	v	θ	H	V	M
Simple roller bearing	✓	×	✓	×	✓	×
Hinge bearing	×	×	✓	✓	✓	×
Translation only bearing	✓	×	×	×	✓	✓
Built-in	×	×	×	✓	✓	✓

Figure 1.33 Positive directions for movements and reactions in Table 1.4.

bridge joint, clearly visible to pedestrians, which allows relative horizontal movement and also rotation but prevents relative vertical movement. The joint has been designed to transfer vertical force in either direction. The two rings act as rollers being free to rotate, but they are held in place by projections which fit into slots in the rings. Recently built bridges might have elastomeric bearing blocks in which natural or synthetic rubber is used as an elastomer and which require less maintenance. Figure 1.32 shows two forms of such bearing blocks (Lee, 1994).

Table 1.4 describes, for four types of support for the end of a beam, the possible movements allowed and the corresponding reactions which are allowed to develop. It should be noticed that reactions are only possible where the corresponding movements (whose nomenclature are shown in Fig. 1.33) are restrained. In graphical representations of structures, idealised depictions of these support conditions are used (those used in this book

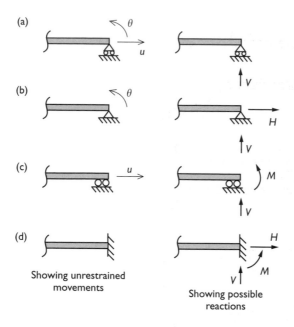

Showing unrestrained
movements

Showing possible
reactions

Figure 1.34 Idealised depiction of beam supports corresponding to those in Table 1.4.

are shown in Fig. 1.34). Although case (c) is not a very practical support condition, this depiction is sometimes required to indicate the allowable displacements at a line of symmetry (as seen in Fig. 6.59).

Ex. 1.15
For the potential structures shown in Fig. 1.35, determine which have possible body movements and which have reactions that can be determined from equilibrium if the geometry and the magnitudes of the loading are specified.

Answer
a   All three body freedoms are restrained by the built-in support and the two non-zero reactions (vertical force and moment) may be obtained from overall equilibrium.
b   Horizontal movement is unrestrained, therefore equilibrium cannot be maintained with the given loading, nor can the reactions be obtained from equilibrium.

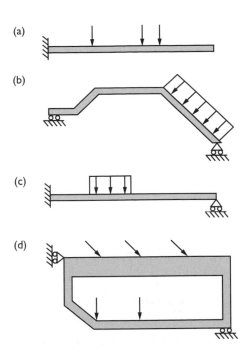

(a)

(b)

(c)

(d)

Figure 1.35 Support conditions for potential structures.

c   All body movements are prevented, but since there are four restraints (three at the built-in end and one at the other end), the three equilibrium equations are insufficient to determine their reactions.

d   Horizontal movement is restrained by the left-hand support and vertical movement and rotation are restrained by the right-hand support. Hence all body freedoms are restrained. Furthermore, the three equilibrium equations are sufficient to determine the magnitudes of the three reactions.

Soft restraints

All of the restraints discussed above may be considered as hard because they are assumed not to move when load is applied. It is also possible for bodies to be held in equilibrium by restraints which are softer than the body itself. In this case, equilibrium is achieved by a small amount of movement occurring such that the reactions are mobilised. The hydrostatic support for the floating pontoon of Ex. 1.10 is an example of a particularly soft restraint.

Ex. 1.16
Illustrate by means of sketches how the ground reaction of a pad footing for a masonry wall may be mobilised when wind forces are added to the gravity forces of the masonry.

Answer
If the wall and footing have a symmetrical cross-section, they may bed down uniformly when under the action of self-weight, so producing a uniform soil pressure as shown in Fig. 1.36(a). However, when the wall is subject to wind pressure, the ground reactions need to change. The pad will press down more on the leeward side and less on the windward side because of the overturning effect of the wind. Also, because of the lateral loading of the wind, the side of the pad is likely to bear on the surrounding ground. However, due to small lateral movement of the pad, ground friction could also be produced on the base of the footing. Hence a qualitative indication of the ground reactions is as shown in Fig. 1.36(b). However predictions of the magnitudes and shapes of these pressure blocks is beyond the scope of this book.

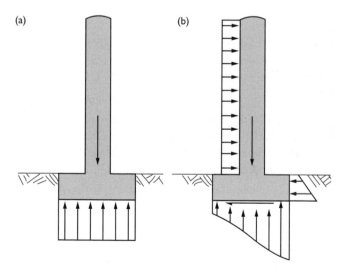

Figure 1.36 Cross-section of a wall and pad footing with possible forces for two loading cases sketched: (a) self-weight only; (b) self-weight and wind pressure.

Restraints in 3D

All of the earlier discussion has been concerned with bodies viewed in two dimensions. If it is necessary to examine equilibrium situations in three dimensions, there are six possible degrees of freedom (three possible translational motions and three possible rotations about each of the x, y and z axes). Hence a minimum of six restraints are required for equilibrium to be possible. This is discussed further in Section 1.13.

Q.1.16
For the potential structures shown in Fig. 1.37, determine which have possible body movements and which have reactions that can be determined from equilibrium, if the geometry and magnitudes of the loading are specified.

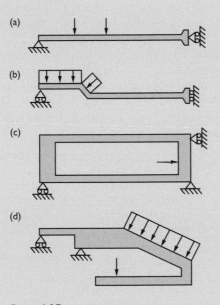

Figure 1.37

Q.1.17   Observation question
Can you identify any other situation in which a structure or body is held in equilibrium by soft restraints? If so, try to sketch the forces that are likely to be acting in this equilibrium state identifying which forces may be classified as reactions.

1.8   Superposition

The principle may be stated as 'The effects of two or more forces or sets of forces applied to a body or system of bodies simultaneously may often be obtained by summing their effects when applied separately.'

It applies to problems in rigid body mechanics provided that geometrical factors do not change significantly. Furthermore, it also applies more widely and will be discussed again in later chapters.

Ex. 1.17
Investigate the applicability of the principle of superposition to the loaded unit of Ex. 1.8.

Answer
The 50 kN/m distributed load gives rise to vertically upwards reactions of 114.3 kN and 85.7 kN at B and C, respectively. The 90 kN load gives rise to vertically upwards reaction of 102.9 kN at C and a reaction at B with components of 90 kN horizontally to the left and 102.9 kN vertically downwards. Adding these two cases gives an overall reaction at C of 188.6 kN vertically upwards. At B the resultant force components are 90 kN horizontally to the left and 11.4 vertically upwards, giving the resultant as 90.7 kN inclined at 7.2° to the horizontal. Thus the theory of superposition yields the same results that are given in Fig. 1.18(c) and is therefore applicable.

Q.1.18
Investigate the applicability of the principle of superposition to (a) Ex. 1.9 and (b) Ex. 1.10.

1.9   Accelerating bodies

The most severe loading cases for many structures occur under conditions of acceleration. Examples are

- aircraft subject to sudden gusts;
- ships in wave action;
- buildings subject to earthquakes;
- road and rail vehicles involved in crashes.

In such cases, structural engineers need to be able to assess the capacity of bodies to withstand the forces imposed on them by invoking d'Alembert's Principle that 'on a body in motion, the external forces are in equilibrium with the inertial forces'.

Consider the case of a dumb-bell comprising two masses, m, on the ends of a light rod of length ℓ which is resting on a smooth horizontal plane. Suppose the rod is subject to a steady lateral force P acting at its centre. The dumb-bell as a whole must develop a lateral acceleration of $P/2m$. However, as far as the two masses are concerned, they will each experience a force of $P/2$ to create their acceleration. Hence the rod must be strong enough to perform the function of force transfer indicated in Fig. 1.38. The structure is in a state of 'pseudo-equilibrium' in which the applied force is reacted by the reversed inertia forces causing acceleration. Such states of pseudo-equilibrium can thus be used to investigate the performance of structures which are not in equilibrium in the conventional sense of the word.

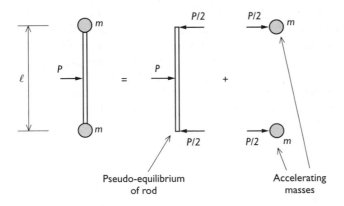

Figure 1.38 An accelerating dumb-bell.

Ex. 1.18
A uniform rod AB of length ℓ and mass, m, lying on a smooth horizontal plane, is subject to a lateral force, P, acting at end A. Determine the state of pseudo-equilibrium of the rod.

Answer
The rod will not only accelerate laterally, but will also develop an angular acceleration. One procedure would be to calculate these and thus calculate the reversed inertia forces. However, a simpler procedure is

to recognise that the local acceleration must vary linearly along the length of the rod. Hence the reversed inertia forces, which must also vary linearly, can be specified in terms of two unknowns say f_A and f_B, the reversed inertia force per unit length at ends A and B respectively. The magnitudes of these variables can then be obtained from the two relevant equations necessary to obtain pseudo-equilibrium. Figure 1.39(a) shows the pseudo-equilibrium condition. The pressure block arising from the reversed inertia forces may be split into two triangular blocks as shown in Fig. 1.39(b). Using this splitting and taking moments about C gives

$$\frac{f_B \ell}{2} \times \frac{\ell}{3} + P \times \frac{\ell}{3} = 0$$

Hence $f_B = -2P/\ell$ and resolving laterally gives $f_A = 4P/\ell$. The forces acting on the rod in the pseudo-equilibrium condition are therefore as shown in Fig. 1.39(c).

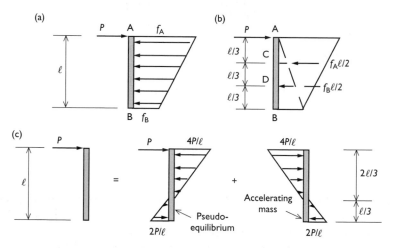

Figure 1.39 A uniform rod with eccentric side load: (a) projected pseudo-equilibrium; (b) showing pressure block resultants; (c) complete solution.

NB   The result shows that the rod pivots about the one-third span position measured from the loose end. This result can be illustrated by applying a sudden force to the end of a pencil or rod when it is lying on a smooth table.

Ex. 1.19
A fairground ride has gondolas which are turned in a vertical circle
by a Ferris wheel at a distance 12 m from its axis. The mass of each
gondola, when laden, is assumed to be 220 kg. If the wheel is rotated
at 2.6 rev/min, determine the magnitudes and lines of actions of the
forces exerted by gondolas in positions A, C, D and E on the Ferris
wheel, as shown in Fig. 1.40, assuming that their centres of gravity are
aligned with their pivot supports.

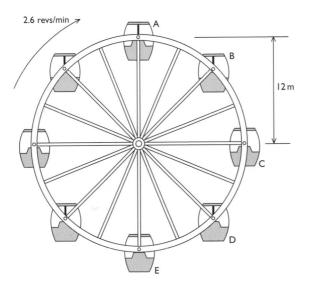

Figure 1.40 A Ferris wheel with gondolas.

Answer
The weight of each gondola is 220 kg × 9.81 m/s^2 which gives
2.158 kN. The angular velocity is 0.2723 rad/s and hence the
centripetal acceleration of each gondola is $\omega^2 r = 0.8896$ m/s^2 giving the centrifugal reversed inertia force as $m\omega^2 r = 196$ kgm/s$^2 =$
0.196 kN. The gondola at A provides a downwards force of
2.158 − 0.196 = 1.962 kN. The gondola at C provides a force of
$(2.158^2 + 0.196^2)^{1/2} = 2.167$ kN inclined outwards from the vertical at tan^{-1}(0.196/2.158) = 5.19°. The gondola at D provides force
components of 0.196 cos 45° = 0.138 kN horizontally and 2.158 +
0.196 cos 45° = 2.297 kN vertically downwards, giving a resultant
force of 2.301 kN at an inclination to the vertical of 3.44°. The gondola
at E provides a downward force of 2.158 + 0.196 = 2.354 kN.

Q.1.19
If, in Ex. 1.18, the centre of gravity of each gondola is below its pivot support, what will the main effect of this misalignment be?

Q.1.20
An aircraft having a total weight of 8.2 kN takes 8 s to perform a 90° turn in a horizontal circular path at a speed of 96 m/s. Determine the magnitude of the aerodynamic lift required to sustain this motion.

1.10   Friction and wheels

Friction

Coulomb's laws of dry friction may be stated as

a   The limiting friction force is proportional to the normal pressure.
b   Once movement takes place, the friction force reduces (by about 25%) from the limiting value when static.

Expressed mathematically, if N and F are the normal and friction forces respectively,

$$|F| < \mu_s N$$

for no relative movement, where μ_s is the coefficient of static friction, and

$$F = \pm \mu_d N$$

for sliding, where μ_d is the coefficient of sliding friction such that $\mu_d < \mu_s$ and the sign depends on the direction of movement.

It is often more convenient, however, to specify the resultants and lines of action of the contact forces at a surface rather than their normal and tangential components. If a resultant is R acting at an angle θ to the direction of the normal, then

$$R = (N^2 + F^2)^{1/2}, \quad \tan\theta = \frac{F}{N}$$

(see Fig. 1.41). Coulomb's laws of friction may thus be expressed in terms of angles of friction, giving $|\theta| < \phi_s$ for static equilibrium where $\tan\phi_s = \mu_s$ and $\theta = \pm \phi_d$ for sliding where $\tan\phi_d = \mu_d$. In Ex. 1.6 involving a block held on an inclined plane, the value of ϕ_s was assumed to be 20°.

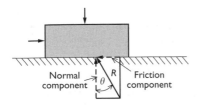

Figure 1.41 Reaction, R, due to surface contact.

Ex. 1.20
For the ladder problem of Ex. 1.11

a   Determine the angle of static ground friction required to prevent slippage.
b   Find the maximum angle of static ground friction required, if the wall could lean up to 30° either forward or backward at B, and the person could be situated anywhere up to 3/4 way up the ladder.

Answer
a   The forces acting on the ladder are shown in Fig. 1.23(a) for which vertical equilibrium yields $V_A = 800\,\text{N}$. Because V_A, the normal component of ground reaction is constant, it follows that the maximum angle of static ground friction will occur when the ground friction force H_A has its maximum value, which has already been established as 144.3 N in Ex. 1.11(a). The minimum angle of friction required to prevent slippage, therefore, is $\tan^{-1}(144.3/800) = 10.2°$.
b   In this question, the normal ground reaction is not constant and hence the case of maximum angle of static ground friction need not necessarily correspond to the case of maximum ground friction. However, the concurrency theorem is particularly effective. To reduce the number of forces acting on the ladder to three, the two vertical loads need to be resolved into one load of 800 N at a distance of $0.375 + 0.75x$ from A (where x is the horizontal distance of the person from A). With $0 \leq x \leq 2.25$, the line of action of this vertical load must be between EH and FG in Fig. 1.42. Also, since the wall reaction at B could lie anywhere within the 60° segment BFG shown in the figure, it follows that these two forces could intersect anywhere within the trapezium EFGH. Since the line of action of the reaction at A must be concurrent with

these two forces, the largest angle of friction is seen to occur when the reaction at A needs to pass through H. Thus to avoid slippage

$$\tan(\phi_s)_{max} \geq 2.0625/(5.196 - 0.9375\tan 30°)$$

giving $(\phi_s)_{max} \geq 23.9°.$

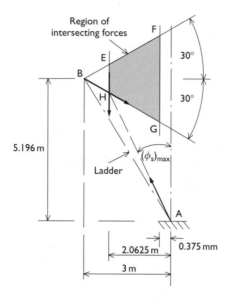

Figure 1.42 Use of the concurrency theorem with the ladder problem, showing the most critical case if the wall inclination can vary by up to ±30°.

Concerning wheels

Wheels are often used in industry and construction in the form of pulleys for hoisting and other applications. Consider a free-wheeling pulley which supports a rope, as shown in Fig. 1.43. The three forces acting are the two rope forces, T_1 and T_2, whose lines of action must be tangential to the rim, and a reaction R which acts through the supporting axle. If everything is at rest, moment equilibrium about the axle gives

$$T_1 = T_2$$

This condition will still apply even when the rope is moving at constant speed. An important result is that the angle of wrap, α, does not have any influence on the force in the rope.

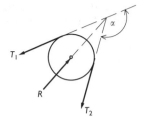

Figure 1.43 Equilibrium of a free-wheeling pulley.

Ex. 1.21
For the hoist shown in Fig. 1.44(a), obtain a formula for the weight which can be lifted in terms of T and α. Also determine the reaction at the axle of the higher pulley when $W = 560\,\text{N}$ and $\alpha = 25°$.

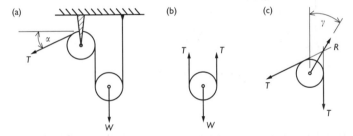

Figure 1.44 Equilibrium of a hoist: (a) the complete hoist; (b) the lower pulley; (c) the upper pulley.

Answer
The force in the rope must be T throughout its length. Vertical equilibrium of the lower pulley (see Fig. 1.44(b)) gives

$$W = 2T$$

Thus the system has a mechanical advantage of two (which is independent of α).

When $\alpha = 25°$, because the line of action of the reaction R of the higher pulley must bisect the angle between the ropes, its inclination to the vertical, γ (shown in Fig. 44(c)) must equal $(90° - \alpha)/2 = 32.5°$. With the force in the rope as $280\,\text{N}$, resolving forces acting on the pulley in the direction of R gives $R = 2 \times 280\cos 32.5° = 472.3\,\text{N}$.

Q.1.21

In considering how megalithic blocks such as those at Stonehenge may have been transported up inclines in ancient times, Parry (2000) compares the forces required for different techniques. These include

a pulling by sled;
b pulling by means of a sled on rollers;
c wrapping the megalith in cradles so that it can be rolled easily, and then pulling it via ropes wound round grooves in the rims of the cradles.

These methods are illustrated in Fig. 1.45. If rolling resistance and weights of sleds and cradles can be neglected, determine formulae for the pull required for each of these techniques as functions of the weight W, inclination, β, to the horizontal and coefficient of sliding friction, μ. Compare these forces for the case of a 400 kN megalith being pulled up a 1 in 10 slope where the coefficient of sliding friction is 0.2.

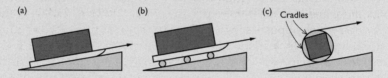

(a)   (b)   (c) Cradles

Figure 1.45 Different methods of moving a megalithic block: (a) by sled; (b) by sled on rollers; (c) by rolling using cradles.

Q.1.22

Devise a hoist using pulleys which will give a mechanical advantage of 4.

Q.1.23

Consider a wheel of negligible weight resting on horizontal ground and supporting a loaded axle. Consider how equilibrium of the wheel may be affected if a horizontal force is applied to the axle. Hence identify what factors would help to prevent the wheel moving or, once moving, would provide resistance to movement.

1.11 Free-body diagrams

In practical applications, the most important stage in solving equilibrium problems is ensuring that the forces contributing to equilibrium have all been correctly identified together with their lines of action (i.e. stage (i) of the four stage process described in Section 1.5). For 2D problems, this can be achieved by specifying a free-body diagram, isolating the body from whatever surrounds it and showing all the forces (including reactions) acting on it.

Ex. 1.22
An anchor block needs to restrain a force of 560 kN acting at 30° upwards from the horizontal. If the block is to rest on a horizontal surface with a coefficient of static friction between the block and the surface being no less than 0.21, determine the weight required for the anchor block.

Answer
The free-body diagram requires the horizontal surface taking away and being replaced by the reaction. Since the minimum weight anchor block must be on the point of slipping, maximum static friction will be present. Hence the reaction R will be inclined to the vertical at the maximum angle of static friction, namely $\tan^{-1} 0.21 = 11.86°$ in the direction which opposes slippage. Because there are three forces acting on the anchor block, these must be concurrent giving the free-body diagram shown in Fig. 1.46. The problem can be solved because there are two unknown forces and two remaining equilibrium conditions to be satisfied (one equilibrium condition has already been used to make the forces concurrent by fixing the line of action of R). Direct solution for W is obtained by resolving perpendicular to the line of action of R, thus

$$560 \cos 18.14° = W \sin 11.86°$$

giving $W = 2589$ kN (well over four times the force to be restrained).

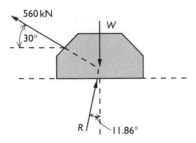

Figure 1.46 Free-body diagram for an anchor block.

A note on anchor blocks

When cables of suspension bridges cannot be anchored to rock, they need to be held by large anchor blocks such as the one shown in Fig. 1.47. The early development of suspension bridges in France was set back by the failure during construction of an anchor block in the 1830s for a bridge over the Seine in Paris designed by Navier. The anchor block as a whole did not slip, but it broke up through being insufficiently strong. The bridge was never completed.

Figure 1.47 An anchor block for the Verrazano Narrows Suspension Bridge, New York.

Ex. 1.23
A door 1.85 m high and 0.84 m wide is supported on two hinges, one 200 mm from the top and the other 300 mm from the bottom. If the door is of weight 74 N, what can be deduced about the reaction forces at the hinges when no other applied forces are acting?

Answer
Assume that the density of the door is uniform such that the resultant weight has a line of action at 420 mm from the line of the hinges. Also, assume that the hinges do not support any moment reactions. If the hinge reactions are each represented by a force of unknown magnitude and line of action, this yields four unknowns, too many to be determined from the three in-plane equilibrium conditions. However, if these reactions are represented by component forces in the horizontal and vertical directions (H_A and V_A at the top and H_B and V_B at the bottom), the two vertical components have the same line of action and hence can be combined into a single force $V = V_A + V_B$. The free-body diagram for the door is therefore as shown in Fig. 1.48 with three unknown forces. Equilibrium analysis gives the result:

$$V = 74\,\text{N}, \quad H_A = -23.02\,\text{N}, \quad H_B = 23.02\,\text{N}$$

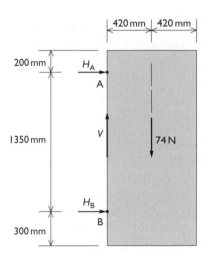

Figure 1.48 Free-body diagram for a door.

If the vertical reaction is equally split between the hinges, $V_A = V_B = 37\,\text{N}$ in which case the resultant hinge forces are both of magnitude $(37^2 + 23.02^2)^{1/2} = 43.58\,\text{N}$. However, one hinge could be carrying significantly more load. For instance, if all the vertical load is taken by the top hinge (as might be the case if this hinge is the first to be secured), the hinge forces will then be $(74^2 + 23.02)^{1/2} = 77.50\,\text{N}$ at the top and $23.02\,\text{N}$ at the bottom.

Q.1.24

Figure 1.49 shows a triangular unit ABCDEF with supports specified according to the convention shown in Fig. 1.34. The position of the lower support, D, is still to be decided and hence it is located at a variable distance x from the corner C.

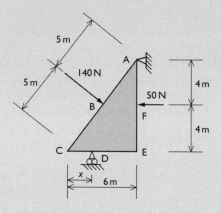

Figure 1.49

a   Specify the free-body diagram for the bracket when it is subject to the two applied forces shown.

b   Determine if there is any value of x for which equilibrium cannot be maintained.

Q.1.25

Figure 1.50 shows a claw hammer being used to extract a nail out of timber. Specify its free-body diagram and determine the force acting

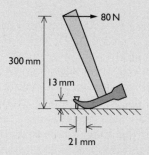

Figure 1.50

on the nail:

a   if the nail is being pulled vertically upwards.
b   if there is negligible friction between the hammer and the timber.

1.12   Compound systems

Mechanical devices and structural systems frequently comprise a number of bodies which are all in equilibrium (or in quasi-equilibrium according to the concept of Section 1.9). The mechanics of such compound systems may be examined from exploded views in which each free-body diagram is separated from its neighbours. In specifying the forces acting, it is important to appreciate that a contact force acting on one body must act also in reverse on the impinging body (as implied by Newton's third law of motion).

If, in a planar equilibrium problem, there are n bodies to be examined, there will be $3n$ independent equilibrium conditions to be satisfied. Hence, under normal circumstances, $3n$ variables will be needed to ensure that equilibrium can be maintained. Whereas it is always possible to determine the necessary equilibrium equations from the individual free-body diagrams, it may in some cases be more convenient instead to use some equilibrium equations derived from two or more bodies together.

Ex. 1.24
a   A block ABCD weighing 500 N, which is resting against a vertical wall, is to be raised by sliding a wedge EFGH of weight 100 N into place beneath it. Determine the horizontal force, P, necessary to start moving the wedge if it slides on a horizontal surface, the interface between the block and the wedge has a slope of 12° and the coefficient of static friction is 0.20 for all surfaces (see Fig. 1.51(a)).
b   Determine the maximum wedge angle that can be used without the force P having to be permanently present to stop the block slipping down again.

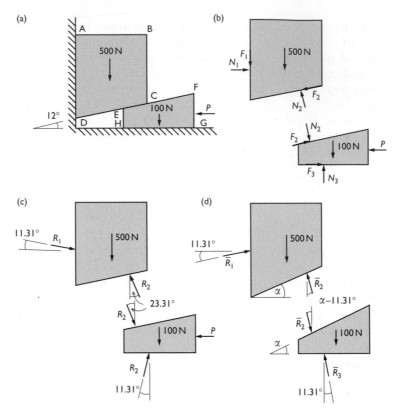

Figure 1.51 The equilibrium of a block raised and held using a wedge: (a) configuration; (b) free-body diagrams; (c) alternative free-body diagrams; (d) free-body diagrams for the block to slip back.

Answer

a   Normal and friction pressures will occur on each of the three surfaces with the friction pressures directed to oppose the intended motion. Representing these pressures by their resultant forces, the free-body diagrams for the block and wedge are as shown in Fig. 1.51(b). (Note that the forces acting on the common interface EC have been specified as equal and opposite.) The total number of unknown forces in Fig. 1.51(b) is seven, but by using the limiting friction condition $F = \mu_s N$ on all three surfaces, this reduces to four. For the two bodies, there are a total of six equilibrium conditions to be satisfied, but the unknown forces may be obtained from the two force equilibrium equations for the block and the two for the wedge without using the moment equilibrium equations.

A direct solution procedure is obtained by replacing the normal and friction forces by their resultants acting at an angle $\tan^{-1} 0.20 = 11.31°$ to the normal at each surface. The modified free-body diagrams are shown in Fig. 1.51(c). In order to determine P, it is first necessary to find R_2. Thus by resolving perpendicular to the line of action of R_1 for the block:

$$R_2 \cos(23.31° + 11.31°) = 500 \cos 11.31°$$

which gives $R_2 = 595.8 \, \text{N}$. Then by resolving perpendicular to the line of action of R_3 for the wedge:

$$P \cos 11.31° = R_2 \sin(23.31° + 11.31°) + 100 \sin 11.31°$$

Substituting for R_2 gives $P = 365.2 \, \text{N}$. Hence to move the block $P > 365.2 \, \text{N}$.

b   With the force P removed, the friction forces reversed and the angle of the wedge/block interface specified as α, the free-body diagrams are as shown in Fig. 1.51(d). By resolving perpendicular to the lines of action of $\bar{R}_3$ and $\bar{R}_1$, respectively,

$$\bar{R}_2 \sin(\alpha - 22.62°) = 100 \sin 11.31°$$

$$\bar{R}_2 \cos(\alpha - 22.62°) = 500 \cos 11.31°$$

Hence $\tan(\alpha - 22.62°) = 0.2 \tan 11.31°$ giving $\alpha = 24.91°$ which is the maximum possible interface angle which could be used without the block slipping back when the force P is removed.

NB   Use of the direct method here makes this difficult problem relatively easy to solve.

Screws

The operation of a screw has similarities with the wedge discussed in the above example. However, the helical path of the ramp of the screw enables more continuous lifting to take place. A screw with too steep a helix would not stay in place under load.

Ex. 1.25

a   A bridge consists of two simply-supported side spans ABC and DEF
which support a central span CD resting on their projecting ends
at C and D as shown in Fig. 1.52(a). If the self-weight of the bridge
is everywhere 240 kN/m span, determine the joint force at C and
the support reactions at A and B when there is no traffic loading.

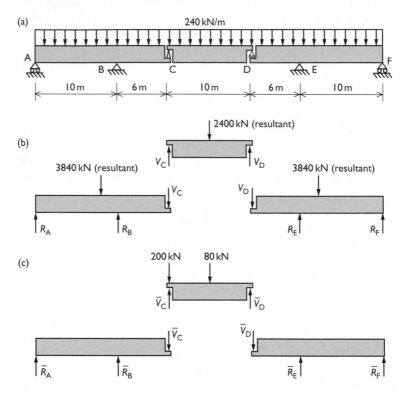

Figure 1.52 Equilibrium of a cantilever bridge with a suspended span: (a) general
arrangement and self-weight; (b) free-body diagrams for self-weight
loading; (c) free-body diagrams for vehicle loading.

b   Determine these same quantities due to a vehicle on the central
span having axle loads of 200 kN and 80 kN at C and the span
mid-point respectively. Also identify the implications of this result
on the type of supports required at A and F.

Answer

a   Figure 1.52(b) shows free-body diagrams for the three segments of the bridge with the unknown forces identified. Three equilibrium equations are required to show that all horizontal reactions will be zero. There are six other equilibrium conditions to satisfy and six unknown forces, hence the problem is likely to have a solution.

   Starting with the central span because that has only two unknowns, taking moments about D gives $V_C = 1200\,kN$. Then, taking moments about B for the left-hand side span, $R_A = 48\,kN$. Vertical equilibrium of the same span gives $R_B = 4992\,kN$. Because of symmetry, there will be similar reactions on the other side of the bridge. Thus nearly all the self-weight is reacted at supports B and E.

b   An analysis for the case of the vehicle load yields the free-body diagrams shown in Fig. 1.52(c) and the following values for the variables $\bar{V}_C = 240\,kN$, $\bar{R}_A = -144\,kN$, $\bar{R}_B = 384\,kN$, $\bar{V}_D = 40\,kN$, $\bar{R}_E = 64\,kN$, $\bar{R}_F = -24\,kN$.

   The principle of superposition can be used to add this result to the earlier result obtained when the bridge was subject to only its self-weight, to indicate that the effect of the vehicle load is to change the reaction at A from 48 kN upwards to 96 kN downwards. Since this could also happen for R_F with the vehicle load near to D, it is important that both of these supports are held down to prevent lift off.

Q.1.26

Two similar uniform circular pipes are placed on horizontal hard ground alongside each other and touching. A third similar pipe with its axis parallel, is gently stacked on top of these two. Determine what the coefficient of friction needs to be between the pipes and also between the pipes and the ground, if the two lower pipes are not to roll or slide apart.

Q.1.27

The two components of a three-pinned arch are hinged together at B and to the abutments at A and C as shown in Fig. 1.53. Hinges A and C are at the same height, distance *s* apart, and hinge B is situated at a height r above the centre point of the line AC.

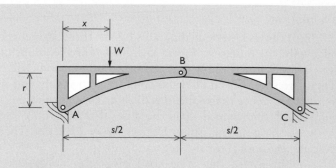

Figure 1.53 A three-pinned arch.

a   If a vertical load, W, is applied to the left-hand segment at a distance x from A, determine expressions for the horizontal and vertical components of the forces carried by all three hinges due to this load acting alone.

b   If $s = 50\,\text{m}$, $r = 7.5\,\text{m}$ and $W = 200\,\text{kN}$, show that the largest magnitudes for five of the six components of hinge force occur when $x = 25\,\text{m}$ and determine their values.

1.13   Equilibrium in 3D

Where force systems are three-dimensional rather than planar, it is still possible to investigate their equilibrium through the four stage process described in Section 1.5. Because of the complication of the extra dimension, however, particular care needs to be taken over specifying forces and reactions. There are six possible independent equilibrium equations for a 3D body corresponding to the three possible translation movements and three possible rotations, with moments being taken about a line rather than a point. Whereas it is useful to construct a perspective diagram of the body on which all the forces are drawn, it can be confusing to develop equilibrium equations directly from it. It is safer to draw suitable projections (usually elevation, end view and plan) and work from these. As with 2D equilibrium problems, it is generally beneficial to look for direct methods of determining required variables.

Ex. 1.26
The platform ABCD shown in Fig. 1.54(a) is supported by hinges at A and B and held in a horizontal position by a chain DE where E is located directly above A. With dimensions as shown and assuming that

the hinges cannot carry any moment, determine the forces in the chain and the minimum values for the hinge forces needed to support a load of 1.1 kN/m^2 acting over the whole area of the platform.

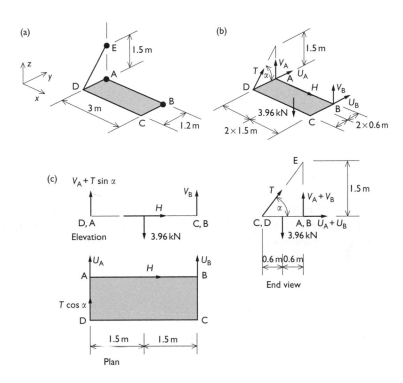

Figure 1.54 3D equilibrium of a platform: (a) general arrangement; (b) 3D free-body diagram; (c) free-body diagram projections.

Answer
Forces acting on the platform are the load of $1.1 \times 1.2 \times 3 = 3.96\,\text{kN}$, whose resultant acts at the centroid, the hinge forces having components H_A, U_A, V_A and H_B, U_B, V_B in the x, y, z directions and the pull T exerted by the chain. However, H_A and H_B both act along the line AB and may be replaced by the single force $H = H_A + H_B$. The full set of forces acting on the platform are thus as shown in Fig. 1.54(b) with the number of unknowns matching the number of independent equilibrium conditions (i.e. six). Figure 1.54(c) shows the projected views. Where a force is orthogonal to the plane of projection, it does not appear in the view. Thus U_A and U_B do not appear in the elevation.

By inspecting the projections, it is easy to form equations which, at each step, involve only one new variable. Thus taking moments about AB (see the end view)

$$T \times 1.2 \sin \alpha = 3.96 \times 0.6$$

in kNm units. But $\alpha = 51.3°$, thus $T = 2.54 \, \text{kN}$. Taking moments about AD (see the elevation)

$$V_B \times 3 = 3.96 \times 1.5$$

in kNm units. Hence $V_B = 1.98 \, \text{kN}$.

Resolving vertically (see the elevation or end view)

$$V_A + V_B + T \sin \alpha = 3.96 \, \text{kN}$$

in kN units. Hence $V_A = 0$.

Can you identify three more equations to determine the remaining forces? The result should be: $U_B = 0$, $U_A = -1.59 \, \text{kN}$, $H = 0$. The minimum possible values of H_A and H_B are thus zero and hence the minimum possible values of the hinge reactions are at A:

$$(H_{A(\text{min})}{}^2 + U_A^2 + V_A^2)^{1/2} = 1.59 \, \text{kN}$$

and at B:

$$(H_{B(\text{min})}{}^2 + U_B^2 + V_B^2)^{1/2} = 1.98 \, \text{kN}$$

Ex. 1.27
The shaft shown in Fig. 1.55(a) is restrained in three directions by the bearing at A and in the two lateral directions by the bearing at B. A motor applies a torque (i.e. a moment about the shaft axis) at A which in turn is driving belts at constant speed through the pulleys at C and D. If the forces in the belts and the dimensions are as shown in the figure, estimate the torque required at the motor and also the bearing forces at A and B. Indicate the direction of the bearing forces on a diagram.

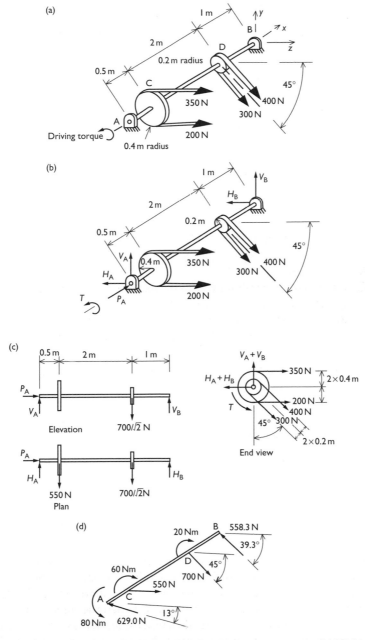

Figure 1.55 3D equilibrium of a driving shaft: (a) general arrangement; (b) 3D free-body diagram; (c) free-body diagram projections; (d) resultant forces acting on shaft.

Answer

The pulley wheels are moving at constant angular velocity and the belts at constant speed. Therefore the only inertia forces which could affect equilibrium relate to the centripetal accelerations of the belts whilst they are travelling round the pulleys. It will be assumed that the masses and speeds of the belts are such that these inertia forces may be neglected. Unknown reactions at the bearings can be specified as P_A, H_A, V_A, H_B, V_B with the directions shown in the free-body diagram, Fig. 1.55(b). With motor torque T also as an unknown, the total number of variables matches the number of equilibrium conditions needing to be satisfied. From equilibrium in the elevation and plan views shown in Fig. 1.55(c), it follows:

$$P_A = 0, \quad V_A = 141.4\,\text{N}, \quad V_B = 353.6\,\text{N},$$
$$H_A = 612.8\,\text{N}, \quad H_B = 432.2\,\text{N}$$

Then taking moments about the shaft axis for the end elevation

$$T = (350 - 200) \times 0.4 + (400 - 300) \times 0.2 = 80\,\text{Nm}$$

The bearing forces are therefore at A:

$$(612.8^2 + 141.4^2)^{1/2} = 629\,\text{N} \quad \text{at } 13.0° \text{ to the horizontal}$$

and at B:

$$(432.2^2 + 353.6^2)^{1/2} = 558\,\text{N} \quad \text{at } 39.3° \text{ to the horizontal.}$$

Hence the equilibrium of the shaft is as shown in Fig. 1.55(d).

Parallel forces in 3D

A special class of 3D equilibrium problem arises when all the applied forces and the reactions are parallel to each other. If a plane at right angles to these forces is considered (e.g. a horizontal plane if all the forces are vertical), then all three in-plane equilibrium conditions will be superfluous leaving only three 'out of plane' independent equilibrium equations to be satisfied. It is also possible to analyse equilibrium by referring to a 2D diagram in which the forces are only seen end on, as seen in the following example.

Ex. 1.28
A rectangular platform ACDF has column supports at B, D and F and
dimensions as shown in Fig. 1.56(a).

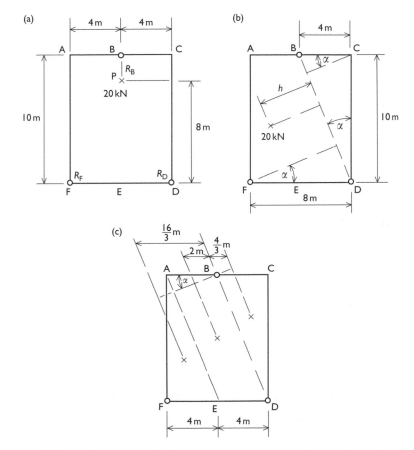

Figure 1.56 Equilibrium of a platform supported on three columns: (a) platform
with a concentrated load; (b) moment arm about BD of a concentrated
load; (c) positions of resultants of distributed loading on platform
segments.

a Determine the reactions due to a concentrated load of 20 kN being
   placed at P whose position is shown in the figure.
b Determine the two places where the weight should be placed to
   yield the maximum and the minimum reaction at F and find those
   reactions.

c If the platform could be subject to a distributed load of $5\,\text{kN/m}^2$ anywhere on its surface, identify where the load needs to be placed to give the maximum and the minimum reactions at F and find those reactions.

Answer

a On Fig. 1.56(a) the downwards 20 kN force at P has been designated by a cross whilst unknown reactions R_B, R_D and R_F (assumed to be positive upwards) have been designated by a circle. The number of unknowns matches the number of equilibrium conditions to be satisfied. Taking moments about the line FD gives $10R_B = 20 \times 8\,\text{kN}$. Hence $R_B = 16\,\text{kN}$. Taking moments about line AF and resolving vertically give $R_D = R_F = 2\,\text{kN}$.

b In order to investigate what affects the reaction R_F, it is helpful to obtain a direct equation for it. This may be obtained by taking moments about the line BD which intersects the lines of action of the other two variables. If the 20 kN weight is placed at a distance h from BD then moment equilibrium gives

$$8R_F \cos \alpha = 20\,h$$

where α is the angle BDC (see Fig. 1.56(b)). However the maximum and minimum values for h are $8 \cos \alpha$ (when the weight is at F) and $-4 \cos \alpha$ (when the weight is at C). Thus

$$-10\,\text{K} \le R_F \le 20\,\text{kN}$$

(The platform is trying to lift-off at F when the weight is placed at C unless other loads also act which counteract this tendency).

c From the previous solution it follows that the sign of its contribution to R_F of any element of loading will depend on the sign of h for the point at which it is applied. For the distributed load, the maximum value of R_F will therefore arise when distributed load only covers the area of the platform with positive h, that is the trapezium ABDF. Similarly the minimum value will occur when only the area with negative h is covered, that is, the triangle BCD (Fig. 1.56(c)). Table 1.5 shows the moment contributions about BD when distributed load covers each of three segments of the platform. Summing the contributions of the first two segments and equating to $8R_F \cos \alpha$ gives $(R_F)_{\text{max}} = 116.7\,\text{kN}$. Similarly the moment contribution from the last segment taken on its own yields $(R_F)_{\text{min}} = -16.7\,\text{kN}$.

Table 1.5 Moment contributions about BD

Segment	Area (m^2)	Total load (kN)	Arm from BD (m)	Moment (kNm)
ABDE	40	200	$2 \cos \alpha$	$400 \cos \alpha$
AEF	20	100	$(16/3) \cos \alpha$	$533 \cos \alpha$
BCD	20	100	$(-4/3) \cos \alpha$	$-133 \cos \alpha$

NB   The direct method of finding R_F has been particularly useful in identifying where on the platform to put distributed loading to obtain the most critical loading cases.

Q.1.28
In Ex. 1.26 why is the vertical reaction at A equal to zero? Is this a result which arises only for one inclination of the chain DE? What changes to the geometry or loading would produce non-zero values for V_A?

Q.1.29
The trailer shown in Fig. 1.57 has points A, B, C, D and E all in the same plane. It is supported by means of two wheels which prevent vertical and lateral movements and a pick-up point which is held in position but cannot resist angular rotations.

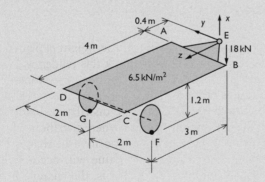

Figure 1.57

a   If there is a uniformly distributed load of 6.5 kN/m² acting over ABCD and a concentrated load of 18 kN at B, determine the vertical reactions at E, G and F for the given geometry.

b   If a horizontal force of 4.8 kN is applied in the direction CD, deter-
    mine from equilibrium the forces acting on the trailer due to this
    load alone, assuming that the wheels cannot lift off.

Q.1.30
A closed door AGED is held in position by two hinges at B and C and
a catch at F and is subject to a concentrated lateral force of 860 N at
E. With the geometry shown in Fig. 1.58 estimate the reactions to this
load, describing any assumptions made.

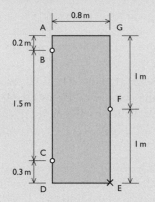

Figure 1.58

Q.1.31
Figure 1.59 shows the plan view of a three-legged circular table having
a diameter of 3 m with legs regularly spaced at distance 1.2 m from the
centre.

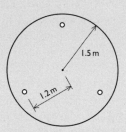

Figure 1.59

a   Determine the support reactions if the self-weight of the table top is $250\,\text{N/m}^2$ and each leg weighs $300\,\text{N}$.

b   If a point load can be placed anywhere on the table, determine its maximum value for there to be no risk of the table toppling over.

1.14   Load paths between components

For a stable structure there must be a continuous flow of force from the loaded point(s) to the reactions, thus creating one or more 'load paths' with equilibrium being satisfied throughout. In this chapter it is only possible to examine load paths as they appear on the interfaces between constituent parts of a composite structure. The fuller picture will emerge in Chapter 3 when forces internal to component parts will be discussed. The concept of load paths is introduced at this stage because of its importance in both analysis and design. Care needs to be taken to ensure that

- equilibrium conditions are always satisfied;
- all loading cases have been accounted for;
- joint action has been examined with particular care.

Ex. 1.29
Discuss the various forces exerted on the components of a pair of scissors when cutting hard material.

Answer
Figure 1.60(a) shows a pair of scissors with typical forces exerted by the user and the material being cut. A pair of scissors comprises three units however: two scissors and a rivet. When each scissor is considered in isolation three primary forces act on it. The additional force supplied by the rivet must act in opposition to the other two as shown in Fig. 1.60(b). When viewed in this way the action of the rivet is very simple. However when viewed from the side, the primary forces exerted on it by each scissor must have some degree of misalignment (Fig. 1.60(c)). To prevent rotation of the rivet, therefore, there must be forces exerted by each scissor on the heads of the rivet. Also, considering the equal and opposite reactions on each scissor, it is necessary for some pressure to exist between the scissors in order to maintain their lateral equilibrium.

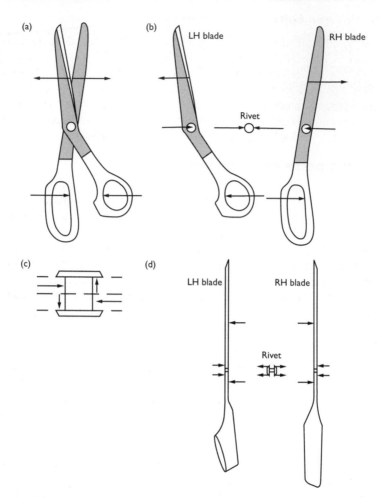

Figure 1.60 Equilibrium of a pair of scissors: (a) primary forces; (b) primary forces on components; (c) side view of rivet subject to primary forces; (d) secondary forces on components.

A pair of scissors generally has blades which are slightly curved in side view so that the cutting blades press together. The outward force thus generated at the cutting edge of each scissor needs to be reacted by an inward force supplied by the rivet and also pressure between the scissors on the handle side of the rivet as shown in Fig. 1.60(d). (If the scissors become slack the user needs to generate the lateral force system by pressing the handles outwards when cutting.)

Rivets, pins and bolts

It would be difficult to predict the magnitudes of the forces described above. These would normally be relatively small and would not be expected to constitute the subject of a structural analysis or design study. However it does illustrate how necessary it is to consider carefully the equilibrium of all components of a compound structure before the complete viability of a force system can be established. Furthermore, the structural action of rivets, pins and bolts is indeed a very important topic for engineers.

Ex. 1.30

A timber shelf ABCD of dimension $b \times \ell$ is nailed to a batten which is attached to a wall along its inner edge CD. Rods AE and BF, which have their ends threaded, hang from a rafter beam to support the outer corners of the shelf. The rods pass through holes in the shelf and also the lower flange of the rafter beam with a washer and two nuts used to secure each end (see Fig. 1.61(a)). What can be deduced about the forces acting on the shelf, rods and rod fittings due to a uniformly distributed load of w/unit area acting over the whole of the shelf?

Answer

The batten can be expected to prevent vertical and horizontal movements but not rotation. Hence it will provide a vertical reaction only. Similarly the rods will provide vertical reactions only. Thus if R_{CD}, R_A and R_B are the reaction forces for the batten and the two rods, the end view of the free-body diagram for the shelf will be as shown in Fig. 1.61(b). Taking moments about AB and CD in turn give $R_{CD} = wb\ell/2$ and $R_A + R_B = wb\ell/2$.

One possibility is that the reaction at the batten is uniform and the rod supports provide equal reactions. If this were so, equilibrium would be as shown in Fig. 1.61(c). Due to tolerances in construction and other factors, however, the reaction at the batten may not be uniformly distributed nor may the rod supports provide equal reactions. Another useful item of information is that the rods cannot give a downwards reaction because a nut has not been placed above the shelf to hold it down. Hence $R_A \geq O$ and $R_B \geq O$ which inserted in the previous equation for these reactions give $R_A \leq wb\ell/2$ and $R_B \leq wb\ell/2$. If the self-weight of the rods and attachments are neglected, the reactions R_E and R_F at the rafters will be equal to R_A and R_B respectively. The rod AE and its attachments are shown in exploded form in Fig. 1.61(d) in which the load path of the force is from shelf → washer → nuts → rod → nuts → washer → rafter beam. Transfer of load between the rod and the nuts is via the screw thread, with the extra

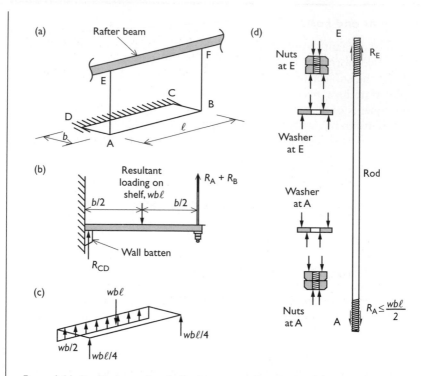

(a) Rafter beam

F

E

C

B

D

b

A

ℓ

(b) Resultant loading on shelf, $wb\ell$

$R_A + R_B$

b/2

b/2

Wall batten

R_{CD}

(c) $wb\ell$

$wb\ell/4$

wb/2

$wb\ell/4$

(d) E

Nuts at E

R_E

Washer at E

Rod

Washer at A

Nuts at A

A

$R_A \leq \dfrac{wb\ell}{2}$

Figure 1.61 Equilibrium of a shelf with two rod supports: (a) general arrangement; (b) end view of free-body diagram of shelf; (c) an unsafe assumption for reactions; (d) exploded view of rod AE and attachments.

lock nut being provided to prevent slippage if the rod goes slack. The washers are present primarily to spread load.

Q.1.32
In the Hyatt Regency Hotel, Kansas City in 1981 three suspended walk ways crossed the foyer. One walkway was suspended independently, but the other two were designed to be suspended one above the other from the same rods (see Fig. 1.62(a)). However, (probably because the need to place nuts below the upper cross-beam would have involved threading a long section of each rod) the design was modified to the arrangement shown in Fig. 1.62(b) in which each single rod ABC was replaced by two shorter ones, AB_1 and B_2C. If each walkway imposes

a load of W on each rod support, determine the forces required to be carried by the rods and all the connections on one side of the walkway for both the configuration as designed and the configuration as built.

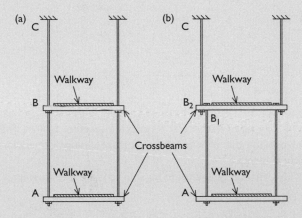

Figure 1.62

NB   The answer to this question explains a crucial factor leading to a major structural disaster. A dance in the foyer of the hotel attracted lots of people onto the walkways. The two walkways which had a common suspension collapsed onto the dancers below. Over 100 people were killed and many others injured.

Chapter 2

Gravity structures

One of the most important secrets of success in building with masonry has been the way gravity forces have been harnessed to provide stability against imposed loads. This aspect will provide the principal theme for examining both ancient and more modern structural forms.

On completion of this chapter you should be able to do the following:

- Examine the stability of simple masonry structures.
- Appreciate some key developments in the structural use of masonry.
- Recognise the value of existing masonry structures, as well as the care required in maintenance, in carrying out modifications or in demolition.

2.1 Masonry as a material

Rock can be immensely strong, as witnessed by many natural arches seen along coastlines and occasionally inland (Fig. 2.1). At times its inherent strength has been utilised in situ (e.g. the rock hewn buildings at Petra, Jordan, or the monolithic churches at Lalibela, Ethiopia (Fig. 2.2)) and is regularly relied on in tunnelling and in forming underground chambers. However, normally, it is required to be cut into blocks in order to be transported from quarry to site, in which case the structural performance of the finished building tends to be dominated by the way the blocks are shaped and put together, rather than by the strength of the original material. A strict definition of masonry would only include those structures built of stone dressed and placed by masons. However in this chapter, a broader definition is used involving also structures built from undressed stone and what might be called artificial stone, comprising bricks, tiles or concrete blocks.

Stonework has been used in construction throughout recorded history and provides some of the most prized historical remains. Bricks (made from clay baked to temperatures of up to 1300°C) have also been produced since ancient times and are now extensively used for domestic buildings

Figure 2.1 Landscape Arch, span 32 m in Arches National Park, Utah.

Figure 2.2 Entrance to Bet Gebriel – Rufael, a church cut from solid rock at Lalibela, Ethiopia, dating from the thirteenth century.

in particular. They are cheap, incombustible, involve relatively low energy consumption for manufacture in comparison with metals and plastics and also allow flexibility in design. Masonry provides thermal and sound insulation, fire protection and weather-proofing as well as acting as a structural material (i.e. maintaining its shape under the application of loads).

Normally masonry blocks are set in mortar, which not only provides load transfer but also fills any gaps, producing a continuous membrane for weather resistance and insulation purposes. Mortars are made from an aggregate, generally sand, mixed with a cementitious material. Traditionally the cementitious material has been lime, burned and crushed to a fine powder. However, with the invention of Portland cement in 1830, small amounts of it have been added to obtain a quicker setting mortar. The two are mixed and water is added on the building site to form a paste. This paste needs to be soft, in order to fill all the interblock spaces, and yet it must set fast enough not to deform when further layers of masonry are added soon afterwards. The layers of mortar should be thick, so as to ensure complete separation of the blocks and prevent large concentrated forces arising through neighbouring blocks touching each other.

How long masonry lasts will depend on how well it is maintained and also on the durability of the primary material in the environment to which it is exposed. For instance, sandstones and limestones tend to have a softer texture than igneous rocks, making them more susceptible to weathering agents (Fig. 2.3).

Whereas masonry structures can generally exhibit high strength when subject to compressive forces, their performance when subject to tensile forces (i.e. those trying to pull the masonry apart) is problematic and unreliable. Cracking may occur due to poor bonding between the blocks and the mortar, due to external effects producing distortion, or due to shrinkage or sag in lintels above window or door openings. For this reason masonry is normally designed as if it can carry only compressive forces. Hendry *et al.* (1997) say 'Direct tensile strength of brickwork is typically about 0.4 N/mm^2 but the variability of this figure has to be kept in mind, and it should only be used in design with great caution'. (This compares with brickwork strengths in compression of the order of 20–80 N/mm^2.) Because jointing between the blocks is the part most likely to have cracks, a lot may be learnt about behaviour and design by considering masonry to be an assemblage of rigid blocks (i.e. ignoring the effects of mortar, even if present).

Q.2.1
Bricks and concrete blocks are frequently manufactured with cavities (as shown in Fig. 2.4). Suggest up to four reasons why this may be advantageous.

Figure 2.3 Sandstone on Lindisfarne Abbey, Northumberland after 800 years of exposure to wind and weather.

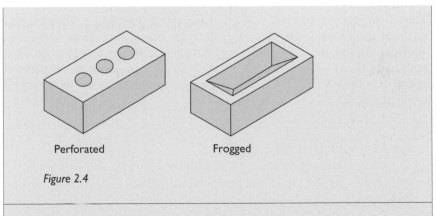

Perforated   Frogged

Figure 2.4

Q.2.2
A half-brick wall is one in which the width of the wall is the width (as opposed to the length) of one brick. Bricks are laid in horizontal

layers with each brick overlapping a joint in the layer below. Suggest two possible benefits of having such overlaps.

Q.2.3
If the external face of a brick wall appears as shown in Fig. 2.5 what might be inferred about it?

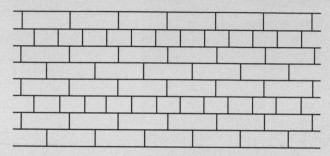

Figure 2.5

2.2 Masonry walls

Free-standing walls

The self-weight of a masonry wall imposes vertical compressive forces within the wall itself and, if the bedding joints are horizontal, there will be no tendency for the blocks to slip. The main concern is its capacity to carry horizontal forces. Two cases will be considered: a uniform lateral pressure acting over the whole of the wall and a horizontal line load acting across the top of the wall.

Ex. 2.1 Uniform lateral pressure
Consider a wall of length 4 m, thickness 210 mm, and self-weight $20 \, \text{kN/m}^3$ subject to a uniform wind pressure of $0.30 \, \text{kN/m}^2$ (Fig. 2.6(a)).

a   What can cause failure of the wall?
b   What would you think would be a safe height to build it?

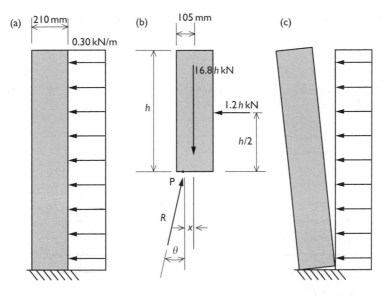

Figure 2.6 A wall subject to lateral pressure: (a) cross-section; (b) equilibrium of top part of wall; (c) collapse mechanism.

Answer

a   Consider the equilibrium of the wall above a bedding joint at depth h in metres below the top (Fig. 2.6(b)). For the resultant reaction R to maintain equilibrium it must have a vertical component of $16.8h$ kN and a horizontal component of $1.2h$ kN. Hence its inclination to the vertical, θ, will be $\tan^{-1}(1.2/16.8) = 4.09°$ (a result which is independent of h). The angle of static friction will be much larger than this. Hence slippage of bedding joints is not expected anywhere in the wall.

   If the reaction intersects the bedding plane at a point P, distance x from the centre-line of the wall, taking moments about P gives

$$16.8hx = 1.2h \times \tfrac{1}{2}h$$

Hence $x = 0.0357h$. As h increases, the reaction point P needs to deviate further from the centre-line of the wall. A purely compressive reaction will not be possible if $x > 0.105$ m which, for the given loading, will occur when $h \geq 2.94$ m. Thus with a wall of height 2.94 m or more, the lateral pressure would be expected to cause the lower bedding plane to open up and so topple it, as shown in Fig. 2.6(c).

b   Even if this is the main criterion for designing the wall, it should not be built 2.94 m high. The safe height may be less than this for the following reasons:

- Dimensions may not be completely accurate.
- Maximum wind pressures may have been underestimated.
- Self-weight may have been over-estimated.
- Strength in compression will not be infinite.
- Additional loads may occur (e.g. a person or object may be blown against the wall).

A standard way of catering for such uncertainties is to use load factors. If , for instance, a factor of 1.6 is used for the wind load, it is inappropriate to use the same factor for the self-weight. Indeed, if this were done, the answer for the safe height of the wall would be unaltered. The wall would be more easily overturned if its self-weight were to be less than anticipated. Since the self-weight is less likely to be as much in error as the wind load, a factor of 0.8 may be appropriate for it. Using these two factors, the forces acting on the wall would be $1.92h$ kN due to wind and $13.44h$ kN due to self-weight, giving a safe height of 1.47 m.

Ex. 2.2 Line load case
Repeat Ex. 2.1, but replacing the distributed lateral load with a line load acting across the top and having a magnitude of 0.7 kN, (Fig. 2.7(a)).

Answer
Equilibrium of a top section of the wall (as shown in Fig. 2.7(b)) gives the inclination, θ, of R to be $\tan^{-1}(0.7/16.8h)$. Also $16.8hx = 0.7h$ giving $x = 0.042$ m.

In contrast to the previous example, since x is independent of h, there is no greater risk of overturning at the bottom of the wall than there is near the top. Indeed for this example, if load factors of 1.6 and 0.8 were again applied to the lateral and self-weight forces respectively, the wall would be deemed safe for any height h. On the other hand, the inclination of R *is* unacceptably high when h is small (see Fig. 2.7(c)). A coefficient of static friction of 0.3 or more is required for h values less than 0.14 m (and that is without imposing any load factors).

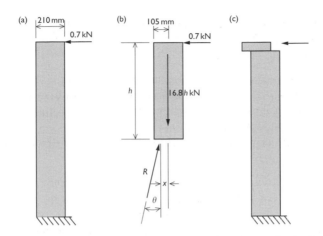

Figure 2.7 A wall subject to a horizontal line load: (a) end view of a 4 m long wall; (b) equilibrium of top part of wall; (c) failure mode.

Comment

A major consideration with free-standing walls is to avoid slippage of the highest bedding joints in cases where lateral forces are applied to the top (as with the previous example). A well-built wall will be topped with heavy coping-stones which perform this function and also protect the interior from moisture (Fig. 2.8).

Figure 2.8 A wall with coping-stones by the old bridge, Stirling.

Walls of churches and public halls

In the nineteenth century in particular, roofs of churches and public halls have frequently been supported on timber arches, which themselves have been supported high up on the walls. However the arches not only generate a downwards force due to the 'dead load' of their own weight and the weight of the roof structure they are carrying, they also push outwards. The tendency of the lateral force high up on the wall to cause it to overturn about its base or to slip at the higher bedding joints needs to be counteracted as follows (see Fig. 2.9):

a   The arches may be supported on 'corbels' projecting on the insides of the walls in order to reduce the moment of the resultant arch force about the base.

b   The walls may be continued above the corbels so that there is sufficient vertical loading on all the bedding joints of the wall.

c   The walls may be buttressed.

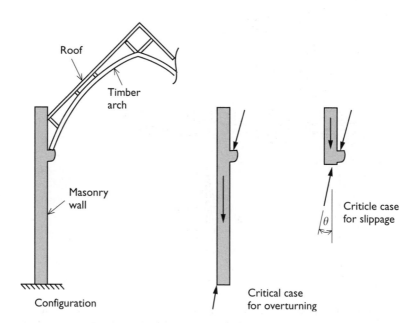

Figure 2.9 A type of construction for public halls and churches showing equilibrium of an unbuttressed masonry wall.

Ex. 2.3 On the use of buttresses
Determine the factor by which a uniform lateral pressure may be increased if a wall of length ℓ, height h, thickness t and density ρ is supported by triangular buttresses of thickness t, depth $2t$ and spacing $5t$ as shown in Fig. 2.10(a).

Answer
Overturning of the wall without buttresses will occur when the moment of the lateral forces about the outer edge of the bottom bedding joint (point A in the cross-section of Fig. 2.10(b)) exceeds the stabilising moment of the self-weight which equals $\rho g \ell h t \times 1/2t$. When buttresses are included, overturning will need to be by rotation about the outer point of the buttresses (point B in the cross-section of Fig. 2.10(c)). Because the moment of the wall self-weight about B is $\rho g \ell h t \times 5t/2$ and that of the buttresses is $\rho g \ell h t / 5 \times 4t/3$, the total moment resisting overturning is increased to $2.77 \rho g \ell h t$ indicating that an increase in lateral loading of more than 5.5 times is possible provided that the masonry bedding joints do not slip.

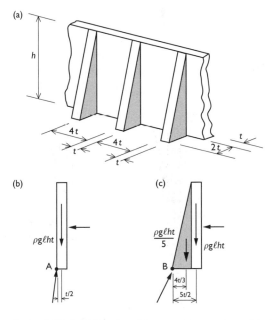

Figure 2.10 Benefit of wall buttresses on the critical overturning condition: (a) dimensions; (b) equilibrium without buttresses; (c) equilibrium with buttresses.

Comment

Even these shallow buttresses have considerably increased the stability of the wall when subject to lateral loading. However, to obtain this improvement, it is essential that the buttresses are well keyed into the wall to prevent slippage of the interface. The buttresses should not be spaced too far apart, otherwise other types of collapse mechanism will occur before the overturning load is reached. Furthermore, the buttresses do little to enhance the resistance of the wall to lateral pressure acting in the reverse direction. Figure 2.27 shows the use of concrete buttresses in dam construction.

Low-rise masonry buildings

Considering masonry to be a material able only to carry compression forces provides insight into some aspects of its use for low-rise buildings (normally one or two storeys).

a   *Stability*

The resistance of a wall to a horizontal force is very much greater if the force is applied in the plane of the wall rather than at right angles to it. Well-designed low-rise masonry buildings make extensive use of this fact to provide greater stability than a free-standing wall would have.

The following are some of the factors which improve stability:

- *Weight of the roof structure*   The walls bear the weight of the roof structure which helps to stabilise them by acting like coping-stones.
- *Cellular construction*   Low-rise masonry buildings are constructed with walls intersecting with each other, normally at right angles. When subject to lateral loading any one wall will tend to be supported by the intersecting walls which act like buttresses (this is particularly true for a box planform when the pressure is external rather than internal).
- *Horizontal linkages*   Roof truss members and floor joists provide horizontal linkages which can improve stability as shown in Fig. 2.11. In both cases the lateral pressure required to cause instability will be greater than if the link was not present.

b   *Cavity walls*

Cavity walls were introduced for domestic buildings in the UK in the 1920s and 1930s and tended to comprise two leaves (e.g. of thickness 114 mm) surrounding a cavity (e.g. of width 52 mm) making a wall which, with the given figures, would be of overall thickness 280 mm. More recently concrete blocks have been used in place of bricks, particularly for the inner leaf of the wall. The cavity controls moisture penetration and increases thermal insulation. It is standard practice to include wall ties of metal strip which link the two leaves across the cavity

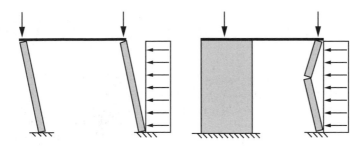

Figure 2.11 Horizontal restraints acting to increase the stability of masonry walls subject to side load.

Figure 2.12 A wall tie with splayed ends to grip the leaves of the wall and a twist to prevent moisture transfer across the cavity.

(Fig. 2.12). The main purpose of these ties is to ensure that lateral forces applied to one leaf can be transmitted to the other so that they share the load.

c *Wind forces*

Wind normally produces suction forces on roofs. The upward components of these forces reduce or cancel the benefit to the walls arising from the self-weight of the roofs. It is important that roofs are heavy enough not to lift off or are well tied down. Also roof structures should transfer any horizontal components of wind forces so that they act along the line of the walls rather than in their weak direction (i.e. at right angles). Furthermore, walls themselves need to be stable enough to resist the lateral wind pressures through the use of cellular construction and horizontal linkages as mentioned earlier.

d *Modifications*

It is important that modifications to low-rise masonry buildings do not compromise their cellular nature. Possible difficulties may arise, for instance, if a terrace of houses is to be converted into a shop or offices. Removal of internal walls may be desirable. However, it should not be carried out unless the structural functions of these walls (involving both support and stabilisation) are catered for. This may require rigid frames to be inserted round any openings.

Q.2.4

Two brick walls of height 3 m and density $1500 \, kg/m^3$ have cross-sectional dimensions as follows:

a   A cavity wall comprising two leaves each of one 100 mm brick thickness with wall ties.
b   A solid wall of thickness 210 mm.

Estimate the uniform lateral pressures required to overturn each of the walls as shown in Fig. 2.13. Hence predict the reduction in stability arising through using the cavity construction.

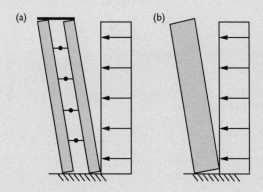

Figure 2.13

Q.2.5

a   It is required to investigate the stability of a masonry wall ABCD of height h, thickness t and self-weight W, which is restrained from horizontal movement at the top and carries a superimposed load of W as shown in Fig. 2.14(a). Consider the possibility of a uniformly distributed side load causing collapse due to the mechanism shown (exaggeratedly) in Fig. 2.14(b). On the point of instability the two blocks ABFE and EFCD will be in equilibrium, with point A carrying the superimposed load and the horizontal reaction, point F carrying the interblock forces and point D carrying the ground reaction. Determine an expression for the magnitude of the uniformly distributed lateral loading which will cause this instability.

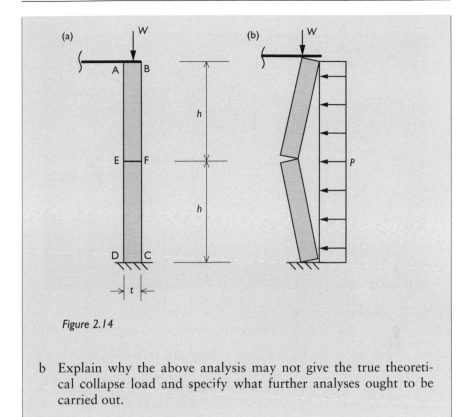

Figure 2.14

b Explain why the above analysis may not give the true theoretical collapse load and specify what further analyses ought to be
carried out.

2.3 Stone upon stone

Building by piling stone upon stone, whether dressed by masons or not, is
a form of construction which has evolved from ancient times and still has
its uses to-day. Such constructions must rely on the forces of gravity due to
self-weight to keep the stones (whether real or artificial) in place.

The Maidum Pyramid

The pyramids of Egypt were monumental constructions built to last, and last
they have – for well over 4000 years to date. Over 100 pyramids were built,
the first one being Zoser's Stepped Pyramid in the twenty-seventh century BC
and the largest being the 140 m high Great Pyramid of Cheops which is the
only one of the seven wonders of the ancient world to be still standing. No
other building reached a similar height till medieval European cathedrals
were built and it was almost 4500 years later, when steel became available,

Figure 2.15 The Maidum Pyramid.

before considerably taller buildings were constructed (Heinle and Leonhardt, 1989).

Some insight into pyramid construction techniques has been gained from the second pyramid to be built, which was at Maidum, 50 km south of Saqqara (the site of the stepped pyramid). Travellers in the eighteenth and nineteenth centuries noted its untypical shape and that it appeared to be built on a large mound (Fig. 2.15). The burial chamber was undecorated and there were signs of the pyramid being abandoned before completion. Furthermore the foundations were discovered below rather than on the top of the mound.

The conclusion from extensive archaeological study is that the present appearance of the pyramid is the result of it partially collapsing when nearing completion (Mendelssohn, 1974). The mound was created from up to 250,000 tons of debris from the collapse. A central core was surrounded by walls sloping inwards with a 75° 'batter' which retained infill material. This infill material has been described by Mendelssohn as 'badly squared stones' (Fig. 2.16). Evidence points to there having been ten such walls with infill between each which would have given the pyramid a stepped profile. Also, like the stepped pyramid, there is evidence that a decision to increase the size was made during construction. Then, unlike its predecessor, a further decision appears to have been made to include a mantle which would provide it with a smooth outer face as shown in Fig. 2.17. The mantle, however, which did not have a good foundation, fell down pulling some of the outer walls

Figure 2.16 Infill of 'badly squared' blocks.

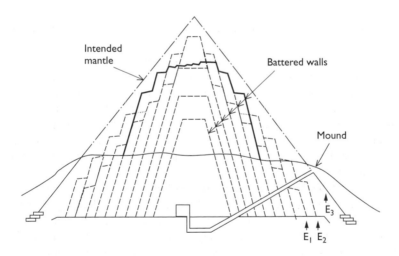

Figure 2.17 North–south section showing the building phases E_1, E_2 and E_3 and the shape of the remaining exposed structure (bold line).

with it. As a result, the construction was abandoned so that the central core and some walls now stand, surrounded by a mound of fallen masonry and wind-blown sand, probably hiding the remains of the ramps which would have been required to haul or roll the masonry blocks up to their required altitude (Fig. 2.15). It is thought that the bands of smooth stone now visible indicate where the external faces were meant to be at different times during the construction.

Lessons learnt at Maidum no doubt influenced subsequent pyramid construction techniques. In particular are the following:

- *Shape* It is thought that the Bent Pyramid at Dahshur, 20 km south of Saqqara (Fig. 2.18), was under construction at the time of the Maidum collapse. Hence the reduction in slope was a precaution to ensure that it did not suffer the same fate. The Red Pyramid (also at Dahshur) was constructed entirely at the reduced slope. After these two were completed successfully, slopes were increased again, indicating a restoration of confidence, probably because of improved techniques.
- *Form of construction* After the Maidum collapse, all stones were dressed and placed in layers which were either horizontal or slightly concave towards the apex.
- *Base* Limestone was replaced by granite for the lowest masonry layers in order to provide a firmer base.

Thus the longevity of the Cheops Pyramid and others can be attributed to the way builders of that era learnt from bitter experience.

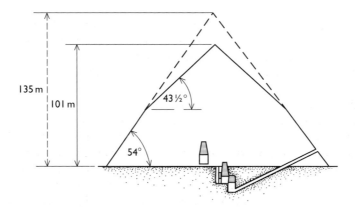

Figure 2.18 Profile of the Bent Pyramid.

Columns of the Parthenon

The Parthenon situated on the Acropolis, a hill dominating Athens, has been heralded as marking the culmination of Ancient Greek architectural development up to that time (Fig. 2.19). It was built as a temple to the god Athena with a plan area of 30.8 m × 69.5 m. The 46 outer columns, whilst acting as supports for the roof, were clearly meant to impress the viewer

Figure 2.19 Columns of the Parthenon.

with the strength, harmony and well-being of the whole edifice. These outer columns were 1.8 m in diameter at the base, 10.4 m tall and were fabricated from 'drums' of Pentelic marble which fitted together perfectly.

Construction would have been easier if the columns were all uniform circular cylinders and equally spaced, in which case a large number of identical drums could have been used. However, optical illusions were appreciated so, in order to give the visual appearance of regularity, uniformity and impressive height, the columns were given both a taper and a small bulge. They were given a small lean inwards towards the inner walls, their diameters were decreased slightly and their spacings increased next to the corners. The columns were also fluted (i.e. they had longitudinal grooves). It is not surprising that the Greeks were exceptional geometricians. They needed to be to achieve this degree of precision in construction.

The columns supported lintel beams which in turn carried the tiled roof structure as well as the ornate gable ends (Fig. 2.20). The weight of these supported items would have given the columns extra stability. Furthermore, because the roof structure was also supported on the internal walls (Fig. 2.21), lateral loading (e.g. due to wind) would have been reacted through the walls rather than via the columns (which may have been toppled by such forces). The Greek architects were not, however, confident enough about their design to rely entirely on gravity forces for stability. They did not use mortar, but they did fit iron cramps to hold adjacent lintels and also

Figure 2.20 End elevation of the Parthenon as built.

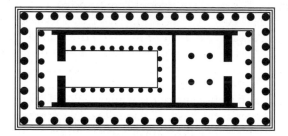

Figure 2.21 Plan of the Parthenon.

adjacent column drums together, and the cramps were covered with lead to avoid corrosion.

The Parthenon later changed use to a Christian church and then to a mosque. In 1687, during a war between Turkey and the Republic of Venice, a shell exploded gunpowder stored in it, which explains the present state of the building. The remaining structure has also been damaged by bursting of stonework where replacement iron cramps have expanded due to rusting. As a result, extensive remedial work has had to be carried out to stabilise the structure and make it safe for visitors. For more information see, for instance, Sprague de Camp (1970) and Coulton (1977).

Smeaton's Eddystone Lighthouse

In the eighteenth century sea trade was increasing substantially. However, sailing the seas was much more hazardous in the days of sailing ships than it is today. Position finding was from the sun and stars (when the sky was clear) and relied also on accurate time keeping. Sea charts were not so well developed and severe storms could drive ships into danger. The number of shipwrecks in the eighteenth century is witness to these dangers. Lighthouses were valuable navigation aids, not only on headlands and at entrances to ports, but also off-shore where reefs constituted particular dangers to shipping.

Smeaton had previously been an instrument maker and had only worked in civil engineering for three years when he was asked to rebuild the Eddystone Lighthouse. The Eddystone rocks, situated 14 miles off Plymouth, are hazards not only to ships entering Plymouth harbour but also to traffic in the English Channel. The first lighthouse was washed away in a storm in 1703 only four years after it was completed. The second lighthouse constructed of timber, but with large stones included for ballast, lasted from 1709 till it was destroyed by fire in 1755.

Since, apart from fire risk, the timber structure had given problems due to worm attack and rot, Smeaton decided that a masonry structure would be more suitable. Furthermore its self-weight would help to provide stability. However, blocks heavier than two tons were too heavy to transport and handle on site, so he faced the problem of ensuring that individual masonry blocks would not be washed away by storms. He could not rely on mortar which may have been washed out before it was properly set and he ruled out the use of cramps because of the time that would be required on site to fix them in place. Smeaton's solution was to use dovetailing to lock each masonry block into its neighbours in the same course of masonry (Fig. 2.22). Furthermore, the lowest courses were dovetailed into the bedrock. Above the level of the bedrock he used marble joggles and treenails (long wooden

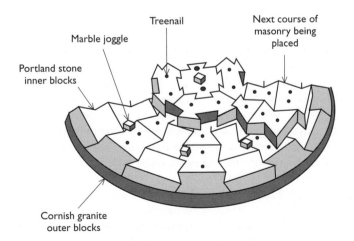

Figure 2.22 Use of dovetailing for the masonry blocks of the Eddystone Lighthouse.

Figure 2.23 Eddystone Lighthouse, 1759 at low water of a spring tide.

pins) to prevent one course of masonry sliding across the one below it. The structure was circular in plan (thus providing no corners to catch the force of the sea and providing uniformity of strength to resist forces from any direction). It was given as wide a base as possible and tapered upwards like a tree trunk (Fig. 2.23). The masonry was topped by a cornice in order to protect the light by deflecting any seas which might have swept up the face of the masonry. Cornish granite was used for the outer blocks of masonry, since the more easily available stone tended to be holed by shellfish when subjected to a marine environment.

The lighthouse which had a total height of 26 m was completed in 1759 and continued in use for over 120 years. When the sea started to undermine the rock on which Smeaton's structure was built, a new taller lighthouse was built on a nearby site. When it was dismantled, Smeaton's masonry was found to be in good condition apart from some of the horizontal joints near the top which had opened up on occasions when upward forces had been exerted by water hitting the cornice. Smeaton's design was so effective (despite his previous lack of experience in this field) that it formed the basis for many subsequent lighthouse designs (Upton, 1975; Skempton, 1981).

NB   John Smeaton 1724–1792 was the first person to call himself a civil engineer (to differentiate from a military engineer) and was a principal co-founder in 1771 of the Society of Civil Engineers. His many other achievements included improving the efficiency of water wheels and windmills.

Gravity dams *(Thomas, 1976)*

Since ancient times, dams have played a vital role in controlling water resources for irrigation, domestic use and power generation. With increasing populations in major cities and a demand for electricity, much larger dams have been constructed in the late nineteenth and the twentieth centuries than previously. A dam is a retaining wall subject to large lateral forces. Structural soundness and water tightness are both vital properties, because any sudden breach causes havoc and devastation as the released surge of water travels downstream (e.g. the rupture of the 60 m high St Francis concrete gravity dam in South California in 1926 resulted in 500 deaths). Where a dam is built in a narrow valley, it may be possible to resist water pressure by curving the dam wall so that it makes a horizontal arch with its abutments founded in the rock of the valley sides. However, the majority of dam sites are unsuitable for such treatment, in which case their stability against sliding and overturning needs to be ensured by making them broad in their base and also heavy enough. Gravity dams and embankment dams both satisfy these criteria.

Traditionally gravity dams were formed from a solid wall of masonry which provided both structural stability and water tightness. The problem has been to identify how to design such structures to safely carry the large forces imposed on them. In 1872 Rankine proposed his 'middle third rule'. To obey this rule, the resultant reaction across the line of any potential rupture needs to lie within the middle third of the particular cross-section (Fig. 2.24). If the resultant acts downstream of the middle third, cracks may open up on the upstream side and admit water under pressure. The uplift pressure of infiltrated water has itself an adverse effect on the stability against both overturning and sliding and is therefore to be avoided. The failure of the Bouzey Dam in France in 1895 (Fig. 2.25) where this rule had not been observed in its design, led to the Rankine proposal being given much more prominence. Although most gravity dams built since 1900 have been of concrete rather than masonry, the Rankine rule is still important because unreinforced concrete, like masonry, is a compression material.

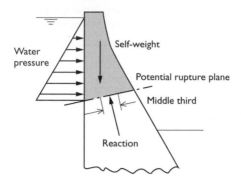

Figure 2.24 Forces acting on part of a dam illustrating the middle third rule.

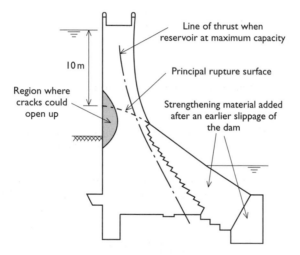

Figure 2.25 Cross-section of the Bouzey Dam at the time of the 1895 failure killing 150 people (Leliavsky, 1958). A 250 m length of the crest, 10 m deep was carried away.

Another problem faced in the design of gravity dams is the need to allow for possible uplift forces due to water seepage underneath. This would increase overturning moment and reduce normal pressures at the base, so increasing the likelihood of slippage. Measures taken to reduce these dangers are as follows:

- Constructing cut-off walls below the dam or grouting into the rock (even if it is impermeable, it may contain fissures).
- Ensuring very good bonding between the base and the underlying rock.

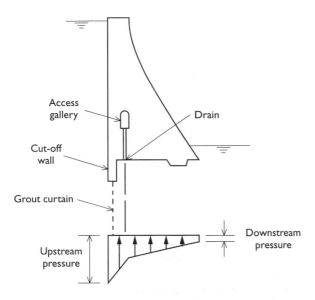

Figure 2.26 Assumption for uplift pressure under a gravity dam.

- Providing drains behind the cut-off wall (if there is one) to relieve water pressure. Drains need to be sufficiently large that they cannot easily get blocked.
- Installing monitoring devices to record pore water pressures under the dam, water flow from drains and other parameters such as angle of tilt of the dam structure. Figure 2.26 shows a typical cross-section and an assumed distribution for uplift pressures. Here most (up to 75%) of the total pressure drop is assumed to occur in front of the drain.

The use of buttresses to support a dam wall can lead to a saving of material (Fig. 2.27). However, because the weight of a dam is important for stability, some buttress dams have been constructed with sloping upstream faces. As a result the water pressure acting on it has a downwards component which helps the self-weight to resist slippage and overturning (Fig. 2.28).

The other principal category is embankment dams. These are made of locally available soil or rock compacted in order to reduce possible settlement. However, no attempt is usually made to make this material water proof. Instead an impermeable barrier is included. With earth dams this has normally been a puddle clay core (first introduced by Thomas Telford in 1820). However with rock fill dams, timber or steel plates, concrete or asphalt layers have been used, with polythene sheeting as a modern addition. For earthfill dams, typical slopes are 1 in 3 or 1 in 4 for the upstream face

Figure 2.27 Buttress Dam at Marmolada, Italy.

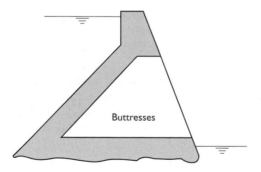

Figure 2.28 Cross-section of a form of Buttress Dam.

and 1 in 2 to 1 in 3 for the downstream face. However for rockfill dams, slopes may be as high as 1 in 1.5. Because embankment dams have a broader base, overturning and slippage of the whole dam on a horizontal plane do not tend to be risks. Yet, there is still the possibility of one or other of the embankments slipping if it is built at too steep a slope for the particular material.

Breakwaters

Inshore waters provide one of the most hostile environments for structures. This is particularly so where there is a long 'fetch' of open sea for a wind to create high waves, which may break and propel water forward with velocity as they move towards the shore. High local pressures will be developed when this water is suddenly halted by a solid object. The high density of water and its relative incompressibility are both factors which contribute to the high pressures. Such pressures can extend cracks and crevices or burst internal chambers. Storm force waves can also pick up stones and hurl them at any structure in the way.

Designers of breakwaters to protect harbours have had to address the problem of providing a structure which resists the onslaught of the waves, coupled also with problems of off-shore construction, where there may be currents and changes in sea level due to tides and where poor weather may frequently disrupt construction activity.

From ancient times (Fig. 2.29) to the present day, rocks have been used extensively in these structures. They have two major advantages:

a   They can be placed quickly either by vehicle, crane or barge when weather conditions are suitable (Fig. 2.30).
b   If heavy enough, their self-weight should prevent them being disturbed by wave action.

Some simple breakwaters may be built as rubble mounds with the heaviest rocks placed as 'armouring' on the seaward face. Where they are to be used also for mooring ships, a masonry or concrete construction is required to provide a quay wall and access to it. Even then, rocks are frequently used to form a foundation and also as armouring on the seaward face.

The selection of rocks suitable for marine works, particularly the heaviest ones required for armouring, tends to be carried out carefully taking account of properties such as density, porosity, angularity, hardness, resistance to marine creatures and presence of fractures. Frequent difficulties of obtaining supplies of suitable rocks near to proposed breakwaters has resulted in the development of many types of concrete blocks. Some of these blocks are meant to interlock in a semi-random manner whilst others are designed to fit together in a regular pattern (Fig. 2.31).

Thus piles of stones (or their substitutes) which rely on gravity forces for stability have been used throughout history to the present day, albeit for very different purposes.

(a)

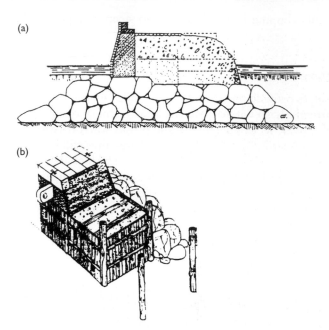

(b)

Figure 2.29 Ancient breakwaters: (a) cross-section of the main breakwater of Portus (Rome) constructed around 50 AD; (b) reconstruction of a Roman cast-in-situ concrete breakwater with rock armouring.

Figure 2.30 End tipping rock to repair a breach in a breakwater.

(a)

(b)

Figure 2.31 Concrete armouring units: (a) tetrapod armouring in Crete; (b) hexagonal 'seabees' armouring at Ardglass, North Ireland.

2.4   Mechanics of masonry arches

Terminology

Figure 2.32 shows a single span masonry arch bridge together with terminology used. There are, however, many variations in forms of construction even of those bridges which do not class as historic. For instance, arch barrels may or may not consist of wedge-shaped stone voussoirs. Furthermore, the barrel may not be constructed of the same material or be as thick as the outward impression given from the external face. The extrados may be stepped over the haunches and the intrados could be ribbed (Fig. 2.33).

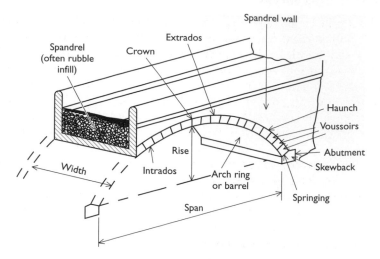

Figure 2.32 Cutaway view of a single span masonry arch.

Figure 2.33 About 600-year-old ribbed arches of Devil's Bridge at Kirby Lonsdale, Cumbria (Barbey, 1981).

The spandrel may or may not be as shown in Fig. 2.32. It could have internal walls or arches which themselves may or may not be surrounded by rubble infill. Sometimes concrete is included with the rubble to increase its strength particularly near to the abutments.

The thrust line concept

Initially it will be assumed that the superstructure which it is supporting imposes only vertical forces on an arch. In this case the forces acting on the ith voussoir will be the imposed and self-weight gravity forces, which may be resolved into a resultant vertical force W_i, plus interblock forces R_i and R_{i+1} at joints i and $i + 1$ respectively. Since only three forces act on the voussoir, they must be collinear (Fig. 2.34(a)). Alternatively, the interblock forces may be resolved into horizontal and vertical components as shown in Fig. 2.34(b), in which case horizontal equilibrium yields $H_i = H_{i+1}$. Since this applies to all voussoirs, the subscripts for H may be discarded. Thus to maintain equilibrium, a large horizontal component of force acts right across the arch pushing outwards against both abutments. If the lines of action of all the interblock forces are represented on one diagram, they form a continuous 'thrust line' stretching between the abutments as shown in Fig. 2.35(a). The thrust line concept helps to resolve two important questions regarding behaviour and design, namely,

a   What is the best shape for an arch?
b   Under what circumstances does an arch become unstable?

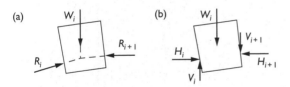

Figure 2.34 Forces acting on a voussoir: (a) in the form of resultants; (b) resolved into components.

Hooke's anagram

The thrust line concept can be attributed to Robert Hooke, who in 1675 published a latin anagram claiming that it solved the problem of the arch. Translated, this anagram reads 'As hangs a flexible chain, so, inverted, stand the touching pieces of an arch' (Hopkins, 1970). Hooke never did find the time, before he died, to elaborate on the problem he was addressing and how it was solved. Since the shape of a uniform chain when hanging is a catenary, it was first assumed to mean that the ideal shape for an arch should be a catenary. However, a flexible chain adopts different shapes for different loading regimes. From other writings of Hooke, it is clear that he really intended the statement to imply that the shape should depend on the loading

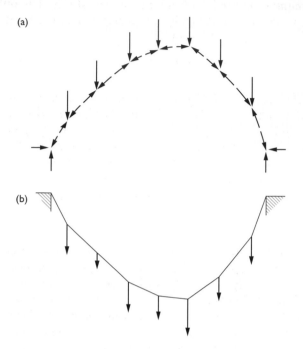

Figure 2.35 A thrust line and its inverse: (a) a thrust line for an arch subject to irregular vertical loading; (b) a 'funicular polygon' formed by a hanging chain subject to the same loading.

distribution to which the arch is subject (Fig. 2.35(b)) shows a hanging chain whose inverted geometry gives the thrust line shown in Fig. 2.35(a).

Determination of a feasible thrust line

For a thrust line to be feasible in the sense that equilibrium is possible:

a   It must intersect each joint in the arch between the intrados and the extrados (so that the joint is only required to carry compressive forces); and

b   Its inclination to the normal at each joint must be less than the angle of static friction (to avoid sliding).

Condition (a) will be satisfied if the thrust line is everywhere contained within the arch ring. (There would be negligible benefit from taking account of individual joint positions.) Condition (b) will normally be satisfied provided that the arch thickness/span ratio is small. It should also be noted that an arch in equilibrium will normally have different feasible thrust lines. An arch not in equilibrium will have no feasible thrust line. The only situation where there is one unique thrust line is when the arch is only just in equilibrium (i.e. it is on the point of instability).

Ex. 2.4

A bridge of span 6.55 m and rise 1.43 m has a segmental arch barrel of thickness 0.22 m and width 3.8 m. The bridge carries a level roadway, such that the average height of superstructure above the springing is 1.75 m. Assuming that all the gravity forces transfer directly to the arch ring and that the weight of the masonry and infill everywhere is $20 \, kN/m^3$, identify whether a feasible thrust line exists for the case of no traffic loading and estimate the forces requiring to be resisted by the abutments.

Answer

A convenient approximate procedure is to divide the bridge into vertical sections for the purpose of determining the gravity forces. Twenty sections are normally sufficient, but by making use of symmetry, only the left-hand side ten will need to be used in this case. Drawing out the profile of the bridge and measuring the depth of the ten vertical sections gives an estimate of the volumes of masonry in each section and hence their weights. Applying these as concentrated forces gives the loading shown in Fig. 2.36. For a thrust line to be close to the centre-line of the arch ring, its rise must be roughly the same as that of the arch itself. If the vertical cross-section of the bridge at the springing is AC and at the crown is DF and the thrust line meets these at B and E respectively, then AB will be set equal to FE with E being 1.43 m above B. It will also be assumed that the thrust line is horizontal at E so that no vertical force acts at this point. (This assumption makes the thrust line on the right-hand side a mirror image of that on the left-hand side.)

Consider equilibrium of the half bridge ACDF subject to the loads shown in Fig. 2.36. Equating moments about B gives directly $H = 158.5 \, kN$. If the height of B above A (and hence also E above F) is assumed to be 0.06 m, the positions of all the points on the thrust line may now be calculated. One way of doing this is to investigate the vertical equilibrium of the forces acting on points P_{10}, P_9, etc. in order to determine the vertical components of the forces in each section of the thrust line. Dividing these by H gives the slopes and hence vertical projections of each straight section of the thrust line, from which the thrust line profile shown in Fig. 2.36 may be derived. Whereas it does not coincide with the centre-line of the arch ring, it lies within its middle third region. The two conditions for stability of the

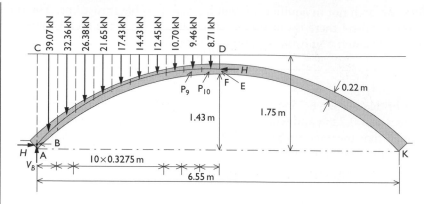

Figure 2.36 Determination of a feasible thrust line for a masonry arch.

arch mentioned above will therefore be met provided that the voussoir joints are at right angles to the arch centre-line and the coefficient of friction is not exceptionally low. Vertical equilibrium of the half bridge ACDF gives $V_B = 192.6\,\text{kN}$ which can be combined with H vectorially to give the abutment reaction as 249.5 kN inclined at 50.55° to the horizontal.

NB   This result is only a possible value for the abutment reaction. It would be necessary to investigate different values of H if it was required to identify the minimum or maximum possible reactions which satisfy equilibrium.

Considerations of shape

If weight distribution happened to be uniform across the span of a bridge, the thrust line would be parabolic. Furthermore, for a relatively shallow arch, the parabola and the circle almost coincide. Hence there is justification for the shape of a relatively shallow arch being the segment of a circle, provided that the weight distribution is approximately uniform. However, most masonry arches are likely to carry a larger weight of super-structure on the haunches than on the crown. In such cases, ideal shapes would be ones in which the curvature increases from the crown to the abutments.

In practice, where the spandrels contain rubble infill, they will exert horizontal as well as vertical pressures on the arch barrel. In that case, the thrust line concept is still useful. However, the horizontal component of force in the arch will no longer be constant across the span, but will reduce

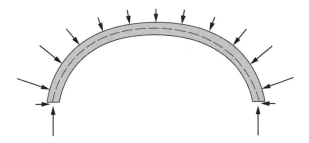

Figure 2.37 Equilibrium of an arch with varying curvature.

towards the abutments. If an arch is subject to a uniform normal pressure all the way round the extrados, a semi-circular shape would be ideal. However, the more usual case will involve pressures increasing from the crown to the abutments, thus making shapes with varying curvature more appropriate (Fig. 2.37). Arches with this type of shape have been called 'basket-handled'.

Instability

Apart from elastic instability which may occur with very shallow and thin arch barrels, there are principally two ways for a well-shaped masonry arch bridge to become unstable. One way is due to movement of an abutment and the other is when changes to the loading (e.g. because of traffic) provoke a collapse mechanism. In both cases, the thrust line concept can be used to investigate when instability will occur.

If one of the abutments of the bridge shown in Fig. 2.36 is unable to carry the horizontal component of force H, the arch will try to adopt a thrust line with a higher rise than the one shown. The limiting equilibrium situation is when the thrust line passes through A and K and also touches the extrados near to L and M with H being approximately 140 kN instead of the previous estimate of 158.5 kN (Fig. 2.38(a)). Where the thrust line touches the edge of the arch ring, joints can open up acting as hinges to form either of the two mechanisms shown in Fig. 2.38(b) or (c) or indeed the symmetrical combination of these shown in Fig. 2.38(d).

Figure 2.39 shows how a thrust line may distort due to the imposition of a large concentrated load to such an extent that the arch is on the point of instability. Also shown is the four hinge collapse mechanism which would be induced. Figure 2.40 shows a similar type of mechanism formed during a load test on an actual masonry arch reported by Page (1993). Normally a failure of this sort would be sudden and catastrophic. This bridge was held in the partially collapsed state because the loading was applied by jacks rather

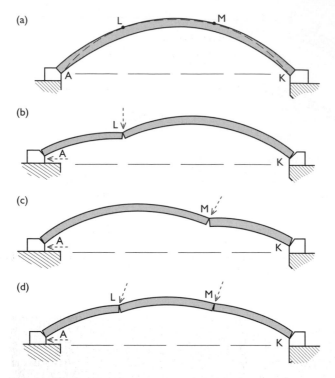

Figure 2.38 Arch instability due to slip of abutment at A: (a) limiting position for the thrust line; (b) a collapse mechanism; (c) an alternative collapse mechanism; (d) a combined collapse mechanism.

Note
Due to symmetry, either or both of the independent mechanisms (b) and (c) or any combination of them can occur at the same load. Arrows indicate the direction of movement of the joints.

than by using heavy weights. Load tests on masonry arches have shown that they are usually considerably stronger than this simple theory would imply.

In complete contrast, Fig. 2.41 shows a 'wobbly arch' in which the voussoirs have curved surfaces which roll against each other. This is a very impractical sort of arch to use in the field but, unlike conventional arches, it displays the actual position of the thrust line which can be identified from the points of contact.

Spandrel walls

Spandrel walls tend to be a particularly vulnerable part of masonry arch bridges. One danger is at road level where impacts by speeding heavy vehicles

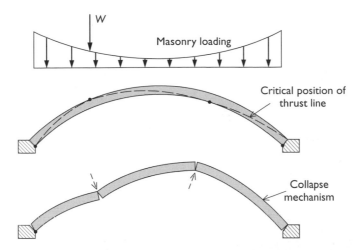

W

Masonry loading

Critical position of
thrust line

Collapse
mechanism

Figure 2.39 Arch instability caused by a large concentrated load. Arrows indicate direction
of movement of the joints in the collapse mechanism.

Figure 2.40 Load test to destruction on the Prestwood brick arch bridge having similar
dimensions to that of Ex. 2.4. A line load has been applied across the width of
the bridge at the 3/4 span position.

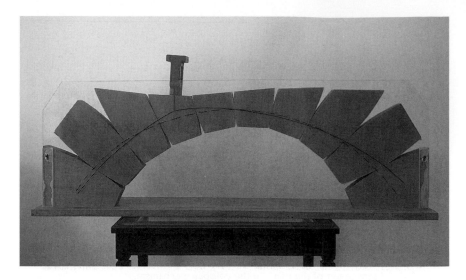

Figure 2.41 A wobbly arch showing the position of the thrust line before loading (dotted line) and after loading (full line).

can do much more damage than the slower moving vehicles of previous centuries. Another more insidious danger (to the bridge itself) emanates from below road level. This is because infill will press outwards on the spandrel walls. It is particularly important that the infill is kept dry where freeze–thaw action in winter can cause it to expand. Expansion of the infill causes the spandrel walls to break away from the main part of the arch barrel. The piers of many old multispan bridges were built up to roadway level above the cutwaters. Whereas this may have been primarily to provide a refuge for pedestrians when horses and carriages were passing by, the masons would, no doubt, also have been aware of the beneficial structural effect it would have had by buttressing the spandrel walls against the outward forces they would be subjected to (Fig. 2.42).

Q.2.6
Figure 2.43 shows a shallow masonry arch comprising two straight pitches rising to a clear height of h above the springing. The depth of masonry is b. Show that, for the arch to sustain a uniformly distributed load acting over the whole span, it is necessary for b to be greater than $h/4$.

Figure 2.42 The 15 span old bridge at Berwick-on-Tweed, Northumberland approximately 400 years old.

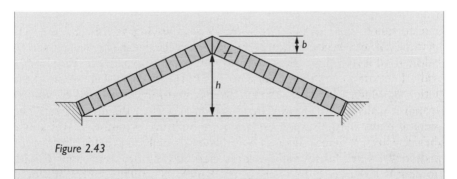

Figure 2.43

Q.2.7
Different equilibrating thrust lines can be drawn for any particular set of parallel forces.

a Identify three independent ways in which the position of the thrust line may be varied.
b Identify a possible collapse mechanism for a flat arch such as the one shown in Fig. 2.44 subject to uniformly distributed loading (flat arches are sometimes seen over windows and doorways in buildings).

Figure 2.44

Q.2.8

The segmental masonry arch shown in Fig. 2.45 has radii for the intrados and extrados of 3 m and 3.4 m respectively and a slope at the abutments of 45°. Determine on which parts of the span a very large concentrated load may be sustained, if the weight of the bridge is negligible compared with that of the concentrated load (it may be assumed that neither do the voussoirs crush, nor do the joints slip).

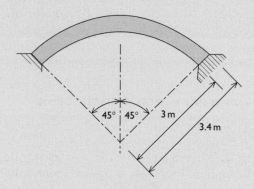

Figure 2.45

Q.2.9

a Assuming that a masonry arch has voussoirs which do not crush and also firm abutments, identify factors which, according to the thrust line theory, improve its capacity to carry heavy traffic loading.

b Where an arch is supporting spandrel walls infilled with rubble, specify reasons why it may be able to carry heavier traffic loads than is indicated by considering the arch on its own.

Q.2.10   Observation question

Inspect a masonry arch bridge if one exists near to you. (If one is not available, find information about one in literature.) Answer the following questions:

- What is the shape of the arch and its approximate dimensions?
- What are the materials of construction, particularly the arch barrel?
- Are there any special features such as a humpback in the road, skew geometry, etc.?
- Are there any signs of deterioration such as cracking or fall out of masonry?
- Are there any weight restrictions or signs of repair, strengthening or modification? If so, can you identify the reasons for these changes?
- Can you find out when it was built?

Hence, if you owned the bridge, decide what you would like to ask a structural engineer about its condition and suitability for its purpose.

2.5 Developments in masonry arch construction

Construction using centring

Unlike corbelled arches (Fig. 2.46), voussoir or 'true' arches require support during construction. A support (normally a timber frame called the 'centring') is first constructed so that the arch barrel can be constructed on it. The centring is not normally removed till both the arch barrel and the superstructure are in place. This process is facilitated by the inclusion of wedges in the supports for the centring. Removal of these wedges cause the centring to drop evenly away allowing the arch barrel to develop the horizontal component of thrust required for the bridge to be self-supporting. Figure 2.47 shows centring with wedges included near to the abutments. It also shows masonry courses between adjacent arches over the piers and a prediction for the thrust line made by Thomas Young in the early nineteenth century.

Roman arches

True arches appear to have originated in the ancient cities between the Tigris and the Euphrates (Shirley-Smith, 1964). Although the names 'extrados' and 'intrados' come from Greek, the ancient Greeks did not favour arches in their monumental buildings. The Romans, on the other hand, appreciated their potential and used them extensively in bridges, aqueducts and

Figure 2.46 A Mayan corbelled arch at Chichenitza, Mexico.

Figure 2.47 Centring for Waterloo Bridge, London.

buildings. They even made triumphal arches. Roman arches were nearly all semi-circular in shape and, where a number of arches were required in an aqueduct or building such as the Coliseum, the ratio of span to pier width was generally between three and four.

The most magnificent remaining example of Roman arch construction is the three tiers of arches forming the 275 m long Pont du Gard (Fig. 2.48). This

Figure 2.48 The Pont du Gard.

aqueduct carried water destined for the nearby city of Nimes in a channel of dimension $1.2\,m \times 1.8\,m$ at a height of $49\,m$ above the valley of the river Gard. Mortar was only used in the top tier (no doubt to provide a water tight conduit). The Romans would have had problems coping with the horizontal component of thrust, which acted on a pier when the centring for the first of two arches it was to support was removed. Particularly when the pier was tall, it would have been difficult to prevent it from toppling, even if the pier was shored with timber or if the centring for the adjacent arch had already been put in place. This is the probable reason why the ratio of span/pier width was never large and also why tall aqueducts like the Pont du Gard were constructed in tiers. The simple geometry of the semi-circular arch resulted in simple layouts, with all the voussoirs being of identical shape, which nineteenth and twentieth century bridge builders also found attractive (Fig. 2.49).

Figure 2.49 Part of the 28 span Royal Border Bridge at Berwick-on-Tweed carrying the east coast main railway line between England and Scotland designed by Robert Stephenson and completed in 1850.

Old London Bridge *(Shirley Smith, 1964)*

In 1176, Peter Colechurch started to construct a stone bridge at London to replace timber bridges which had experienced problems due to flooding and fire. The bridge, having 19 pointed arches and a drawbridge, took 33 years to complete, building one arch at a time. The idea of the pointed arch came from Persian, Moslem and Byzantine influence. Less precision was required in dressing the stones and the horizontal component of thrust acting on the abutments was reduced. One of Colechurch's main problems was the founding of piers in the tidal waters of the Thames estuary. A plan of the bridge in the fifteenth century (Fig. 2.50) shows broad piers with cutwaters to facilitate water flow, surrounded by 'starlings' (i.e. platforms constructed by driving timber piles into the river-bed). These piles not only formed coffer-dams within which the pier foundations were constructed, but also were retained and backfilled to protect the piers from scour. During refurbishments, further piles were sometimes placed round the existing ones making the starlings even larger. The blockage in the river due to the piers alone was 44%, but when the water level was below the top of the starlings, the blockage was 75% (Fig. 2.51). This had the effect of producing differences in water level between upstream and downstream of as much as 1.5 m. As a result, there was a saying that 'wise men go over and only fools go under'. Furthermore

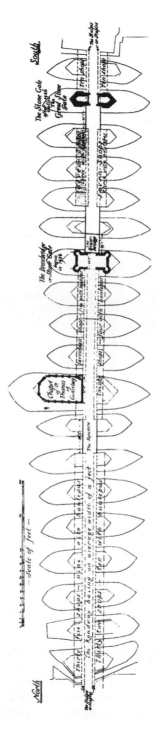

Figure 2.50 A plan of Old London Bridge based on measurements made in the fifteenth century.

Figure 2.51 The chapel of St Thomas of Canterbury on Old London Bridge from a reconstruction in Home (1931).

the tidal range was so reduced upstream that the Thames sometimes froze over in winter. Even so, the bridge remained in use for over 600 years.

The Ponte Vecchio

In 1333 Florence, the most advanced manufacturing city in Europe at the time, urgently needed a bridge after all those over the river Arno had been washed away by a flood. Taddeo Gaddi is the person attributed with the bold step of designing the Ponte Vecchio, a masonry bridge having segmental arches with rise to span ratios of up to 1 : 7.5 instead of the more conventional 1 : 2 for semi-circular arches (Fig. 2.52). It is clear that the larger horizontal component of thrust (whose magnitude would have been unknown at the time) was adequately catered for. Indeed it has survived for over 600 years

Figure 2.52 The central arch of the Ponte Vecchio.

despite (or maybe because of) supporting shops as well as traffic loading. Shallow arches have the particular advantage of enabling larger spans to be used. The Ponte Vecchio has just three spans of approximately 26 m each, which makes it less of a blockage to floodwaters than previous bridges were. Shallow arches have also enabled the roadway to be set at a lower level relative to the clearance height required below the arches.

The honour of building the first shallow arch has been attributed to the Chinese, on account of the An Ji Bridge in Zhao Xian, Hebei Province which has an amazing single 37 m span masonry segmental arch with a rise of 7 m dating back to the seventh century (Brown, 1993). This Chinese bridge may not have been known about in Florence, and in any case, it would not have helped Taddeo Gaddi to design the piers in the river which needed to carry large horizontal forces. After the success of the Ponte Vecchio, innovation with different shallow arch shapes started to occur in Italy and elsewhere, but for more than 400 years, none were built with a lower rise/span ratio than 1 : 7.5.

Jean-Rodolphe Perronet

Perronet, who in 1747 was appointed as director of the newly created Ecole des Ponts et Chaussées (the first engineering school in the world), has been regarded as the father of modern bridge building. Hopkins (1970) states 'Having investigated a project thoroughly, he was uncompromising in his reliance on theory and thus freed arch design and construction from the empirical bonds which had so far prevented their full development'. He strove to make bridges more slender and shallower than others would dare. However, his greatest innovation was achieved with a crossing of the Seine at Neuilly using five arches each of span 43.6 m and piers of width just 4.43 m. He achieved such slender piers by erecting all the arches simultaneously. Then, in a spectacular event attended by King Louis XV, he had all the centring fall away into the river at the same time. By making all the arches similar to each other and take up load at the same time, he had ensured that the slender piers were not subjected to any out of balance horizontal components of thrust during construction.

It is interesting to note that, unlike earlier bridges, his design was no longer fail-safe. If one arch later collapsed (e.g. due to flood damage or warfare), piers would be subjected to horizontal loading which they had not been designed for and progressive collapse of the whole bridge could ensue. The advantage, apart from speeding up the construction, was that narrowing of the piers meant that the bridge offered less obstruction to flood water, thus reducing the risk of collapse.

Another feature of the bridge was the use of 'cornes de vache' or cow's horns, rarely used previously, to speed the flow of flood water through the bridge by avoiding 90° angles in the masonry of the haunches (Fig. 2.53).

Figure 2.53 Cornes de vache on the Pont de Neuilly.

Trial and error

In parallel with the work of Perronet, William Edwards was attempting to bridge the river Taff in Wales. His contract required him to provide a bridge that would stand for seven years. Because his first multi-arch bridge was washed away by flood after less than three years, he resolved to create a single span crossing. However, the centring of the second bridge was carried away by floodwaters when it was under construction. For his third attempt, he used stronger centring, but when it had been removed and superstructure was being placed on the haunches, collapse occurred with the haunches moving inwards and the crown rising. Observing this Edwards was able, for his fourth attempt, to correct the weight distribution by using open spandrels. The resulting bridge which still stands at Pontypridd has an elegant segmental span of 42.7 m with a rise/span ratio of 1 : 4 and a thin arch barrel (Fig. 2.54). These setbacks illustrate some of the pitfalls which innovators faced with masonry bridges. The final solution would have interested Robert Hooke greatly, had he still been alive.

An arch never sleeps

The three span St John's Vale arch bridge, built in 1849, which was made entirely of brick except for rubble infill, required to be demolished to make way for station platform extensions (Parker, 1993). Because explosives were ruled out for safety reasons, a scheme was devised in which excavators would lower the superstructure to extrados level and then break out the arches starting with the central one. Because the time for possession of the railway (54 hours over one weekend) was short, the schedule involved men placing sleeper mats under the arches to protect the railway lines whilst the

Figure 2.54 Edward's 1756 bridge at Pontypridd (from *Proc. ICE* Vol. 5, 1846 after p. 476).

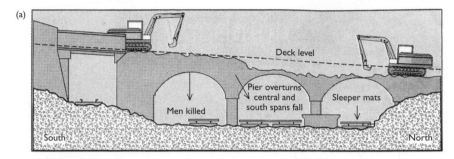

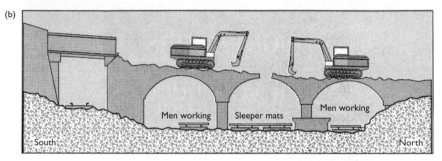

Figure 2.55 A flawed 1992 demolition procedure: (a) situation when collapse occurred; (b) proposed scheme.

excavators were at work. Figure 2.55(a) shows how the 1992 demolition was proceeding when an unanticipated collapse killed two men. In fact, if the excavators had been working to the intended schedule, the central arch would have been broken whilst men were under both side arches, as shown in Fig. 2.55(b), making the tragedy even worse.

If the people who devised the schedule had been aware of the historical development of masonry arch bridges, they must surely have considered whether the arches were built on the Perronet principle and were supporting each other. So, even if warnings about progressive collapse in British Standard 6187, Codes of Practice for Demolition, and also guidance notes from the Health and Safety Executive had been overlooked, they should have been alive to the danger involved with their schedule. The old proverb 'An arch never sleeps' indicates that arches should always be treated with respect.

The masonry legacy

This chapter has given just a small glimpse of the achievements of masons and builders over the centuries. Although masonry has been largely replaced as the primary building material for large structures, it is still in use for domestic

Figure 2.56 A section of a bay in Amiens Cathedral erected in the 1230s showing high masonry vaulting and flying buttresses (Sandström, 1970).

buildings and for wall cladding of framed buildings, for instance. Furthermore, there is a need to maintain and sometimes modify masonry structures which are vital to society. Great efforts should be made to preserve those which are truly historic. Not least amongst these are the high Gothic cathedrals (Fig. 2.56). They are marvellous examples of what could be achieved building with a primary material suitable only to carry compression forces.

Chapter 3

Internal forces

The discussion of equilibrium is extended to discover what can be found out about the forces and load paths within bodies. Knowledge of the internal forces is required whenever strength requirements are under consideration, and also when predicting deflections and their influence on performance.

On completion of this chapter you should be able to do the following:

- Use the method of sections to determine internal forces of statically determinate systems both in 2D and 3D.
- Appreciate the significance of principal stresses and how load paths may form within a structure.
- Determine bending moment diagrams for three-pinned arches and portal frames.
- Determine stresses in cylindrical pressure vessels.
- Appreciate the importance of cross-sectional shape on torsional characteristics of members, being able to estimate stresses in some cases.

3.1 The method of sections

Existence of internal forces

In order to investigate what is happening inside a body in equilibrium, consider a part of the body separated from the rest by a cutting line known as a 'section'. It is evident that the body part is not accelerating and so, like the complete body, must be in equilibrium. To achieve this equilibrium, it is normally necessary for additional forces to act on the body part from the adjoining part(s) across the section. When distributed across the interface, the internal forces are called 'stresses'. They consist of normal components, 'direct stresses' and tangential components, 'shear stresses'. Whereas equilibrium conditions alone will not enable stress distributions to be determined, they can be used to determine or provide information on stress resultants, otherwise known as internal forces.

Ex. 3.1

For the bracket shown in Fig. 3.1, determine the internal forces acting across the section AB having a thickness 20 mm, and show that these forces maintain equilibrium of both left-hand and right-hand parts of the bracket.

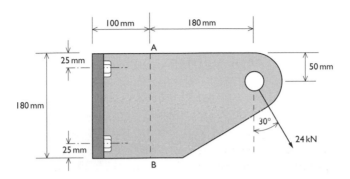

Figure 3.1 A loaded bracket.

Answer

Consider the equilibrium of the right-hand part of the bracket with internal forces P, S and M specified as acting at the section AB as shown in Fig. 3.2. Resolving horizontally gives $P = 12\,\text{kN}$, resolving vertically gives $S = -20.78\,\text{kN}$ and taking moments about C gives $M = -4221\,\text{Nm}$.

To consider equilibrium of the left-hand part of the bracket, it is first necessary to estimate the external reactions. Assume that the upper bolt carries a horizontal reaction, H_D, to stop the bracket pulling away from its support. Also, assume that the reaction to stop the bracket moving to the left, being bearing pressure, has a resultant, H_E, situated 5 mm above the bottom of the bracket. Vertical reactions

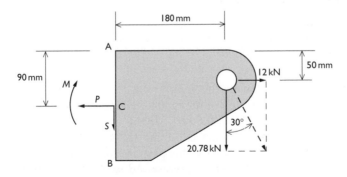

Figure 3.2 Forces acting on right-hand part of bracket.

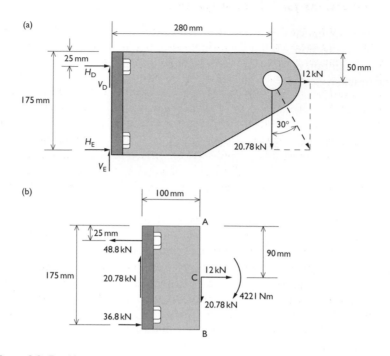

Figure 3.3 Equilibrium investigation for left-hand part of bracket: (a) determination of external reactions from overall equilibrium; (b) forces acting on left-hand part of bracket.

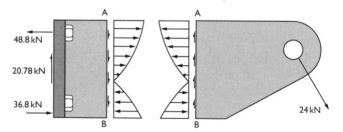

Figure 3.4 Exploded view of bracket showing possible stress distribution at section AB.

V_D and V_E will occur at the support face as shown in Fig. 3.3(a). Hence, overall equilibrium of the bracket yields $V_D + V_E = 20.78$ kN, $H_D = -8.80$ kN, $H_E = 36.80$ kN. If the internal forces acting on the right-hand part are reacted on the left-hand part as shown in Fig. 3.3(b), the forces acting on this part are found also to be in equilibrium. The stresses on AB to correspond with these internal forces could, for instance, have a distribution of the form shown in Fig. 3.4.

Sign conventions for 2D force systems

Sign conventions are necessary for forces (and moments), for instance, in order to specify external loading when inputting to a computer or for specifying reactions when the direction is initially unknown. For 2D analyses in which horizontal and vertical force components are to be used, the convention could be as shown in Fig. 3.5(a). Such a convention would be meaningless, however, for specifying internal forces. Each internal force is, in practice, a pair of equal and opposite forces acting on opposite faces of a section. Alternatively, if two closely spaced sections are made, the opposing forces can be considered to be applied to an incremental length of a structure. The sign conventions used in this book are shown in this way in Fig. 3.5(b). (the unknown external and internal forces in Figs 3.2 and 3.3(a) use this convention).

Care has to be taken when analysing structures or members of structures which are at different inclinations to the horizontal. Rotation of Fig. 3.5(b) either clockwise or anticlockwise through 90° in order to represent a vertical member would give the same conventions for axial and shear forces, but the bending moment convention would be different. Ambiguity is only avoided by giving the axis of a member a positive direction. With the axis direction to the right as shown in Fig. 3.5(b), the tension side due to a positive bending moment will always be to the right (starboard) to someone facing in the positive direction.

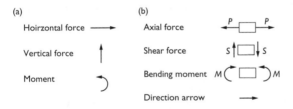

Figure 3.5 Sign conventions for 2D problems: (a) forces; (b) internal forces.

Resultants of distributed loading

It is important that resultants of distributed loading are only determined after sectioning has taken place, so that forces are not allocated to the wrong part of the divided structure.

Ex. 3.2
Neglecting the thickness of the members of the bent cantilever shown in Fig. 3.6, determine the internal forces acting at B and C in the

horizontal member and at D and E in the inclined member. Check the equilibrium of joint C/D.

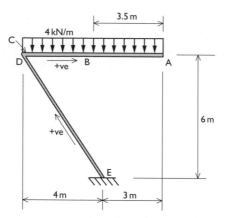

Figure 3.6 A bent cantilever with positive member direction indicated.

Answer
Figure 3.7 shows outboard parts of the bent cantilever with sections at B, C, D and E respectively and including the internal force variables specified in conformity with the sign convention and specified member direction arrows. All of the internal forces may be obtained using direct methods applied to the appropriate equilibrium diagram. Thus for

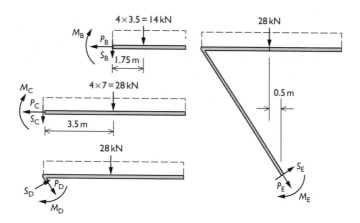

Figure 3.7 Forces acting on parts of the bent cantilever.

the equilibrium of part AB, resolving forces perpendicular to S_B gives $P_B = 0$, and perpendicular to P_B gives $S_B = -14\,\text{kN}$. Taking moments about the intersection of S_B and P_B gives $M_B = -24.5\,\text{kNm}$. Similarly for the other sections: $P_C = 0$, $S_C = -28\,\text{kN}$, $M_C = -98\,\text{kNm}$; $P_D = -28 \sin 56.31° = -23.30\,\text{kN}$, $S_D = 28 \cos 56.31° = 15.53\,\text{kN}$, $M_D = -98\,\text{kNm}$; $P_E = -23.30\,\text{kN}$, $S_E = 15.53\,\text{kN}$, $M_E = 14\,\text{kNm}$. The internal forces acting on joint B/C, shown in Fig. 3.8, are found to be in equilibrium.

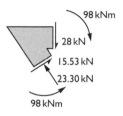

98 kNm

28 kN

15.53 kN

23.30 kN

98 kNm

Figure 3.8 Equilibrium of joint C/D of the bent cantilever.

Plane statically determinate bar type structures

It has been possible to obtain any required internal forces using the method of sections for both of the given examples because they are 'statically determinate'. A structure is externally statically determinate if there are just sufficient restraints to prevent body movements (see Section 1.7). In that case, all the reactions can be obtained from equilibrium of the whole structure. If, therefore, the structure is sectioned into two parts, the only unknowns will be the internal forces at the particular section. Where a statically determinate bar type structure does not contain loops, it may be sectioned anywhere in such a way that the three unknown internal forces may be obtained from equilibrium of one or other of the two parts (see Fig. 3.9). On the other hand, if there are more than sufficient restraints or if a structure contains one or more loops, it is not statically determinate (see Fig. 3.10).

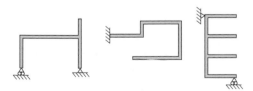

Figure 3.9 Examples of statically determinate bar structures.

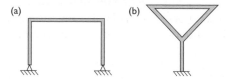

Figure 3.10 Examples of redundant structures: (a) with one external redundant force; (b) with three internal redundant forces.

Figure 3.11 Statically determinate bar structures which include internal hinges.

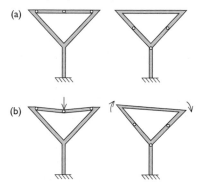

Figure 3.12 Structures with hinges placed wrongly for statical determinacy: (a) position of hinges; (b) mechanisms created showing direction of movement.

The number of unknown forces in excess of the number which can be obtained from equilibrium are called 'redundancies'. However, the presence of a hinge within a member of a structure provides an extra equation, namely that the bending moment at the hinge position must be zero. Therefore, the introduction of sufficient internal hinges to match the number of forces that would otherwise be redundant, will produce a statically determinate structure provided that they are placed suitably. Figures 3.11 and 3.12(a) show examples of suitably and unsuitably placed hinges, respectively. An unsuitably placed hinge creates a mechanism (allowing the structure to move without distorting the members) (see Fig. 3.12(b)).

Q.3.1

Determine the internal forces acting at the sections A and B for the crane hook shown in Fig. 3.13 and indicate their directions of action on a diagram.

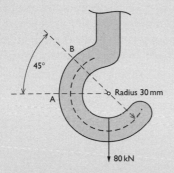

45°

B

A

Radius 30 mm

80 kN

Figure 3.13

Q.3.2

For the structure shown in Fig. 3.14, determine the internal forces acting at section B, C, D and E using the standard sign convention with

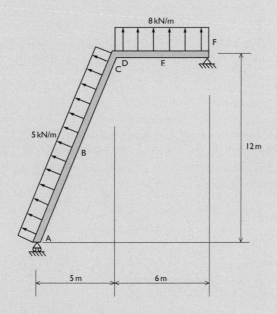

8 kN/m

F

D
C   E

5 kN/m

B

12 m

A

5 m   6 m

Figure 3.14

member directions A→C and D→F (B and E are the centre sections of members AC and DF, respectively).

Q.3.3

For the canopy structure shown in Fig. 3.15, determine the support reactions at F. Hence determine the internal forces acting at B, C and E, adjacent to the central joint, from the equilibrium of the three members AB, CD and EF. Use the results to check that the central joint is in equilibrium.

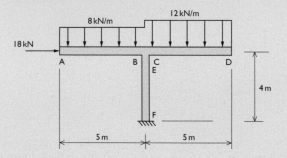

Figure 3.15

Q.3.4

Which of the structures shown in Fig. 3.16 can be classified as statically determinate? For the others, indicate a way of making them statically determinate by introducing or moving internal hinges.

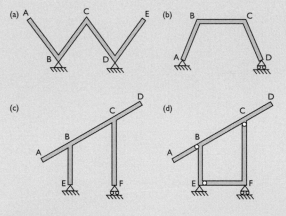

Figure 3.16

3.2   Stresses acting on an incremental element

Complementary shear stresses

Whereas the method of sections provides information on what happens with internal forces on a macroscopic scale, it is also possible to use equilibrium to study what happens microscopically. Consider a rectangle of material of dimensions $dx \times dy$, with unit thickness, subject to the 2D stress system shown in Fig. 3.17. Here the conventional terminology of σ for direct stress and τ for shear stress has been used. Equilibrium of the element in the x and y directions yield no useful information because incremental changes to stresses have been ignored. However, moment equilibrium of the element gives

$$(\tau_{xy} + \tau_{yx})\, dx\, dy = 0$$

The result $\tau_{yx} = -\tau_{xy}$ implies that shear stresses on planes at right angles are complementary to each other.

Rotation of planes

To consider what happens on other planes, consider the equilibrium of a triangle of material formed from half of the rectangle shown in Fig. 3.17. Let ds be the length of the diagonal section and θ be the angle between the diagonal and the y axis. With σ and τ as the stresses acting on the diagonal

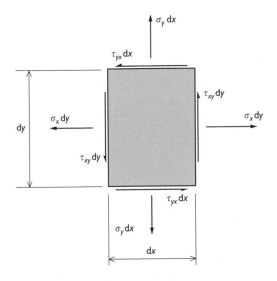

Figure 3.17 Forces acting on an incremental rectangle.

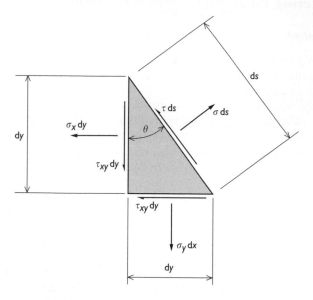

Figure 3.18 Forces acting on an incremental triangle.

face, the forces acting on the triangle are as shown in Fig. 3.18. Specifying direct equations for σ and τ, substituting $dx = \sin \theta \, ds$ and $dy = \cos \theta \, ds$ and cancelling ds gives

$$\left.\begin{aligned}
\sigma &= \sigma_x \cos^2 \theta + \sigma_y \sin^2 \theta + 2\tau_{xy} \sin \theta \cos \theta \\
\tau &= (\sigma_y - \sigma_x) \sin \theta \cos \theta + \tau_{xy}(\cos^2 \theta - \sin^2 \theta)
\end{aligned}\right\} \tag{3.1}$$

Readers who are adept at trigonometry might like to see how these equations can be transformed by using double angle formulae and the substitutions $\tau_{xy} = R \sin 2\alpha$ and $\frac{1}{2}(\sigma_x - \sigma_y) = R \cos 2\alpha$ into

$$\sigma = \tfrac{1}{2}(\sigma_x + \sigma_y) + R \cos 2(\alpha - \theta) \quad \text{and} \quad \tau = R \sin 2(\alpha - \theta) \tag{3.2}$$

where

$$\left.\begin{aligned}
&\tan 2\alpha = 2\tau_{xy}/(\sigma_x - \sigma_y) \\
&\text{and, in order to make } \sin^2 2\alpha + \cos^2 2\alpha = 1, \\
&R^2 = \left[\tfrac{1}{2}\left(\sigma_x - \sigma_y\right)\right]^2 + \tau_{xy}^2
\end{aligned}\right\} \tag{3.3}$$

When θ is varied through $180°$, the maximum and minimum direct and shear stresses occur as specified in Table 3.1.

Table 3.1 Planes on which critical stresses occur

Critical stress	$\alpha - \theta$	σ	τ
σ_{max}	0° or 180°	$\frac{1}{2}(\sigma_x + \sigma_y) + R$	0
τ_{max}	45° or 225°	$\frac{1}{2}(\sigma_x + \sigma_y)$	R
σ_{min}	90° or 270°	$\frac{1}{2}(\sigma_x + \sigma_y) - R$	0
τ_{min}	135° or 315°	$\frac{1}{2}(\sigma_x + \sigma_y)$	$-R$

Ex. 3.3
Stresses of $\sigma_x = 110$, $\sigma_y = -45$ and $\tau_{xy} = 50$, in N/mm^2 units, occur at a point in a material. Determine the maximum and minimum direct and shear stresses and the planes on which they act.

Answer
Using N/mm^2 units throughout, $\frac{1}{2}(\sigma_x + \sigma_y) = 32.5$, $\frac{1}{2}(\sigma_x - \sigma_y) = 77.5$, $R^2 = 77.5^2 + 50^2$ giving $R = 92.23$. Hence, from Eq. (3.2), the critical stresses are

$$\sigma_{max}(= \sigma_1) = 32.5 + 92.23 = 124.73$$

$$\sigma_{min}(= \sigma_2) = 32.5 - 92.23 = -59.73$$

$$\tau_{max, min} = \pm 92.23$$

Furthermore, from Eq. 3.3, $\tan 2\alpha = 50/77.5$ giving $\alpha = 16.41°$, hence using Table 3.1, the equivalent stress systems shown in Fig. 3.19 are obtained.

Principal stresses and planes

The following are important rules which arise from the above study:

- The maximum and minimum direct stresses, called 'principal stresses', occur on planes at right angles to each other. These planes are called 'principal planes'.
- No shear stresses act on the principal planes.
- Maximum shear stresses occur on planes at 45° to the principal planes.
- The maximum shear stress is half of the difference between the principal stresses.

For some materials, particularly those that are brittle, the maximum tensile stress is the one which is of most concern. However, for ductile materials,

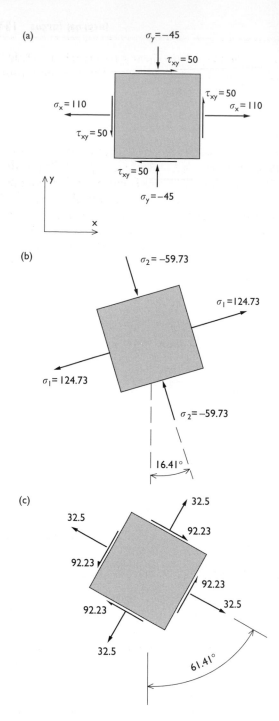

Figure 3.19 Equivalent stress systems (units $= N/mm^2$): (a) stresses on x and y planes; (b) principal stresses; (c) planes for maximum shear stresses.

distortion starts to occur when the maximum shear stress reaches a critical value. Hence, equivalent stress systems such as in Fig. 3.19(c) are important in predicting load carrying capabilities of structures.

Mohr's circle of stress

In the late nineteenth century, Otto Mohr developed a convenient graphical technique for solving problems involving stress systems at a point. By plotting a graph of σ against τ as θ varies, he discovered that a circle is mapped out with centre at $\sigma = \frac{1}{2}(\sigma_x + \sigma_y)$ and radius R as specified in Eq. (3.3). Furthermore, the angle between the radius containing (σ, τ) and the σ axis is $2(\alpha - \theta)$ where α is defined in Eq. (3.3). As the angle of the plane is rotated, the point representing its stresses rotates round the Mohr's circle in the opposite sense at twice the angular rotation. Figure 3.20 shows a Mohr's circle representing Eqs (3.2) and (3.3).

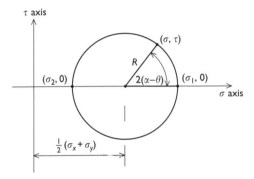

Figure 3.20 The Mohr's circle of stress.

Ex. 3.4
Repeat the investigation of Ex. 3.3 using the Mohr's circle concept.

Answer
Two known points on the Mohr's circle are $(\sigma_x, \tau_{xy}) = (110, 50)$ and $(-\sigma_y, -\tau_{xy}) = (-45, -50)$. Note that shear stresses in the Mohr's circle have been defined as positive anticlockwise, so the shear stress in the y plane must be given a negative value. Furthermore because the x and y planes which these points represent are 90° apart, the points representing them on the Mohr's circle must be 180° apart at opposite ends of a diagonal. Therefore, plotting their position on the graph and joining them identifies a diagonal of the circle, and hence the

circle itself may be drawn on this diameter as shown in Fig. 3.21. The maximum and minimum direct stresses, σ_1 and σ_2, and the maximum shear stress can then be obtained from the graph, together with the angles at which they act.

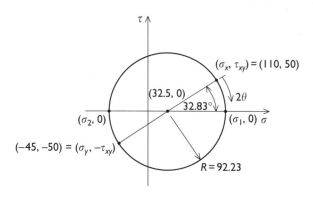

Figure 3.21 The Mohr's circle for Ex. 3.3.

Present day value of the Mohr's circle

Otto Mohr did not have a pocket calculator or computer on hand. In modern times, people find the Mohr's circle useful just as an aide memoir as to how to perform the calculation. From a sketch of the Mohr's circle for a particular problem, it is easy to recognise what calculations need to be performed. For Ex. 3.3, these are

a   The σ coordinate for the centre of the circle is found from $\sigma_c = \frac{1}{2}(110 - 45) = 32.5$.
b   From Pythagoras, the radius may be computed as $R = 92.23 (= \tau_{max})$.
c   Hence σ_1 and σ_2 can be computed by addition and subtraction.
d   Angles can be determined by trigonometry if required.

The nett result is that Eqs (3.2) and (3.3) have been solved without the formulae being remembered or graphical measurements being made.

Principal stresses as an eigenvalue problem

Equations defining the principal stresses and planes in 2D can be specified directly by resolving forces in the x and y directions for Fig. 3.18 while omitting the τ term (because $\tau = 0$ on the principal planes). This gives

$$\sigma_x \, dy + \tau_{xy} \, dx = \sigma \cos \theta \, ds \quad \text{and} \quad \tau_{xy} \, dy + \sigma_y dx = \sigma \sin \theta \, ds$$

Hence substituting $dx = \sin \theta \, ds$ and $dy = \cos \theta \, ds$ gives

$$\begin{bmatrix} \sigma_x & \tau_{xy} \\ \tau_{xy} & \sigma_y \end{bmatrix} \begin{bmatrix} \cos \theta \\ \sin \theta \end{bmatrix} = \sigma \begin{bmatrix} \cos \theta \\ \sin \theta \end{bmatrix} \tag{3.4}$$

This is the standard eigenvalue form

matrix × vector = scalar × same vector

with the two principal stresses being the two eigenvalues of the symmetric matrix

$$\begin{bmatrix} \sigma_x & \tau_{xy} \\ \tau_{xy} & \sigma_y \end{bmatrix}$$

and the principal planes defined by the corresponding eigenvectors.

A note on eigenvalues

Principal stress analysis is one of four types of problem mentioned in this book which give rise to eigenvalue problems. When only two variables are involved, as here, they reduce to the solution of quadratic equations. When more variables are present hand solutions become difficult, in which case computer packages such as MATLAB become particularly useful.

Q.3.5
The following plane stresses act at a point: $\sigma_x = 100 \, \text{N/mm}^2$, $\sigma_y = 60 \, \text{N/mm}^2$, $\tau_{xy} = -90 \, \text{N/mm}^2$ with the negative sign implying a clockwise shear stress. Determine the principal stresses, the maximum shear stress and specify on a diagram the orientations of the planes on which they act.

Q.3.6
For Ex. 3.3 construct a matrix whose eigensolution will yield the principal stresses and planes. Show that the eigenvectors describing the

principal planes, namely,

$$\begin{bmatrix} 0.9592 \\ 0.2826 \end{bmatrix} \text{ and } \begin{bmatrix} -0.2826 \\ 0.9592 \end{bmatrix}$$

both satisfy the eigenvalue equation of this matrix and hence determine the principal stresses.

3.3   Internal load paths

Significance of the Lamé–Maxwell equations

The Lamé–Maxwell equations, derived here using partial differential calculus, are not usually included in undergraduate structures courses. However, they do provide some useful insight into internal load paths, applicable whatever material properties exist. Whether or not the mathematical derivation is followed, knowing the result is an aid to understanding structural behaviour.

Stress field equilibrium constraints

The lines of principal stress within a material provide a way of visualising internal load paths. To identify how the stress field is constrained by equilibrium considerations, consider an element $dp \times dq$ of material in which p and q are coordinates aligned to the local principal axes. Although the principal axes will be everywhere at right angles to each other, the principal stresses may be varying with p and q, and so may be their angle α (measured positive anticlockwise) as shown in Fig. 3.22. In the following analysis, only first order small quantities will be taken into account.

The force on face AB of Fig. 3.22 is $\sigma_p \, dq$ acting in the direction $-p$. On the face CD, the stress will have changed to $\sigma_p + (\partial\sigma_p/\partial p)dp$ and the width of the face will be $dq+(\partial\alpha/\partial q) \, dp \, dq$. Multiplying these gives the force on this face as

$$F_{\text{CD}} = \sigma_p dq + (\partial\sigma_p/\partial p + \sigma_p\partial\alpha/\partial q) \, dp \, dq$$

This force will act at an anticlockwise angle $\phi = (\partial\alpha/\partial p) \, dp$ to the direction p. However, because ϕ is small, to the first order of small quantities, $\cos\phi = 1$ and $\sin\phi = \phi$ making components of this force in the p and q directions equal to F_{CD} and $\sigma_p(\partial\alpha/\partial p) \, dp \, dq$, respectively. Figure 3.22 shows the corresponding forces on faces AD and BC including their directions. In the absence of self-weight or other body forces, equilibrium of forces

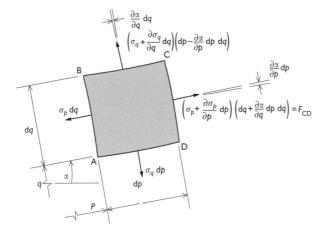

Figure 3.22 Element equilibrium using principal planes.

in the p and q directions gives

$$\left.\begin{array}{l} \dfrac{\partial \sigma_p}{\partial p} = (\sigma_q - \sigma_p)\dfrac{\partial \alpha}{\partial q} \\[3mm] \dfrac{\partial \sigma_q}{\partial q} = (\sigma_q - \sigma_p)\dfrac{\partial \alpha}{\partial p} \end{array}\right\} \tag{3.5}$$

These are known as the Lamé–Maxwell equilibrium equations (see, for instance, Frocht, 1941). However, $\partial\alpha/\partial p$ and $\partial\alpha/\partial q$ are the curvatures of the lines of principal stress in the p and q direction, respectively. Hence these equations state that the rate of change of a principal stress equals the product of the difference in the principal stresses times the curvature of the other principal stress.

Stress field properties

A 2D stress field within a body must satisfy the following conditions:

- Principal stress lines must be continuous.
- They must always intersect each other at right angles.
- In the absence of body forces, the curvature of the principal stress lines in one direction and the rate of change of the other principal stress are related through the Lamé–Maxwell equilibrium equations.
- The boundary conditions must conform with the external pressures. In particular, if there are no friction forces at a boundary, it must act as one of the principal planes.
- Crowding of stress lines occur at stress concentrations.

Figure 3.23 illustrates the situation at a boundary where no friction forces are present. Figure 3.24 shows a constriction in a compression member causing a stress concentration. Here the Lamé–Maxwell equilibrium equations can be used to show that, in the outer uniformly tapering sections, the compressive stresses will be increasing towards the constriction. If there are no external pressures on the top and bottom faces, there will be no stresses acting in the orthogonal directions. However in the centre, where the primary stress field is curving, secondary compressive stresses will be required in the orthogonal directions in order to maintain equilibrium.

The Lamé–Maxwell equations are necessary, but not sufficient, to determine a stress field. They are useful for checking the feasibility of derived or computed solutions and for visualising qualitatively what kind of stress may be feasible.

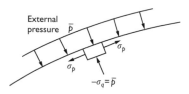

Figure 3.23 Stresses at a frictionless boundary.

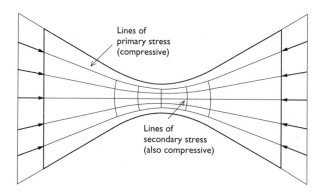

Figure 3.24 A possible stress field at a stress concentration.

Hoop stress in a circular cylinder

The formula for hoop stress in a circular cylinder is normally obtained by considering equilibrium of a unit length of cylinder sectioned across a diameter. If the cylinder has an internal radius r, thickness t, the internal fluid pressure is $\bar{p}$ and the hoop stress, assumed to be constant across the thickness, is σ_h, the forces acting on the half cylinder and its associated fluid are as shown in Fig. 3.25. Equilibrium of the system gives

$$\sigma_h = \bar{p}\frac{r}{t} \qquad (3.6)$$

Examining the equilibrium of an element of the cylinder (Fig. 3.26), it is seen, from the Lamé–Maxwell equations, that the curvature of the hoop stress needs to be balanced by a gradient in the radial stress which will be positive. (In practice, this will be compressive stress decreasing from the inner to the outer face). Since the hoop stress is constant and α does not vary in the radial direction, the other Lamé–Maxwell equation is automatically satisfied.

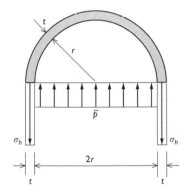

Figure 3.25 Forces acting on a sectional cylinder.

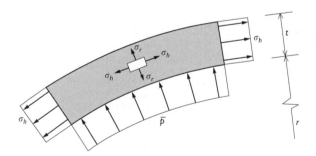

Figure 3.26 Forces acting on an element of a cylinder.

The Boussinesq stress field

Another important classical solution is for the elastic stresses in an infinite half plane of elastic material when a concentrated load is applied normal to

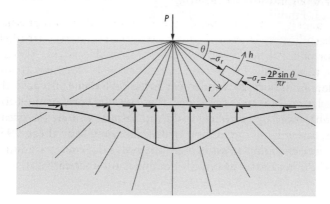

Figure 3.27 The Boussinesq stress field showing the internal stress distribution across a horizontal plane given by elastic theory.

the surface. Boussinesq derived the stress distribution as

$$\sigma_r = -2P \sin \theta / \pi r, \quad \sigma_h = 0$$

where P, r, h and θ are as shown in Fig. 3.27 (see for instance, Timoshenko and Goodier, 1970).

Consider the equilibrium of the element shown in Fig. 3.27. Unlike the previous example, the primary (radial) stress field is not curved. However, because it is diverging away from the load application point, $\partial \alpha / \partial h$ is positive and since also $\sigma_h - \sigma_r$ is also positive, the Lamé–Maxwell equations are seen to require $\partial \sigma_r / \partial r$ to be positive (i.e. with compressive stress decreasing as r increases).

Figure 3.27 also shows the distribution of stress, split into vertical and horizontal components, acting on the part above a horizontal section. The Boussinesq stress field and derivations from it are useful in various applications such as listed here:

- Determining how a concentrated load affects local stresses if it is applied directly to a structure such as a beam.
- Assessing the performance of a bearing plate placed between a concentrated load and the surface of a structure.
- Considering the behaviour of road pavements subject to wheel loading.

In none of these applications will Boussinesq's assumptions completely apply. The material will not stretch to infinity, it may not be properly elastic or uniform and the load will not be concentrated at a point. However, it provides a benchmark against which any variations in behaviour may be compared.

A solid cylinder with diametral loading

A solid cylinder acting as a roller bearing will be loaded in compression between opposite ends of a diameter. Primary compressive stresses are likely to spread out approximately as shown in Fig. 3.28. Since these stress lines are curved outwards, to satisfy the Lamé–Maxwell equations, secondary stresses must occur having a positive gradient towards the centre-line. Because the secondary stresses must be zero at the boundary of the cylinder, it is possible to deduce, without knowing the precise stress distribution, that the greatest tensile stresses must occur across the loaded diameter. Indeed the 1983 British Standard Institution's concrete test BS 1881-117 involves loading a cylinder across its diameter to measure its tensile splitting strength.

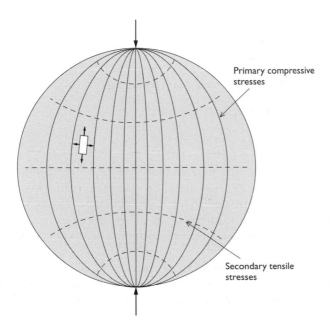

Primary compressive stresses

Secondary tensile stresses

Figure 3.28 A solid cylinder loaded across a diameter.

Q.3.7

A circular cylinder with radius r, thickness t and internal pressure $\bar{p}$ carries a uniform hoop stress of $\bar{p}r/t$. Assuming that t is negligible compared with r and therefore that radial stresses are negligible compared with the hoop stress, show, from the Lamé–Maxwell equations, that the radial stress gradient is approximately $\bar{p}/t$.

Q.3.8

A long uniform bar with a rectangular cross-section, pulled in tension, is expected to develop a uniform longitudinal tension field with no compression stresses. However, suppose that it contains a notch as shown in Fig. 3.29. The Lamé–Maxwell equations will need to be satisfied and also the internal forces acting on the reduced central cross-section will need to be in equilibrium with the external loading. Using this information, sketch a possible stress field indicating which principal stresses will be tensile and which will be compressive.

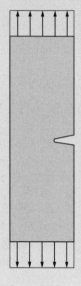

Figure 3.29

3.4   Statically determinate arches and portal frames

Bending moments in arches

Arches and portal frames made out of steel, reinforced concrete or other materials, which have high tensile as well as compressive strength, are used widely in many contexts. Although not many have three pin-joints (the number required to make them statically determinate), quite a few have two, a well-known example being the Sydney Harbour Bridge. The study in this section should help in the appreciation of their behaviour without introducing complications in analysis caused through the presence of redundancies (the effect of redundancies will be discussed at greater length in Chapters 6 and 8).

In these types of structure, one of the most important considerations is the bending moments which they need to carry. It is normal to plot a graph of bending moments alongside the profile of the structure itself, so that the bending moment at a cross-section can be read directly from the adjacent graph. There is a convention for the graph always to be plotted on the tension side of a member, which eliminates any possible confusion over signs.

Ex. 3.5
A segmental arch bridge has the configuration shown in Fig. 3.30(a). The arch itself is three-pinned and is to be analysed for the loading case shown in Fig. 3.30(b).

a Determine all the internal forces acting at cross-section C.
b Determine the bending moments at all the labelled points.
c Plot the bending moment diagram for the arch radially on the profile of the arch.

Answer
a The support reactions shown in Fig. 3.30(b) already have the horizontal equilibrium equation satisfied. The other two overall equilibrium conditions yield $V_A = 837.5\,\text{kN}$ and $V_J = 462.5\,\text{kN}$. Because the bending moment at F must be zero (on account of the hinge), taking moments about F for the right-hand arch segment yields, in kNm units,

$$9.238H = 462.5 \times 16 - 100(4 + 8 + 12)$$

Hence $H = 541.2\,\text{kN}$.

The internal forces at C may be obtained from the equilibrium of the segment AC of the arch shown in Fig. 3.30(c). Resolving forces in the direction of P_C and S_C and taking moments about C gives

$$P_C = -541.2 \cos 30° + (400 - 837.5) \cos 60°$$
$$= -687.4\,\text{kN}$$
$$S_C = 541.2 \cos 60° + (400 - 837.5) \cos 30°$$
$$= -108.3\,\text{kN}$$
$$M_C = 837.5 \times 6.762 - 400 \times 2.762 - 541.2 \times 6.762$$
$$= 898.8\,\text{kNm}$$

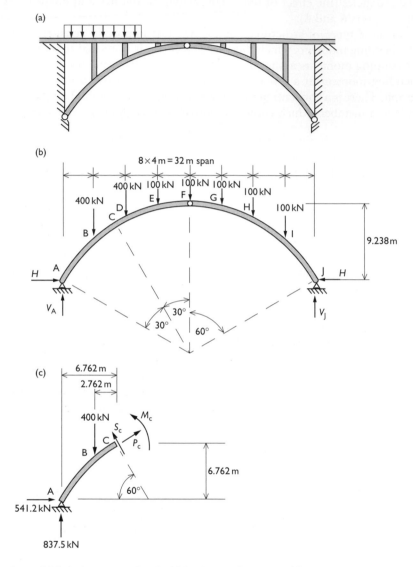

Figure 3.30 A three-pinned arch: (a) bridge configuration; (b) arch geometry and loading; (c) forces acting on part AC.

b   Bending moments for other cross-sections derived in a similar way are, in kNm units, $M_B = 747$, $M_D = 1087$, $M_E = 488$, $M_G = -412$, $M_H = -713$ and $M_I = -753$.

c   To determine the full bending moment diagram, intermediate bending moments between each load application point need to be evaluated. The resulting diagram (plotted on the tension side) is shown in Fig. 3.31(a).

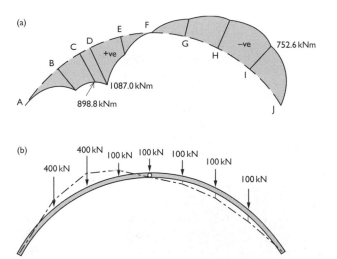

Figure 3.31 A comparison between the bending moment diagram and thrust line for the three-pinned arch: (a) bending moment diagram plotted radially; (b) the thrust line.

Recall of the thrust line concept

The bending moment diagram in Fig. 3.31(a) may seem bizarre. However, where the loading on an arch is entirely vertical, the thrust line concept can be recalled to help with interpretation. It can be shown that the bending moment at a cross-section of the arch equals the vertical distance between the thrust line and the axis of the arch at the particular cross-section, multiplied by H. Figure 3.31(b) shows the position of the thrust line, which necessarily passes through the three hinges. The sudden changes of slope of the bending moment diagram correspond to the sudden changes of slope in the thrust line and the segments in between are curved because the arch itself is curved. Indeed, the bending moment diagram would resemble the thrust line more closely if it was plotted on the compression side rather than the tension side of the arch.

Modern arches

Arches able to withstand large tensile as well as compressive stresses do not need to be designed in such a way that the thrust line remains within the arch thickness. Therefore bridges they support do not need to be heavy in order to provide stability. Thinner arch rings, lighter superstructures and much larger spans are thus possible than with masonry arches (Fig. 3.32).

Figure 3.32 Coldspring arch, California.

Statically determinate portal frames

Portal frames have much similarity with arches, particularly when subject just to gravity loading. They are really arches having an angular rather than curved profile. Thus

- like arches they develop a horizontal component of thrust;
- the reactions and internal forces can be calculated in a similar way to that of an arch with similar hinge positions;
- the thrust line concept is similarly useful in visualising bending moment distributions for cases of gravity loading.

Portal frames are extensively used for single storey buildings such as factories, warehouses and retail stores. They provide a complete load path for forces down to ground level, and their columns can be integrated with the walls of the building. Figures 3.33 and 3.34 show two types of portal frame, the second of which more closely resembles an arch and hence is more likely to have lower bending moments when subject to distributed gravity loading.

To illustrate this consider the portal frame with the dimensions and loading shown in Fig. 3.34. From overall equilibrium, $V_A = V_E = ws/2$ and by

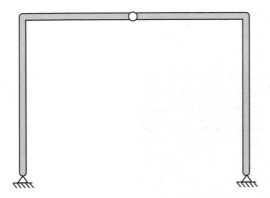

Figure 3.33 A simple portal frame with pin-joints.

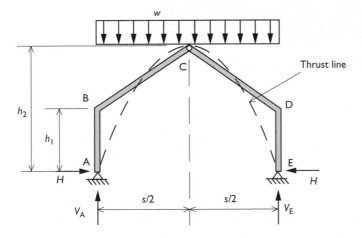

Figure 3.34 A pitched roof portal frame with pin-joints showing the thrust line for a uniformly distributed load.

taking moments about C for the LHS:

$$\frac{V_{AS}}{2} - \frac{ws^2}{8} = Hh_2$$

Hence $H = ws^2/8h_2$, showing that the horizontal component of force is reduced as h_2 is increased. Because the loading is uniform, the thrust line is a parabola passing through the three hinge positions, as shown in Fig. 3.34. From this figure, it can be seen that, unless the roof pitch is very steep, the largest bending moments will occur at the eaves joints, B and D. Equilibrium of the column AB gives

$$M_B = Hh_1 = \frac{ws^2h_1}{8h_2}$$

Thus the eaves joint bending moment, M_B, will be significantly less for the pitched roof portal frame than for the equivalent simple portal frame (where $h_2 = h_1$).

Ex. 3.6

a   Determine the bending moment diagram for the pitched roof portal frame shown in Fig. 3.35(a) when subject to a vertical loading of 240 kN distributed uniformly across the rafter members.

b   Also, determine the bending moment diagram for a single horizontal load of 60 kN applied to the left at B.

c   If the members are tapered such that the bending strength varies linearly along the members with that at B and D being four times that at A, C and E, determine the minimum acceptable bending strength at B and D (not allowing for load factors).

Answer

a   The distributed loading is 15 kN/m. With reactions as shown in Fig. 3.35(a), because of symmetry $V_A = V_E$. Vertical equilibrium gives $V_A = V_E = 120$ kN. Taking moments about C for one half of the frame gives

$$M_C = -9H + 120 \times 8 - 120 \times 4 = 0$$

Hence $H = 53.3$ kN.

(a)

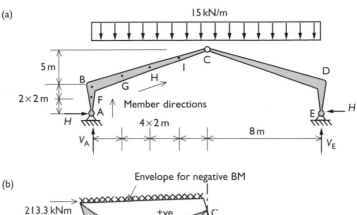

(b)

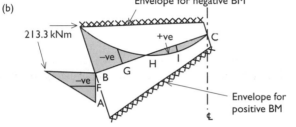

Figure 3.35 A portal frame with tapered members: (a) geometry and loading; (b) bending moment diagram for LHS (RHS being similar through symmetry).

Making a section at F and taking moments about F for length AF of the frame gives $M_F = -2H = -106.7$ kNm. Making a section at G and taking moments about G for length AG of the frame gives

$$M_G = 120 \times 2 - \tfrac{1}{2} \times 15 \times 2^2 - 53.3 \times 5.25 = -70.0 \text{ kNm}$$

Similarly for other named sections, $M_B = -213.3$, $M_H = 13.4$ and $M_I = 36.7$ all in kNm units. Plotting the bending moment diagram (Fig. 3.35(b)) for the LHS (the RHS being similar through symmetry), it may be noted that the largest bending moment occurs at B with a magnitude of 213.3 kNm.

b   For the second loading case (Fig. 3.36(a)), taking moments about E for the whole frame gives $V_A = 15$ kN and hence from vertical equilibrium $V_E = -15$ kN. Taking moments about C for part CDE of the frame, $H_E = 8V_E/9 = 13.33$ kN. Hence from horizontal equilibrium $H_A = 46.7$ kN. Taking moments about B and D for the parts AB and DE, respectively, gives the bending moments at the eaves joints as $M_B = -186.7$ kNm and $M_D = 53.3$ kNm,

respectively. If more bending moments are evaluated, it will be discovered that the bending moment diagram is linear between joints
as shown in Fig. 3.36(b).

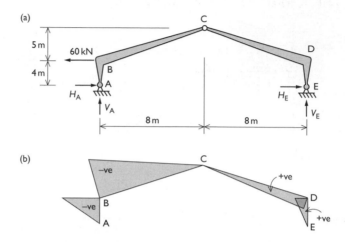

Figure 3.36 The portal frame with side loading: (a) loading; (b) bending moment
        diagram.

c   The largest overall bending moment is 213.3 kNm for the first load-
    ing case. If this is taken as the strength in bending at B, the strength
    of the neighbouring members will reduce linearly to a value of
    213.3/4= 53.3 N/m at both A and C. An envelope indicating the
    bending strength along member BC has been added to Fig. 3.35(b),
    assuming that the member is equally strong in positive and negative
    bending. This shows that the bending moment capacity of member
    BC, like other members, is not exceeded if the bending strength
    at B and D is made equal to 213.3 kN/m. Note that the thrust
    line concept can be used to verify the bending moment diagram of
    Fig. 3.35(b), but not Fig. 3.36(b).

Pin-joints and lack of them in portal frames

Often pin-joints are assumed to be present at the supports of portal frames
for design purposes even though they may not be there in practice. This is
because, where the fixity against rotation of the ground supports is unknown,
it is better to assume that the structure is weaker rather than stronger than
it really is. In that case, maximum predicted bending moments will usually
tend to be more than for the actual structure, leading to an oversafe rather
than an unsafe design.

Q.3.9
For the loaded frame shown in Fig. 3.37, determine the bending
moments at the named points. Draw the bending moment diagram
and justify its shape by sketching the thrust line.

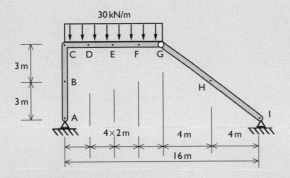

Figure 3.37

Q.3.10
Sketch thrust lines and bending moment diagrams for the four frames
shown in Fig. 3.38 indicating, without performing any calculations,
approximately where you expect the largest bending moments to
occur.

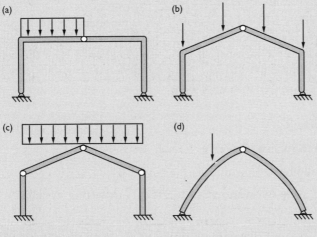

Figure 3.38

Q.3.11

Figure 3.39 shows a form of gantry crane which runs on tracks on either side of a ship building dock. With the left-hand column pinned at the top, the gantry is a three-pinned portal structure. Discuss its structural action under vertical loading, comparing it to symmetrical three-pinned arches and portal frames.

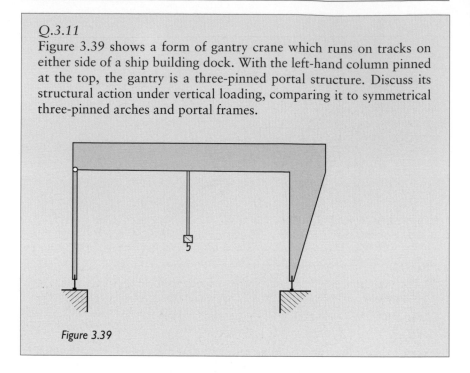

Figure 3.39

3.5   Stress systems in 3D

Internal forces for bar structures

If a section is taken of a bar type structure, the resultants of the stresses acting across the interface will have, in general, six components (to correspond with the six possible body freedoms in 3D). These will normally be related to the centroidal axis of the member. If the x axis is aligned with the centroidal axis and the y and z axes form a right-handed system with it, the internal forces may be designated as follows:

Axial force: P

Shear forces: S_y and S_z

Bending moments: M_y and M_z

Torque T (otherwise known as torsional or twisting moment)

Positive directions for these are indicated in Fig. 3.40. Where the bar structure is statically determinate, the internal forces for any cross-section can be obtained using the method of sections.

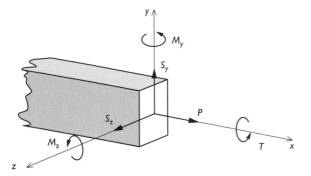

Figure 3.40 Internal forces for 3D sections.

Ex. 3.7
A semi-circular rail ABCDE carries a trolley which provides a single concentrated vertical load of 600 N anywhere along its length. The cantilever support CF is to be designed (see Fig. 3.41).

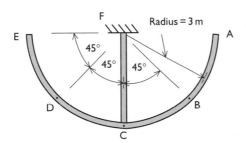

Figure 3.41 A semi-circular rail with cantilever support shown in plan view.

a   Ignoring self-weight of the structure, determine all the internal forces at F, the root of the cantilever due to the trolley being placed at B.
b   Indicate where the worst case of combined bending and torque of the cantilever is located and where the trolley needs to be to produce this.

Answer
a   Figure 3.42 shows the structure sectioned at F with all forces acting. Such a 3D representation is often confusing, so it is easier

and safer to work with projected views as shown in Fig. 3.43. From the side-view it can be deduced that $P = 0$, $S_y = 600\,\text{N}$, $M_z = -600 \times 2.121 = -1272\,\text{Nm}$, from the elevation $S_z = 0$, $T = -1272\,\text{Nm}$ and from the plan (not shown) $M_y = 0$.

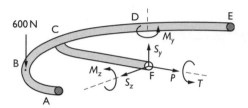

Figure 3.42 Forces acting on rail and support when sectioned at F.

Note
The cantilever is shown below the rail.

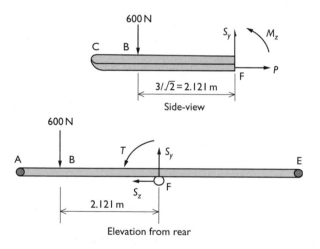

Side-view

Elevation from rear

Figure 3.43 Projected views taken from Fig. 3.42.

b   Consider the internal forces at the root, F. The maximum value of M_z (positive or negative) occurs when the trolley is at C giving a lever arm of 3 m. However, T is zero in this case. The maximum T is when the trolley is at A or E with a lever arm of 3 m, but then $M_z = 0$.

Consider the internal forces at the end, C, of the cantilever. When the trolley is at A or E, $M_z = 600 \times 3 = 1800\,\text{Nm}$ and

$T = \pm 1800$ Nm. Thus, the maximum bending moment and torque occur together and this is the worst combined case (see Fig. 3.44).

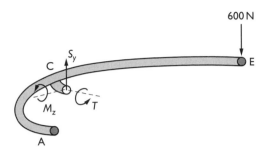

Figure 3.44 Internal forces acting at end C of cantilever with trolley at E.

On stress distribution

For every set of internal forces, there are different possible stress distributions. Factors influencing these distributions will be the characteristics under load of the material, the magnitudes of the applied loads, the previous history of loading, etc. Two of the more common assumptions are that

a   stresses conform to elastic theory (which is generally the case when stresses are small);
b   stresses conform to plastic theory (which is sometimes the situation at failure for ductile materials).

Internal forces may be divided into two types. Axial force P and bending moments M_y and M_z are the resultants of normal stresses whilst torque T and shear forces S_y and S_z are the resultants of shear stresses. The following notes refer to internal forces carried by long uniform members.

Stresses corresponding to P

When an axial load is applied, the usual assumption is that the normal stress distribution is uniform. This accords with both elastic and plastic theory. However, even in this simple case, complications can arise if there are stresses present caused in manufacture (e.g. through rolling, extruding or cooling). Initial stresses could, for instance, be tension next to the outer face and compression elsewhere. A uniform stress distribution will still be correct for plastic theory. But for elastic theory, a uniform distribution is only correct as far as the additional stress due to loading is concerned and only then if the elastic limit for the material is not exceeded once this stress is added.

Stresses corresponding to M_z and M_y

The elastic assumption is that the normal stress is proportional to the distance from the neutral axis (the neutral axis, being the line of zero stress, must pass through the centroid). Where M_z only is applied, the stress is usually proportional to y (Fig. 3.45). However, this is not always the case. For instance, a stress distribution proportional to y for the angle section shown in Fig. 3.46 has resultant moments of both M_y and M_z. Particular care should be taken with cross-sections which are not symmetric about the y or z axis (this is discussed further in Section 4.11).

The plastic assumption is that all the material has reached its yield stress either in tension or compression. This leads to the stress being in blocks as shown in Fig. 3.45. As with the elastic assumption, special consideration needs to be given to unsymmetric cross-sections.

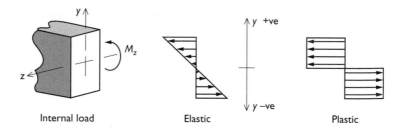

Figure 3.45 Assumed bending stresses for symmetric cross-sections.

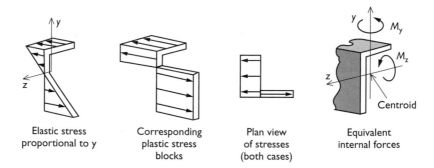

Figure 3.46 Bending of an unsymmetric cross-section.

Stresses corresponding to P and M$_z$

According to elastic theory, when P and M_z are both present, the stresses are the sum of those occurring when the internal forces act separately (i.e. the theory of superposition holds in this case). Provided that the cross-section is symmetrical about the y or z axis, the stress distribution will be linear in y. The neutral axis will not pass through the centroid, however. Indeed, it may not intersect the cross-section at all (see Fig. 3.47(a)).

An important variation is where a material (such as masonry) is considered as unable to carry tensile stresses. If compressive stresses conform to elastic theory and the cross-section is rectangular with depth d, the effect of increasing the ratio $e = -M_z/P$, is shown in Fig. 3.47(b). This ratio, called the eccentricity, has limiting values of $\pm\frac{1}{2}d$ when the compressive stress tends to infinity (this is consistent with the thrust line remaining within the cross-section for masonry walls and arches). Furthermore, it is only when $|e| < d/6$ (i.e. the thrust line is within the middle third) that the whole cross-section is under compressive stress. When $|e| > d/6$, the part of the cross-section which would ordinarily be taking tension is considered to be unloaded because any cracks there would start to open up.

The plastic assumption requires the stress to reverse from its positive to negative yield values at the neutral axis. In this case, the neutral axis must always intersect the cross-section whenever $M_z \neq 0$ as shown in Fig. 3.47(c).

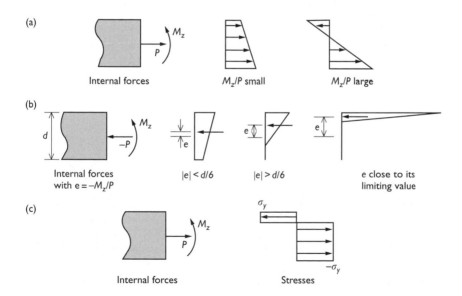

Figure 3.47 Some stress distributions for combined axial force and bending moment: (a) elastic; (b) elastic, compression only; (c) plastic.

Stresses corresponding to S_y and S_z

Where cross-sections are thick, the shear stresses are likely to be small and therefore having a negligible effect on performance. However, where cross-sections are thin, important deductions can often be made about shear stress distributions just from equilibrium.

Because shear stresses cannot occur across the plane of a surface (unless friction is present), feasible shear stresses in thin-walled sections must be parallel to the neighbouring surfaces. Thus for a doubly symmetric I section, shear forces in the horizontal 'flanges' must be virtually horizontal and, in the vertical 'web', they must be virtually vertical. Thus with a horizontal shear force of S_z applied, the web will not provide a horizontal shear force component and the flanges will therefore carry $\frac{1}{2}S_z$ each (see Fig. 3.48(a)). Similarly, a vertical shear force of S_y will be carried virtually entirely by the web (Fig. 3.48(b)). Shear stresses arising due to shear force should not be expected to be uniform, however. For the I section, flange stresses will not be uniformly distributed when S_z is applied and, when S_y is applied, there will normally be shear stresses present in the flanges (as shown in Fig. 3.48(b)) even when the resultant shear force for each flange is zero. Shear stress distributions due to shear force need to be related to bending stress distributions and those which conform to elastic theory are discussed in Section 4.6.

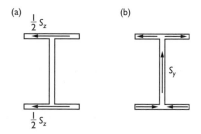

Figure 3.48 Shear forces in an I section: (a) due to S_z; (b) due to S_y.

Stresses corresponding to T

Solid circular shafts resist torque by means of axi-symmetric shear stresses as shown in Fig. 3.49. The capacity of thin-walled sections to carry torque is an important topic which will be discussed further in the next section.

Principal stresses and planes

An incremental element of material of dimension $dx \times dy \times dz$ may be subject to six independent stress components $\sigma_x, \sigma_y, \sigma_z, \tau_{xy}, \tau_{yz}$ and τ_{zx}. Figure 3.50

(a)   (b)

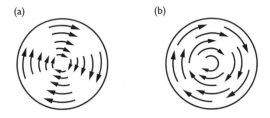

Figure 3.49 Shear stresses in a circular shaft subject to torque: (a) elastic − τ proportional to distance from centre; (b) plastic − τ equal to yield value throughout.

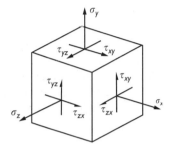

Figure 3.50 Stresses acting on a 3D element.

shows these drawn with a right-handed set of axes and including the complementary shear stresses. Consider a principal stress σ, acting on an inclined plane, which intersects the axes at P, Q and R as shown in Fig. 3.51. If ℓ_x, ℓ_y and ℓ_z are the direction cosines of the normal to the plane PQR, the components of σ in the x, y and z directions will be $\sigma\ell_x, \sigma\ell_y$ and $\sigma\ell_z$. Furthermore, with A as the area of PQR, the areas of OQR, ORP and OPQ are $A\ell_x, A\ell_y$ and $A\ell_z$, respectively. Thus by resolution of forces in the axis directions:

$$\begin{bmatrix} \sigma_x & \tau_{xy} & \tau_{zx} \\ \tau_{xy} & \sigma_y & \tau_{yz} \\ \tau_{zx} & \tau_{yz} & \sigma_z \end{bmatrix} \begin{bmatrix} \ell_x \\ \ell_y \\ \ell_z \end{bmatrix} = \sigma \begin{bmatrix} \ell_x \\ \ell_y \\ \ell_z \end{bmatrix}$$

Because this is a 3×3 eigenvalue problem, it is known that there will be precisely three eigenvalues defining the principal stresses. Furthermore, because the matrix is symmetric, it is known that the corresponding eigenvectors defining the principal planes will be orthogonal to each other (i.e. at right angles). Fortunately there is rarely a need to perform 3×3 eigensolutions.

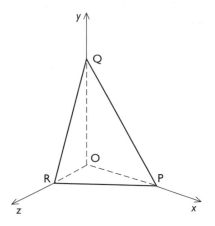

Figure 3.51 Inclined plane geometry.

Ex. 3.8

The propeller shaft of a deep-sea submersible craft is subject to a torque which gives rise to a maximum shear stress in the plane of the surface of $32 \, \text{N/mm}^2$. The shaft is also subject to a bending moment which produces a maximum longitudinal stress in the top fibre of $36 \, \text{N/mm}^2$.

a   If no other stresses act at this point, determine the maximum shear stress acting here.

b   When submerged at depth, the shaft is subjected to an additional hydrostatic pressure (i.e. in all directions) of $10 \, \text{N/mm}^2$. What effect will this have on the maximum shear stress?

Answer

a   With the x axis aligned with the shaft axis and the y axis vertical, $\sigma_x = 36$, $\sigma_y = \sigma_z = 0$, $\tau_x = 32$ in N/mm^2 (Fig. 3.52). The Mohr's circle in the x–z plane has a centre at $(18, 0)$ and a radius of 36.71. Hence the principal stresses are $\sigma_1 = 18 + 36.71 = 54.71$, $\sigma_2 = 0$ (the unaltered stress in the y direction) and $\sigma_3 = 18 - 36.71 = -18.71$ all in N/mm^2 units. The maximum shear stress in the material occurs on the largest Mohr's circle, that is the one based on the principal stresses of 54.71 and -18.71 and is thus $36.71 \, \text{N/mm}^2$.

b   The principal stresses will be changed to $\sigma_1 = 44.71$, $\sigma_2 = -10$ and $\sigma_3 = -28.71$. However, the largest difference between any two principal stresses will still be 73.42, giving the same maximum shear stress of 36.71 N/mm².

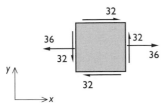

Figure 3.52 Surface stresses in N/mm².

Ex. 3.9
A pressure vessel comprises a circular cylinder closed off at both ends by hemispheres. The internal radius is r and the thickness is t.

a Determine an expression for the maximum shear stress anywhere in the material of the cylinder due to an internal pressure of p and specify a plane on which this acts.
b If r = 200 mm and p = 14 N/mm², what thickness, t, should be used to ensure that the shear stress in the material of the cylinder nowhere exceeds 120 N/mm²?

Answer
The hoop stress in a cylinder subject to internal pressure p was evaluated in Section 3.3 as

$$\sigma_h = pr/t$$

If the cylinder is sectioned at right angles to its axis, equilibrium of one part (as shown in Fig. 3.53(a)) gives

$$\pi[(r+t)^2 - r^2]\sigma_\ell = \pi r^2 p$$

where σ_ℓ is the longitudinal stress which is assumed to be uniform. Hence

$$\sigma_\ell = r\sigma_h/(2r + t)$$

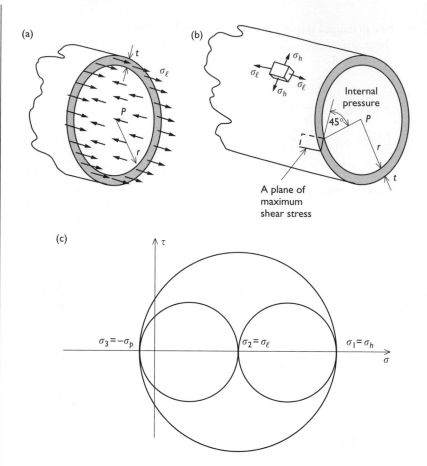

Figure 3.53 Stresses in a cylinder of a pressure vessel: (a) forces acting at a cross-section; (b) principal stresses; (c) Mohr's circles for inner face.

which is less than σ_h. The principal stresses at the outer surface of the cylinder will thus be $\sigma_1 = \sigma_h$, $\sigma_2 = \sigma_\ell$ and $\sigma_3 = 0$ and, at the inner surface $\sigma_1 = \sigma_h$, $\sigma_2 = \sigma_\ell$ and $\sigma_3 = -p$ (see Fig. 3.53(b)). Hence the maximum difference in principal stresses are recorded at the inner surface where the maximum shear stress in the material equals the radius of the largest Mohr's circle, $\frac{1}{2}(\sigma_1 - \sigma_3)$, as shown in Fig. 3.53(c). Thus

$$\tau_{\max} = p(r + t)/2t$$

which acts on planes at 45° rotation between the circumference and the radius as illustrated in Fig. 3.53(b).

b   For the shear stress not to be exceeded

$$14(200t^{-1} + 1)/2 < 120$$

Hence $t > 12.4$ mm.

Q.3.12
A horizontal semi-circular rail of radius r has points A, B, C, D and E spaced at $45°$ intervals in an anticlockwise direction, with simple supports at A, C and E. Determine expressions for all the reactions (assumed to be vertical) and also the bending moments and torques at B, C and D when a load of W is placed at B.

Q.3.13
The following stresses occur on orthogonal axes at a point (in N/mm^2 units): $\sigma_x = 226$, $\sigma_y = -82$, $\sigma_z = 34$, $\tau_{xy} = 0$, $\tau_{yz} = 100$, $\tau_{zx} = 0$. Determine the principal stresses at this point and also the maximum value of shear stress.

Q.3.14
A spherical housing for an underwater camera is to be designed to descend to 2000 m depth of water.

a   If its internal diameter is to be 180 mm, determine the thickness required of the shell if the maximum allowable stress for the material in compression is 500 N/mm^2 and there is to be a load factor of 2.
b   A circular access hole in the sphere, having diameter 30 mm, is to have a reinforcing ring round its edge in order to carry the forces that the removed part of the shell would otherwise carry. Estimate the axial force that this ring needs to carry and hence its required cross-sectional area if made of the same material as the sphere.

3.6   Torsion of thin-walled sections

Torque induced shear stresses

An applied torque needs to be resisted by shear stresses within the cross-section. If a bar is thin-walled, shear stresses can only form on planes parallel to the neighbouring surfaces, thus limiting the number of possibilities for shear patterns. Figure 3.54 shows the three main ways that different types of cross-sections resist torque. Of these the 'closed' cross-section provides by far the most effective resistance to torque for long members.

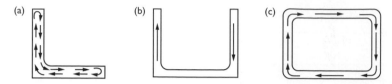

Figure 3.54 Different ways of resisting torque: (a) an open cross-section showing a basic stress system; (b) an open cross-section with the possibility of differential bending; (c) a closed cross-section with Bredt–Batho shear flow.

Bredt–Batho shear flow

When the cross-section is of 'closed' hollow box form, a uniform 'shear flow' tends to develop round the section, where shear flow, q, is defined as the product of the local shear stress and thickness. If there are thinner parts to the box, these parts will experience higher shear stresses. The significance of the shear flow being uniform is that complementary shear forces acting in the longitudinal direction are all in equilibrium (see Fig. 3.55).

To determine the relationship between torque and shear flow for an arbitrary shaped closed section, take moments about any point O. If s is a coordinate measured round the perimeter and h is the normal distance between the tangent at s and the point O, the moment due to an element of

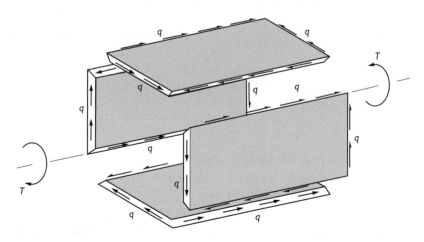

Figure 3.55 Exploded view of a Bredt–Batho shear flow.

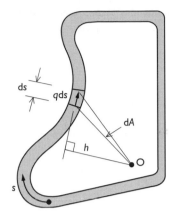

Figure 3.56 An increment ds of Bredt–Batho shear.

shear flow $q\,ds$ is $qh\,ds$ (see Fig. 3.56). Hence the torque is

$$T = \oint qh\,ds$$

where integration involves s completing one circuit. However, $h\,ds = 2\,dA$ where dA is the area of the triangle made by ds as base and O as vertex. Therefore, since q is constant,

$$T = 2q \oint dA = 2qA \tag{3.7}$$

with A being the total enclosed area (measured to the centre-line of the thickness).

Factors which ensure that a section has good torsional characteristics are as follows:

a The cross-section is closed rather than open.
b The enclosed area is as large as possible.
c No part of the cross-section is very thin (thus avoiding high local shear stresses).

Ex. 3.10
Figure 3.57 shows the cross-section of the outer 'skin' of a steel box girder bridge which constitutes a torsion tube (Longitudinal

stiffeners, although present, make no significant difference to torsional characteristics, except that they delay buckling of the skin). Estimate the shear stresses in the skin arising from a torque of 14,000 kNm assuming that it is resisted entirely by Bredt–Batho shear flow.

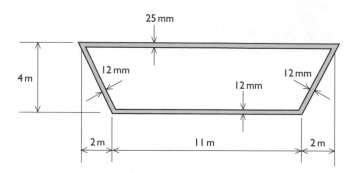

Figure 3.57 Torsion tube of a steel box girder bridge.

Answer
The enclosed area $A = 4 \times 13 = 52\,\text{m}^2$. Hence $q = 14,000/2 \times 52 = 134.6$ in units of kN/m which is equivalent to N/mm. Thus for the deck $\tau = 134.5/25 = 5.4\,\text{N/mm}^2$ and for the rest of the cross-section $\tau = 134.5/12 = 11.2\,\text{N/mm}^2$.

NB   Despite the large twisting moment, the resulting shear stresses are quite small for steel.

Open cross-sections: the basic stress system

For an angle section there is no possibility of unidirectional shear forces in the two legs reacting an applied torque. The only type of stress system which can develop to resist torque involves loops of shear flow within the thickness of the cross-section as shown in Fig. 3.54(a). Not only must shear stresses be relatively large because of the small enclosed area within any possible loop, but also large twisting deflections need to occur before such a stress system is mobilised.

If elastic theory holds, the stress will vary linearly through the thickness from $\bar{\tau}$ at one side to $-\bar{\tau}$ at the other. If also the thickness is constant, it may be assumed that the stress distribution is unvarying. The two triangular stress blocks, as shown in Fig. 3.58, each give a shear flow of $q = \bar{\tau}t/4$ and are spaced $2t/3$ apart. Hence, in this case, the Bredt–Batho formula

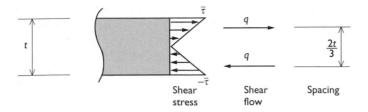

Figure 3.58 Circulating shear flow within a thin section.

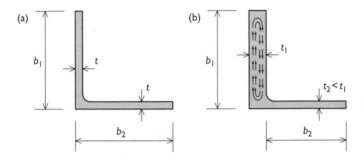

Figure 3.59 Use of Bredt–Batho shear flow with thin open sections: (a) constant thickness, $b = b_1 + b_2$; (b) varying thickness $b = b_1$, $t = t_1$ for a very approximate answer.

could be used with q as defined here and $A = 2tb/3$ where b is the sum of the breadths of the components which make up the cross-section (as shown in Fig. 3.59(a)). If, however, the thickness varies the shear flow as well as the spacing of the forces will increase where thickness is greater. The Bredt–Batho formula can then only give a very approximate indication of maximum shear stresses by ignoring thinner (and thus weaker) components (see Fig. 3.59(b)).

Differential bending

For the channel shown in Fig. 3.54(b), there is the possibility that opposing unidirectional shear forces may develop in the parallel walls of the section in such a way that torque is resisted. However, these shear forces do not form a continuous flow loop and hence the complementary shear flows do not balance. The only way this type of stress system can arise is if there are axial stresses which vary in the longitudinal direction. For instance, if a channel section is simply supported at both ends and subject to a torque at centre span as shown in Fig. 3.60(a), longitudinal stresses will develop in the tops and bottoms of the walls of the channel which balance the complementary shear stresses as shown in Fig. 3.60(b). This stress regime involves the walls bending in opposite directions as shown (magnified) in Fig. 3.60(c).

On the other hand, if a channel section is subject to unidirectional torsion, as shown in Fig. 3.61(a), axial stresses will not develop unless at least one of the ends have a diaphragm which is so stiff that it adequately restrains out of plane distortion (called warping) of the type shown (magnified) in Fig. 3.61(b).

Thus, whether the torsional characteristics of an open section will be improved by differential bending depends on the type of loading and end

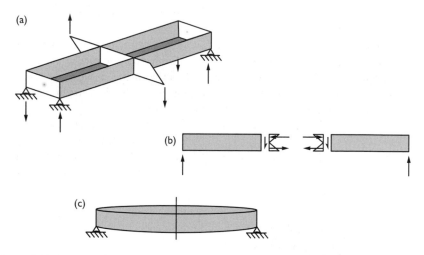

Figure 3.60 A channel section subject to differential bending: (a) geometry and loading; (b) exploded view of forces acting on front wall; (c) elevation showing differential bending deflections.

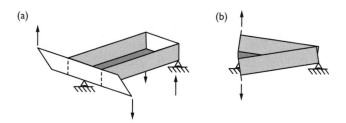

Figure 3.61 A channel section unable to develop differential bending: (a) geometry and loading; (b) elevation showing twisting and also warping of end sections.

conditions. If no such benefit is possible, only the basic stress system of Fig. 3.54(a) will develop to resist torque.

Q.3.15   Observation question
Obtain a matchbox (or make a large one out of stiff card) and separate the tray from the sleeve. Observe the following to check the performance of the different stress regimes shown in Fig. 3.54:

a   The ease with which the tray can be twisted (regime (a)).
b   The difficulty of preventing warping at one end of the tray (see Fig. 3.61).
c   With all four corners of the tray supported, the resistance of the central section of the tray to applied torque (differential bending, regime (b) as per Fig. 3.60).
d   In the above test, the advantage that would be gained from having a diaphragm at the centre to avoid distortion of the cross-section.
e   The way in which the cross-sectional shape of the sleeve distorts if a torque is applied by means of vertical forces at the corners (thus indicating the benefit of having end diaphragms).
f   The torsional strength of the sleeve if it is loaded in such a way as not to distort the end cross-sections, thus illustrating regime (c) of Fig. 3.54 (this can be done by pressing four corners with forces directed parallel to end diagonals as shown in Fig. 3.62 or by including end diaphragms).

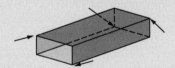

Figure 3.62

Q.3.16
For the cross-sections shown in Fig. 3.63, identify which regimes of Fig. 3.54 apply. In each case, indicate if you think that there is any part of the cross-section which is unlikely to be important as far as torsional strength is concerned.

NB   in some cases, you may need to say (a) or (b) depending on the circumstances.

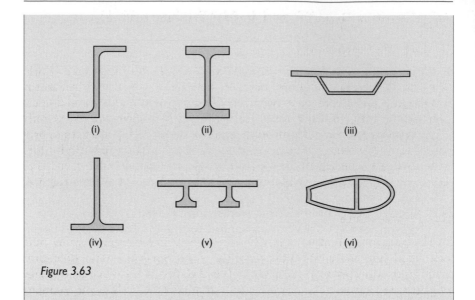

(i)   (ii)   (iii)

(iv)   (v)   (vi)

Figure 3.63

Q.3.17
A member is to be fabricated from two channel sections of outer dimensions 120×60 and thicknesses 6 for the webs and 12 for the flanges (in mm units). There is a question as to which of the two ways shown in Fig. 3.64 they should be welded together. Estimate the shear stresses induced in the two proposed members by a torque of 880 Nm and compare their maximum values (elastic stress distributions should be assumed where relevant ignoring the possibility of differential bending).

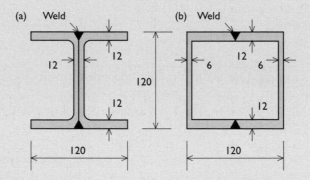

Figure 3.64

3.7 Comets G-ALYP and G-ALYY (Cohen, 1955)

Cutting edge technology

In order to restart the manufacture of civil aircraft after the Second World War, the de Haviland aircraft company designed a new aircraft called the 'Comet' to take advantage of what, at the time, were recent technological developments. Key to the project was the use of jet engines previously only used for military planes. They were capable of better performance than propeller driven aircraft, particularly in the rarefied atmosphere of high altitudes where lower drag made higher speeds and longer ranges possible. Crucial to the performance of such an aircraft would be the lightness of the structure.

Design procedure

Particular care was taken over all the aspects of the design. Stresses were calculated, but these could not be obtained as accurately as was possible some years later once computer methods had been developed. An extensive range of structural tests were carried out on components as well as on complete sections of air frame with more reliance being put on these results than on theoretical calculations.

If a maximum load of P was expected on the structure, the design requirements were that a load of $1.3P$ should be carried without permanent deformation and failure should not occur at less than $2P$. Even so de Haviland aimed at a design pressure of $5P/2$ for the cabin. Their estimated maximum design stress levels were $193\,N/mm^2$ compared with an ultimate material strength of $448\,N/mm^2$. These design stress levels were often averages of stresses occurring over $50\text{--}60\,mm$ because of the difficulty of obtaining more accurate information either theoretically or experimentally. They thought that the additional factor over and above the conventional requirement would allow for inaccuracies in the prediction of stresses and also provide a fatigue life of at least 10 years.

Fatigue

Metals tend to have microscopic cracks present which do not cause damage unless they grow. It is well known that high stress concentrations should be avoided and that repeated loading is more dangerous than steady loading because, each time a tensile load is applied or increased in magnitude, a highly stressed crack will extend a little further. Once a crack becomes of significant size in relation to the cross-sectional area of a loaded member, so much extra load is thrown onto the remaining cross-section that the crack growth accelerates with devastating effect. Aluminium alloys, particularly the higher strength ones developed after the war, were susceptible to fatigue failure. Fatigue was (and still is) a particularly awkward problem to deal with because apparently similar items may vary considerably in the number of loading cycles they endure, and also because of the difficulty of creating tests to match properly the conditions to be met by aircraft in service. de Haviland's considerable experience indicated that the wings of an aircraft were the parts most susceptible to fatigue, being exposed to

a   stress reversals on landing and take-off;
b   stress variations due to gust loading;
c   stress variations caused by aircraft manoeuvres.

Pioneering flights

On 2 May 1952, having then flown a total of 339 hours, the first Comet aircraft entered scheduled passenger service. It was the first jet aircraft in the world to do so and operated at a cruising altitude of over 10,000 m, double that of other airliners. When military aircraft had flown to these altitudes, crew had been provided with oxygen masks. The solution for the Comet was to pressurise the whole cabin so that no-one would require masks. Every time the plane reached cruising altitude, the cabin would approximate to a large balloon carrying an internal pressure of $56 \, kN/m^2$. Indeed, the need to sustain internal pressure was the reason why the central part of the fuselage was a circular cylinder. Tension stresses arising due to this internal pressure could be calculated from formulae given in Sections 3.3 and 3.5 modified to allow for some of these forces being carried by longitudinal 'stringers' and circular fuselage frames. Where there were openings for doors and windows, extra structure was provided to enable stress to flow round them.

Disaster

On a flight from Rome to London on 10 January 1954, Comet G-ALYP (code-named Yoke Peter) fell from the sky in pieces when it was reaching its cruising altitude. The Comet fleet was grounded, safety reviews were carried out and some modifications were made. Debris recovery from the sea was slow, but eventually about 70% of the structure was retrieved. Having no definite evidence as to the cause (black boxes had not been invented), flights were resumed in March. However, on 8 April of the same year, Comet G-ALYY (Yoke Yoke) suffered the same fate on a flight from Rome to Cairo. It was significant that it was also just reaching its cruising altitude (i.e. when the cabin pressure was reaching its maximum).

Investigation

After this similar accident, the evidence pointed more to fatigue in the fuse-lage than the wings and so a full-scale fatigue test was carried out on the fuselage of G-ALYU (Yoke Uncle). However, to do this by pumping in air would have been dangerous because of the energy stored in the compressed air. Failure would cause an explosion equivalent to a 250 kg bomb which could have done a lot of damage including destroying the evidence of what had initiated failure. Instead water was used, but in order to avoid hydrostatic forces which would not be present in practice, the fuselage was placed in a tank of water so that internal and external pressures on the fuselage balanced before pressurisation took place. Because water is almost incompressible, the internal pressure fell away quickly once a fatigue failure occurred. Figure 3.65 shows the failure which was obtained after the equivalent of 3060 flights. Detailed calculations showed that higher stresses than anticipated occurred at the corners of the almost square windows (approximately $315 \, N/mm^2$ of which about 94% arose through internal cabin pressure). Strain gauge measurements on test specimens also confirmed this result.

These much higher stresses than previously envisaged, were consistent with the lower fatigue life obtained in the tests. Although Yoke Peter had made 1290 flights and Yoke Yoke only 900 before their accidents, variability in fatigue performance between different specimens and also differences between test configurations and flight conditions could easily explain these differences (see Fig. 3.66).

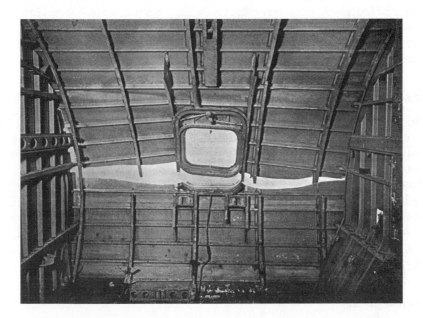

Figure 3.65 Fatigue failure of forward escape hatch of Comet G-ALYU tested in a water tank.

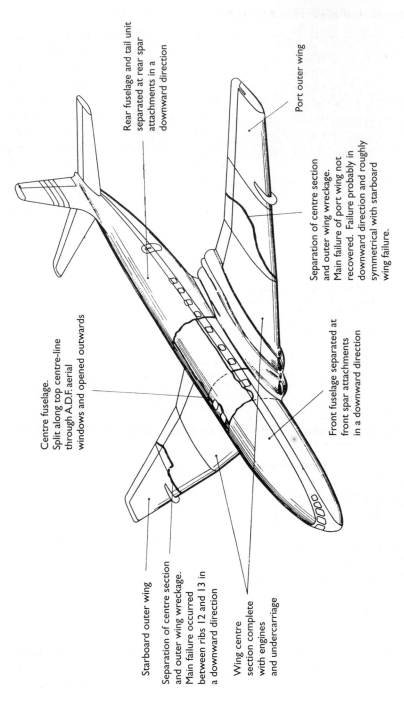

Rear fuselage and tail unit separated at rear spar attachments in a downward direction

Port outer wing

Separation of centre section and outer wing wreckage. Main failure of port wing not recovered. Failure probably in downward direction and roughly symmetrical with starboard wing failure.

Centre fuselage.
Split along top centre-line through A.D.F. aerial windows and opened outwards

Front fuselage separated at front spar attachments in a downward direction

Starboard outer wing

Separation of centre section and outer wing wreckage. Main failure occurred between ribs 12 and 13 in a downward direction

Wing centre section complete with engines and undercarriage

Figure 3.66 Location of fractures for Comet G-ALYP.

Openings and their reinforcement

Where a hole is made in a membrane carrying a 2D tension field, reinforcement round the edge of the opening is generally needed to carry the forces otherwise carried by the removed material. The ideal shape for the opening is one in which the forces in the reinforcement can be entirely axial. In an aircraft cylindrical fuselage, the hoop stress/longitudinal stress ratio is likely to be close to two, in which case the ideal shape is found to be an ellipse with a major axis/minor axis ratio equal to $\sqrt{2}$ as shown in Fig. 3.67. Because the windows on the Comet were almost square, the reinforcement would be subject to bending moments as well as axial loads. The critical points would be at the corners and top and bottom centre where maximum bending stresses occur (see Fig. 3.68). Fatigue fracture started at the inside of a corner where tensile stresses from both axial load and bending moment in the reinforcement add together.

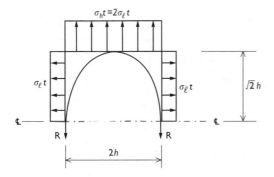

Figure 3.67 Ideal elliptical shape for a reinforced opening in a pressurised cylinder stress field.

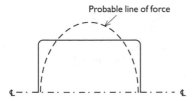

Figure 3.68 Line of force compared with an almost square reinforced opening.

Consequences

- The detailed theoretical and experimental investigations increased the understanding of structural behaviour, fatigue life and their prediction.
- In the UK, a requirement was introduced that two complete airframes should be tested for each new aircraft design: one for static strength and the other for fatigue life.

- The Comet airliner did not fly again with fare paying passengers, though a later version with thicker fuselage skins, smaller window openings and more powerful engines operated in a military role for many years.
- The UK lost its lead in commercial airliner development to the US.

Q.3.18

Consider a reinforced opening in a pressurised cylinder in which the hoop stress is twice the longitudinal stress.

a   Show that an elliptical opening with the major axis/minor axis ratio equal to $\sqrt{2}$, as shown in Fig. 3.67, avoids the reinforcing ring being subject to bending moment.
b   Show that the axial load is not constant round the ring.

Q.3.19

Consider the above stress field, but with a square opening. Figure 3.69 shows the forces exerted on a half-frame ABCDE. Because the force line will cross AE at right angles, there will be no horizontal reaction at A and E. In the figure, the reactions at A and E have been displaced sideways a distance e to account for the line of force not passing exactly through these points.

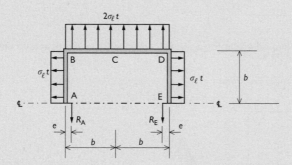

Figure 3.69

a   Determine expressions for the bending moments at B and C when $e = 0$.
b   Show that, if e has a small positive value, the bending moment at the corner B will be greater in magnitude than that at C.

Chapter 4

Beams

Beams are the most universal of structural forms. They are present in almost every structure and, in some cases, the whole structure may be considered to be a beam or an assembly of beams. This is the case with ships or aircraft and also many bridges. Because of their many uses, they take on a large variety of cross-sectional shapes. It is thus important to understand as thoroughly as possible their structural action, and to know when rule of thumb techniques may be adequate for analysis and when particular care needs to be taken.

On completion of this chapter you should be able to do the following:

- Derive and check shear force and bending moment diagrams.
- Recognise some ways of keeping critical bending moments as low as possible.
- Determine section properties and their influence on elastic stresses.
- Understand how equilibrium principles may be applied to tapered beams, unsymmetric beams and beams with diaphragms.
- Appreciate the reasons behind the shapes of some commonly used types of beam.

NB   This chapter does not deal with statically indeterminate beams or composite beams, nor is it concerned with estimating beam deflections. These will be discussed in Chapter 6.

4.1   Bending moment and shear force diagrams

2D analysis of statically determinate beams

Beams are straight bars which carry transverse loading by means of bending action. The determination of internal forces present in beams is simpler than for arches, not only because of their straight line geometry, but also because their supports are usually designed so that no axial forces are present. If beams are symmetrical about the plane of the loading, a 2D analysis is possible both to determine internal forces and also to investigate possible stress distributions. Statically determinate beams are generally either 'cantilevered',

that is, having one end built-in, or 'simply-supported', in which one support
is pinned and another has a pin and roller. The two relevant internal forces
are conveniently represented graphically in the form of shear force diagrams
(SFDs) and bending moment diagrams (BMDs).

Ex. 4.1

Draw the SFD and BMD for the loaded cantilever shown in Fig. 4.1.

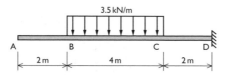

Figure 4.1 A cantilever beam.

Answer

Consider parts AE, AF and AG of the cantilever, where E, F and G are
situated along AB, BC and CD, respectively. With x measured to the
right of A, the internal forces at E, F and G shown in Fig. 4.2 can be
determined from equilibrium (using kN and m units) as

a $S_E = M_E = 0$

b $S_F = 3.5(x - 2)$, $M_F = -1.75(x - 2)^2$ where $2 \leq x \leq 6$

c $S_G = 14$, $M_G = -14(x - 4)$ where $6 \leq x \leq 8$

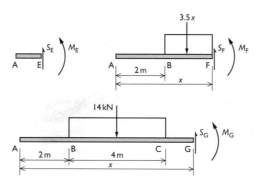

Figure 4.2 Forces acting on portions of the cantilever beam (with x measured
from A).

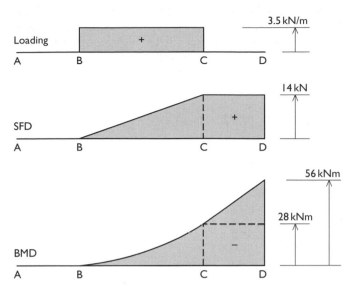

Figure 4.3 Internal forces for the cantilever beam.

which yield the diagrams shown in Fig. 4.3. Bending moments have been drawn negative upwards to correspond with the tension side of the beam (as specified in Section 3.4).

Ex. 4.2
Draw the SFD and BMD for the loaded simply-supported beam shown in Fig. 4.4.

Answer
From overall equilibrium $R_A = 9\,\text{kN}$ and $R_D = 15\,\text{kN}$.

With F situated on BC at distance x from A, equilibrium of part AF gives in kN and m units:

$$S_F = 3, \quad M_F = 9x - 12(x - 4) = 48 - 3x$$

Using similar analyses for other parts of the beam enables the SFD and BMD shown in Fig. 4.4 to be drawn.

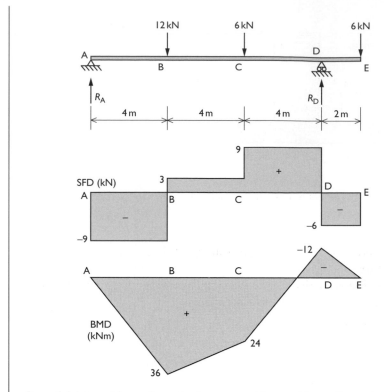

Figure 4.4 Internal forces for a simply-supported beam with cantilevered end.

Relationships between load, shear force and bending moment

Do you notice anything interesting about the SFD and BMDs of Figs 4.3 and 4.4? In both cases the SFD describes the slope of the BMD with the sign reversed and the loading diagram describes the slope of the SFD.

The calculus of variations can be used to show that these are general rules. Consider a beam which has internal forces varying from S and M at spanwise coordinate x to $S + dS$ and $M + dM$ at coordinate $x + dx$ where dx is an incremental length (Fig. 4.5). Equilibrating forces and dividing by dx gives

$$w = \frac{dS}{dx} \tag{4.1}$$

Similarly for moment equilibrium

$$S = -\frac{dM}{dx} \tag{4.2}$$

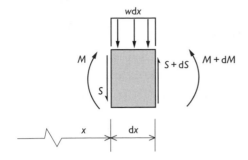

Figure 4.5 Forces acting on an incremental beam element.

The corresponding integral relationships to these are $S = \int w \, dx$ and $M = -\int S \, dx$.

If integration is carried out between spanwise coordinates x_1 and x_2

$$S_2 - S_1 = \int_{x_1}^{x_2} w \, dx \tag{4.3}$$

and

$$M_2 - M_1 = -\int_{x_1}^{x_2} S \, dx \tag{4.4}$$

These integral relationships are also seen to be satisfied by the graphs of Figs 4.3 and 4.4. Where the SFD has an infinite slope the differential relationship appears just to indicate that there is an infinite pressure (i.e. a concentrated load). However, from the corresponding integral relationship it is evident that the magnitude of the concentrated load equals the step change in shear force. Figure 4.6 illustrates these relationships for a further example.

Consequences of variational relationships

Two consequences of the variational relationship between shear force and bending moment for beams, which do not have an externally applied moment, are as listed here:

(i) Local maxima and minima on the BMD must correspond to zero positions on the SFD.

In Fig. 4.4, B and D have zero values on the SFD corresponding to the largest positive and negative bending moments. In Fig. 4.6 zero shear

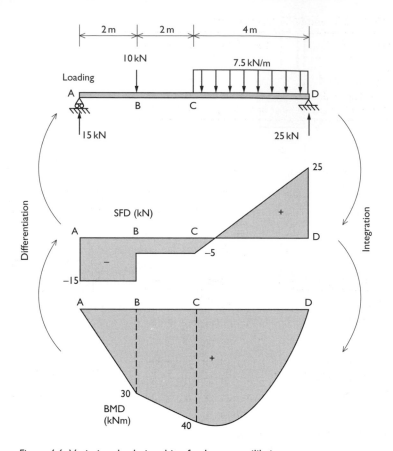

Figure 4.6 Variational relationships for beam equilibrium.

force occurs at 0.667 m to the right of C, thus enabling the maximum bending moment to be easily calculated. Figure 4.3 is a special case where there is a moment applied directly to the beam at D and that is where the maximum bending moment occurs.

(ii) Between any two points that have zero bending moment, the area of the SFD must be zero (from the integral relationship 4.4).

This is useful because most beams (a cantilever beam being an exception) have zero bending moment at both ends. In such cases, the total area of the SFD is zero.

In addition, the characteristics shown in Table 4.1 derive from the variational relationships. Figure 4.7 shows three simple frequently occurring loading cases.

Table 4.1 SFD and BMD characteristics

Loading	Shear force	Bending moment
None	Constant	Linear
Uniform	Linear	Quadratic
Linear (e.g. hydrostatic)	Quadratic	Cubic
Concentrated	Step change	Change in slope

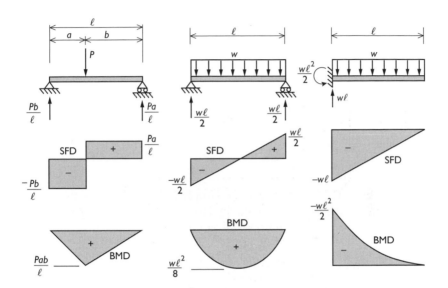

Figure 4.7 Some standard loading cases.

Q.4.1

Figure 4.8 shows a beam loaded on its overhanging portion. Determine the SFD and BMD for this beam and check the validity of (i) and (ii) above.

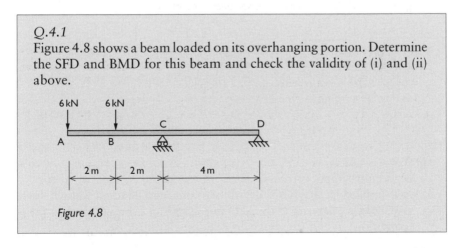

Figure 4.8

Q.4.2
Determine the maximum bending moment for the loaded beam shown in Fig. 4.9.

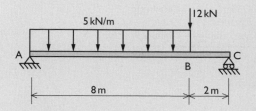

Figure 4.9

Q.4.3
Draw the complete SFD and BMD for each of the two loading cases acting on the three span bridge of Ex. 1.25.

Q.4.4
Sketch the SFD and BMD for the simply-supported beam shown in Fig. 4.10 which is subject to a hydrostatic loading. Determine an expression for the maximum bending moment in terms of w and ℓ.

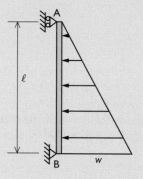

Figure 4.10

Q.4.5
a A simply-supported beam ABC is loaded by means of a couple applied at B. With the dimensions and loading shown in Fig. 4.11(a), determine the SFD and BMD and show that the two conditions (i) and (ii) above do not apply in this case.

b   Determine how the SFD and BMD change if the couple is replaced
by two vertical forces spaced at 0.2 m apart as shown in Fig. 4.11(b)
and show that conditions (i) and (ii) are now satisfied.

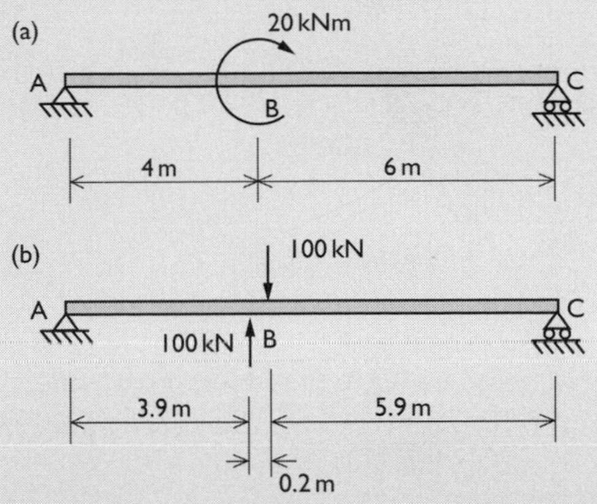

Figure 4.11

4.2   Control of bending moments

End-supported beams

Where a horizontal beam is simply-supported at its ends and is subject to vertical downwards loading, the bending moment will be positive along its entire
length. Whereas shear forces will be maximum adjacent to the supports,
bending moments will be largest across the central regions. The maximum
bending moment is often at centre span. For instance, Fig. 4.12 shows BMDs
corresponding to a single concentrated load placed at various spanwise
positions. The envelope, which is parabolic, shows the locus of largest bending moments occurring at different cross-sections, with the largest overall
bending moment recorded at centre span when the load is placed there.

It is interesting to note, however, that if two loads, spaced a fixed distance apart, travel slowly across such a beam, the maximum overall bending
moment does not necessarily occur at centre span. For the case shown in
Fig. 4.13, the maximum overall bending moment on an 8 m span beam
occurs 1 m from the centre when one of the loads is placed there and the
other is 1 m from the distant support. Thus for small span beam bridges, the
axle spacing as well as the axle loading of heavy goods vehicles is important
in determining not only the overall maximum bending moment to be resisted,

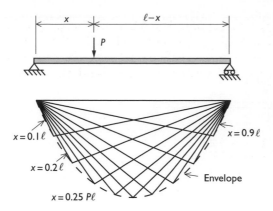

Figure 4.12 BMDs for a travelling load applied to an end-supported beam.

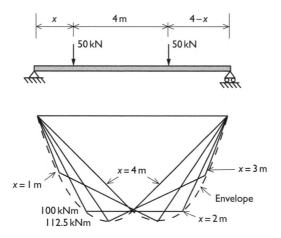

Figure 4.13 BMDs for two travelling loads applied at a fixed distance apart to an end-supported beam.

but also the shape of the bending moment envelope. For this type of beam, it is not possible to control the magnitude of the bending moments except by changing the span length or modifying the loads.

Optimisation of support positions

Where it is possible to move the supports of a beam inwards from the ends, large reductions in the magnitude of the overall maximum bending moment are possible.

Ex. 4.3
A beam ABCDE of length ℓ has supports B and D at distance β from the ends with C being the centre span, as shown in Fig. 4.14. If the beam is to carry a uniformly distributed load of w/unit length, determine the β value which gives the largest reduction in overall bending moment.

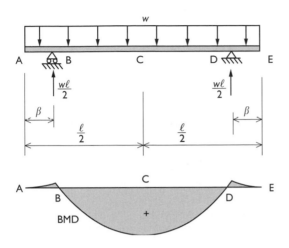

Figure 4.14 A beam with moveable supports.

Answer
From overall equilibrium, the support reactions are each $w\ell/2$ and from the method of sections:

$$M_B = -\frac{1}{2}w\beta^2$$

$$M_C = \frac{1}{2}w\ell\left(\frac{1}{2}\ell - \beta\right) - \frac{w\ell^2}{8}$$

With M_D (which is equal to M_B) these constitute the largest negative (hogging) and the largest positive (sagging) bending moments (Fig. 4.14). When β is increased from zero, M_C is decreased in magnitude and there is continuing benefit whilst it remains the largest overall bending moment. However, benefit ceases when M_B (which is increasing in magnitude with β) takes over as the largest bending moment. Therefore the optimum β is when $M_C = -M_B$. From the above formulae, this occurs when

$$4\beta^2 + 4\ell\beta - \ell^2 = 0$$

the relevant solution being $\beta = 1/2(\sqrt{2} - 1)\ell = 0.207\ell$. Choosing this value for β, the maximum overall bending moment is reduced to $\pm (3 - 2\sqrt{2})w\ell^2/8$ which is only 17.2 % of that for the end-supported beam (Fig. 4.15).

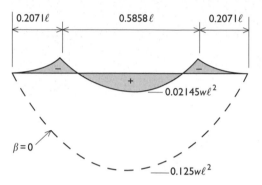

Figure 4.15 BMD for optimum support configuration compared with that for an end-supported beam ($\beta = 0$).

Beam extensions

More frequently the span between supports of a beam is fixed, but it is possible to extend the beam outwards beyond the supports by making it longer. If the span between supports is ℓ and a uniformly distributed load is to be applied over the whole of the beam, the optimum extension length is $\ell/2\sqrt{2} = 0.354\ell$, in which case the overall maximum bending moment is reduced to one half of its magnitude when unextended. However these benefits do not arise with live loading. A single concentrated load, P, will give a central sagging bending moment of $0.25P\ell$ when placed there, whether or not the beam has extensions. Furthermore, if extensions have a length greater than 0.25ℓ, a larger hogging bending moment than $0.25P\ell$ will be recorded over a support when the concentrated load is placed at the nearest overhanging end of the beam.

One use of beam extensions has been with timber housing where upper storeys were frequently corbelled out as shown in Fig. 4.16. The weight of the walls acting on the ends of the first floor beams gave them a negative bending moment which would help to counteract the positive bending moments arising from both dead and live loading acting between the supports.

(a)

(b)

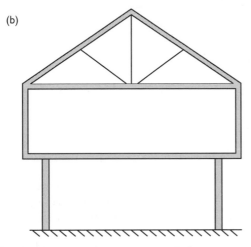

Figure 4.16 Use of corbelled floor beams in timber house construction: (a) a half-timbered house in Wiltshire, England; (b) cross-section.

Use of cantilevered side spans

An alternative technique for spanning a gap using beams is for side spans to cantilever into the main span gap, with a simple beam spanning between the cantilever ends. To assess the structural action, the whole structure may be considered to be a beam on four supports with two internal hinges. For the case of a uniformly distributed load and using the distances between supports already adopted for Fig. 1.52 and Q.4.3, Fig. 4.17 shows the effect on the BMD of moving the hinge positions. The choice of hinge position is

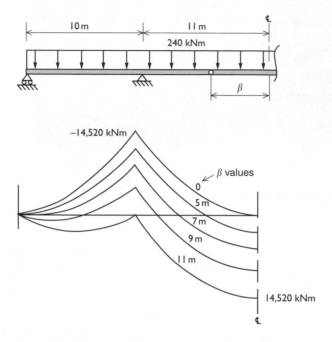

Figure 4.17 BMDs for different hinge positions for the bridge shown in Fig. 1.52.

seen to provide a choice to designers from all hogging bending moment at one extreme to all sagging at the other (for this loading case at least).

Q.4.6
A beam of length ℓ may be subject to a live load of w/unit span anywhere along its length. Find the position of simple supports which minimise the maximum possible overall bending moment and compare the maximum bending moment with that of the end supported case.

Q.4.7
The beam shown in Fig. 4.18 has three supports and one internal hinge. With the loading and dimensions shown, determine expressions for the maximum positive and negative bending moments on the span. Determine the optimum position of the hinge which minimises the maximum overall bending moment.

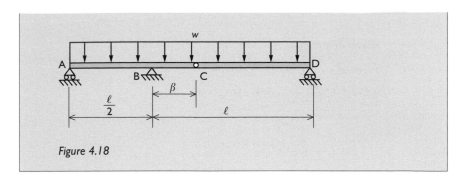

Figure 4.18

4.3   Elastic bending stresses

Engineers theory of bending

When a beam with uniform material properties is subject to positive bending moment, M, in the vertical plane, without axial load being present, top fibres will be in compression and bottom fibres in tension causing the beam to develop a 'sagging' curvature. Conversely with a negative bending moment, the outer fibre stresses will be reversed in which case the curvature is said to be 'hogging'. The principal assumption of engineer's theory of bending derives from elastic theory (discussed further in Chapter 6). It is that bending stresses vary linearly within any cross-section. If the beam being analysed has a cross-section which is symmetrical about either the y or z axis (i.e. types I, II or III of Fig. 4.19, but not type IV), stress will be proportional to the vertical distance, y, above its centroid, that is,

$$\sigma = ky$$

where k is a constant. Hence an incremental element dA of the cross-section at distance y above the centroid carries a force of $kydA$ (Fig. 4.20).

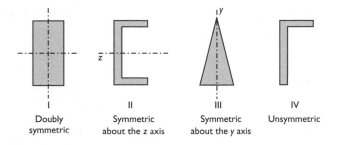

I	II	III	IV
Doubly symmetric	Symmetric about the z axis	Symmetric about the y axis	Unsymmetric

Figure 4.19 Types of cross-section.

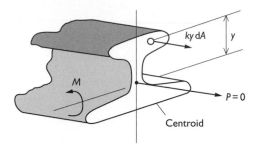

Figure 4.20 Force due to bending acting on an incremental area of the cross-section.

Integrating the forces and moments for the whole cross-section gives

$$P = k \int y \, dA$$

$$M = -k \int y^2 \, dA$$

However $\int y \, dA = 0$ because y is measured from the centroid. If $\int y^2 \, dA = I$, the above equations yield

$$\frac{\sigma}{y} = -\frac{M}{I} (= k) \tag{4.5}$$

Second moment of area

The function I is known as 'second moment of area'. This is a preferable name to the frequently used 'moment of inertia' because it has nothing to do with mass. Equation (4.5) indicates its key role in relating the magnitudes of elastic bending stresses at a cross-section to that of the applied bending moment.

The value of I must always be positive. Evaluation of I by integration is usually performed using horizontal strips taking advantage of y being constant for each strip. Thus, for a rectangular beam of breadth b and depth d as shown in Fig. 4.21,

$$I = b \int_{-d/2}^{d/2} y^2 \, dy = \frac{bd^3}{12} \tag{4.6}$$

Figure 4.22 shows this result alongside three other results obtained by integration.

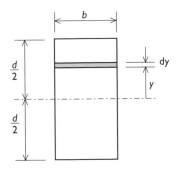

Figure 4.21 Evaluation of I for a rectangular cross-section.

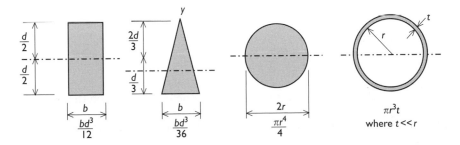

Figure 4.22 I values for some simple cross-sections.

The parallel axis theorem

In practice, mathematical integration rarely needs to be used for finding second moments of area. Instead more complex geometries can be investigated by considering them as combinations of rectangular components or other simple shapes. However, in order to do this, it is necessary to evaluate the I contribution of a component, not about its own local centroidal axis, but about the centroidal axis of the whole cross-section.

If the second moment of area is required of a component about an axis $\bar{z}\,\bar{z}$ separated by a distance s from its own centroidal axis zz (see Fig. 4.23), the required I value contribution is

$$\bar{I} = \int (y+s)^2 \, dA = \int y^2 \, dA + 2s \int y \, dA + s^2 \int dA$$

However, $\int y^2 \, dA = I$, $\int y \, dA = 0$ and $\int dA = A$, which yields the parallel axis theorem as

$$\bar{I} = I + As^2$$

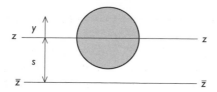

Figure 4.23 Parallel axes.

Note that, after shifting from the centroidal axis of the area itself to a parallel axis, the I value is always increased. Furthermore, the sign of s does not affect the result.

Methods of computing I

The contribution of I of an individual component of a cross-section can be determined by one or a combination of the following techniques:

a   Use of $bd^3/12$ when a rectangular component is centred on the centroid of the cross-section.
b   Use of $bd^3/3$ when the top or bottom of a rectangular component aligns with the centroid of the cross-section.
c   Use of the parallel axis theorem or integration when the component is not rectangular.
d   Use of subtraction as well as addition of areas to build up the complete cross-section.

In addition, it is important to note that, with cross-sections of types III and IV in Fig. 4.19, it is necessary to first determine the position of the centroid.

Elastic modulus and critical stresses

Once I has been determined, the maximum elastic bending stress occurring on a cross-section may be determined from Eq. (4.5) by ignoring signs of M and y as

$$\sigma_{max} = \frac{My_{max}}{I}$$

However, y_{max}, like I, is a property of the cross-section geometry. Hence $Z = I/y_{max}$, called the 'elastic modulus', contains all the information about the cross-section required to compute the maximum elastic bending stress at a cross-section given the bending moment.

Ex. 4.4

a For the six cross-sections shown in Fig. 4.24, determine, in each case, the *I* value for bending in a vertical plane.

b For all of those classified as types I, II or III in Fig. 4.19, determine the elastic modulus, *Z*, and the maximum elastic bending stress when the cross-section is subjected to a bending moment of 100 kNm.

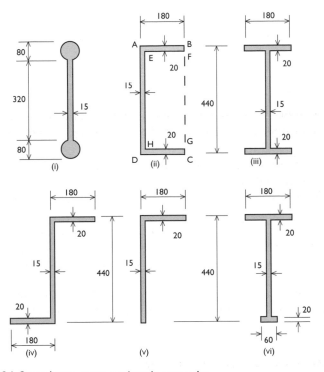

Figure 4.24 Some beam cross-sections in mm units.

Answer

a Using mm units

(i) On account of symmetry, the centroid is half way up the central web, thus making its contribution $15 \times 320^3/12$. For each circular bar $A = 5027$, $I = 2.011 \times 10^6$ and $s = 200$. Hence, from the parallel axis theorem, one bar contributes $2.011 \times 10^6 + 5027 \times 200^2 = 203.1 \times 10^6$ to the overall I value. Thus for the whole cross-section:

$$I = (2 \times 203.1 + 41.0) \times 10^6 = 447.2 \times 10^6 \text{ mm}^4$$

(ii)   Consider subtracting the contribution of EFGH from that of ABCD. Since the centroidal axes of both areas coincide with that of the total cross-section,

$$I = \frac{1}{12} \times 180 \times 440^3 - \frac{1}{12} \times 165 \times 400^3$$

$$= 397.8 \times 10^6 \text{ mm}^4$$

(iii)   The component areas correspond to those of (ii) except for shifts in the z direction which have no effect on the overall I value.

(iv)   This also corresponds to (ii)

(v)   First the height of the centroid needs to be determined, which is found to be 290 mm from IJ as shown in Fig. 4.25(a). One procedure would then be to split the cross-section into two rectangles (say ABEC and CDJI) and employ the parallel axis theorem. An alternative method is to add the contributions of areas ABHF and FGJI and then subtract the contribution of area DEHG. This gives

$$I = \frac{1}{3}(180 \times 150^3 + 15 \times 290^3 - 165 \times 130^3)$$

$$= 203.6 \times 10^6 \text{ mm}^4$$

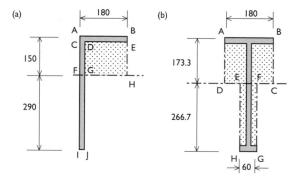

Figure 4.25 Use of subtraction for calculating I values (mm units).

(vi)   The centroid is found to be 266.7 mm above the base. By adding the contributions of areas ABCD and EFGH shown in Fig. 4.25(b) and subtracting the contribution of negative areas gives

$$I = \frac{1}{3}(180 \times 173.3^3 + 60 \times 266.7^3$$
$$- 165 \times 153.3^3 - 45 \times 246.7^3)$$
$$= 268.2 \times 10^6 \, \text{mm}^4$$

NB   When using the subtraction method, it is important to carry sufficient significant figures in the calculation to allow for loss of accuracy during subtraction.

b   (i) $Z = 447.2 \times 10^6/240 = 1.863 \times 10^6 \, \text{mm}^3$
Thus $\sigma_{max} = 100/1.863 = 53.7 \, \text{N/mm}^2$.

(ii) $Z = 397.8 \times 10^6/220 = 1.808 \times 10^6 \, \text{mm}^3$
Thus $\sigma_{max} = 100/1.808 = 55.3 \, \text{N/mm}^2$.

(iii)   As (ii) before.

(iv)   This is an unsymmetric section (type IV).

(v)   This is an unsymmetric section (type IV).

(vi)   The outer fibres at the bottom are further from the centroid than the outer fibres at the top. Therefore using $y_{max} = 266.7$ mm, $Z = 268.2 \times 10^6/266.7 = 1.006 \times 10^6 \, \text{mm}^3$. Thus $\sigma_{max} = 100/1.006 = 99.4 \, \text{N/mm}^2$. This will be a tensile stress if the bending moment is positive (sagging).

Note that type III cross-sections are more complicated than types I and II as far as computing bending stresses are concerned because the centroid height first needs to be identified and also because the largest tensile and compressive bending stresses will not normally be the same.

Q.4.8

a   For the five cross-sections having dimensions shown in mm in Fig. 4.26, determine, in each case, the I value for bending in a vertical plane.

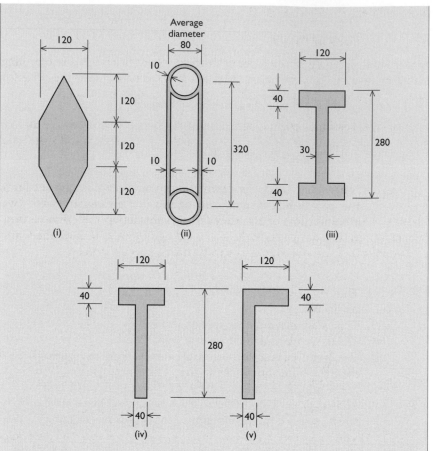

Figure 4.26

b   For all of those classified as types I, II or III in Fig. 4.19, determine the maximum elastic bending stress when the cross-section is
subject to a bending moment of 100 kNm.

Q.4.9
A rectangular hollow section of dimension 180 mm × 80 mm has
thin walls of uniform thickness, t. If the section needs to resist a
bending moment of 56 kNm about its major axis, determine the
thickness t required to ensure that elastic bending stresses nowhere
exceed 360 N/mm^2.

4.4   Cross-section shape

The importance of shape

If asked to support a plastic beaker of water by means of a banknote spanning a 130 mm gap, it is worth trying, particularly if the banknote is the prize for success! Lessons to be learnt from attempting this party trick are

- To withstand bending moments, a beam cross-section needs depth
- It is important that the cross-sectional shape does not distort (see Fig. 4.27).

To be efficient in bending in a vertical plane, a beam needs to have a large I value and this can be achieved by placing most of the material into two 'flanges' one at the top and the other at the bottom of the cross-section. Where a beam is required to withstand bending moments in both the y and z directions, the cross-section needs to be broad as well as deep.

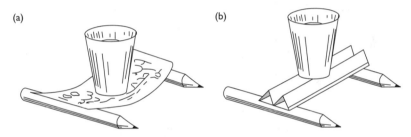

Figure 4.27 The bank note trick: (a) how not to support a beaker of water? (b) the jackpot solution?

Hot-rolled steel sections

Figure 4.28 shows some typical hot-rolled sections in structural steel used in the construction industry. Of these, the first three are designed to resist one-way bending and therefore have most of the material in two flanges set well apart. Although columns carry axial compression, they also need to have good two-way bending characteristics. The broad flanges of the universal column satisfy this requirement. The tee and angle sections tend to be used more for axially loaded members of frames where performance in bending is not so important. They were used more extensively before welding became common, because of the ease of riveting and bolting other members to them. Hollow sections, being of uniform thickness, are not quite as efficient in one-way bending as universal beams, joists and channels of

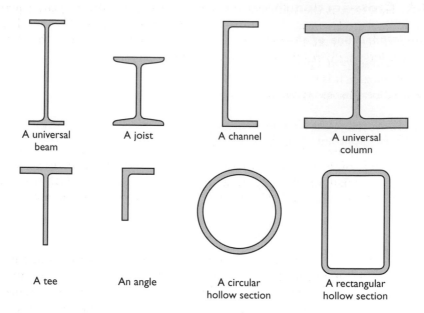

Figure 4.28 Some typical rolled steel sections.

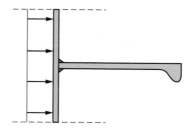

Figure 4.29 A bulb flat stiffener.

the same depth. However, they are particularly efficient in two-way bending and have a number of other advantages.

The steel plates of a ship's hull are stiffened against water pressure by stiffeners welded to their inner faces in order to act as beams spanning between adjacent frames or bulkheads. A common form of such stiffener is a bulb flat (Fig. 4.29). One flange is in the form of a bulb so that it does not impede access to welders during fabrication nor does it have concave corners which are difficult to access. The other end of the stiffener does not have a flange itself, but instead uses a section of the plate it is welded to as a flange. It is difficult to determine how much of the plate can be relied on to act in this way. However, if the stiffeners are close together, a width of plate equal to the spacing between the stiffeners can be assumed to be effective. The stiffener does not require to have a large torsional stiffness because the plate itself is torsionally restrained by neighbouring stiffeners.

Two other shapes of beams

A trestle table

Figure 4.30 shows a typical trestle table with retractable legs near to each end. Because of the length of such tables, it is necessary to secure the surface material to longitudinal members (generally wood or metal) which will act with it to form a longitudinal beam resisting the sagging bending moment.

A tape measure

Retractable steel tape measures flatten out so that they can be rolled up. However, when extended, they have a curved cross-section and are thus able to remain reasonably straight making them easier to use. For instance, a tape measure with the cross-section shown in Fig. 4.31 was able to extend unsupported a distance of 2.4 m.

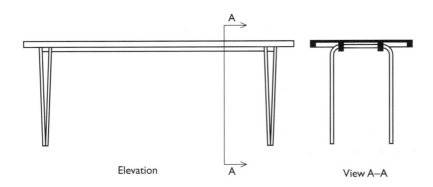

Elevation   View A–A

Figure 4.30 A trestle table (the beam cross-section is the dark area in view A–A).

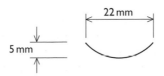

22 mm

5 mm

Figure 4.31 The cross-section of a retractable steel measuring tape.

Q.4.10
Specify advantages and disadvantages of using hollow (closed) as opposed to open sections for different types of beam applications.

Q.4.11 Observation question
What other examples of beams can you identify? For each indicate, if possible, what are the materials of construction, the principal types of loading and the factors influencing cross-sectional shape.

Q.4.12 Observation question
Obtain a retractable steel tape measure and determine how far it can extend without support (a) with the scale upright; (b) inverted. Hence determine the ratio of the two bending moment capacities and suggest why these differ so much.

4.5 The elementary beam

The concept of an elementary beam is suggested here in order to help with the understanding of beam action and clarify internal load paths by giving each component only one simple structural function. Figure 4.32 shows an elementary beam of span ℓ, simply-supported at the ends and carrying

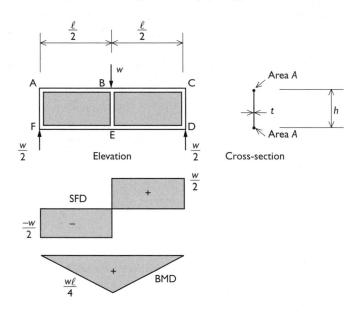

Figure 4.32 An elementary beam.

a central concentrated load W. It consists of rods ABC and DEF as flanges and AF, BE and CD as posts, which carry only axial loads, and web panels ABEF and BCDE, which carry only shear. The height of the cross-section is h, the flange cross-sectional areas are A and the web thickness is t.

Figure 4.33 shows the forces acting on the elements of this beam which split into four actions as follows:

(i) *The central post (or diaphragm)* transfers the applied load to the shear webs. The post will be in compression with the force varying linearly from W at the top to zero at the bottom. Note, however, that the internal forces depend on how the external force is applied. If instead, a load of W is hung from the bottom of the beam, the force in this post would be tensile rather than compressive, though nothing else in the beam would be different.

(ii) *The webs* transfer the load from the central post to the end posts. To do this, the vertical components of the shear force are uniform across each panel, but opposite in sign between the two panels, thus matching the SFD. It is important to note which directions the shear forces act in and also their equivalent representation using principal stresses shown in Fig. 4.34. Tensile stresses occur in the 45° direction closest to the diagonals AE and EC (as would be the situation if a cable was supporting the load) and compressive stresses occur at right angles to this (the thrust line for an arch subject to this loading would follow the diagonals FB and BD).

(iii) *The end posts (or diaphragms)* convey the load from the shear webs to the reaction points. As with the central posts, the internal forces in the posts depend on the nature of the external forces (they would be in tension rather than compression if the beam were to be hung from A and C rather than resting on supports at D and F).

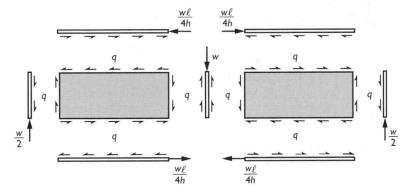

Figure 4.33 An exploded view showing equilibrium of components (note $q = w/2h$).

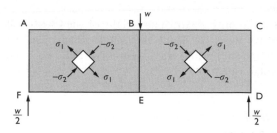

Figure 4.34 Principal stresses in the web of an elementary beam ($\sigma_1 = -\sigma_2 = q/t = W/2ht$).

(iv) *The flange rods* react with the horizontal complementary shear forces in the webs by developing axial forces. By using the method of sections, it is possible to show that the spanwise distribution of axial force in either flange rod is the bending moment diagram divided by the distance between the flanges. Maximum longitudinal stresses occur at B and E with magnitude $|\sigma|_{\max} = W\ell/4hA$.

The behaviour of the elementary beam is closest to the behaviour of beams which have most of their cross-sectional area in two flanges (e.g. the first four sections of Fig. 4.28) and can be used to roughly estimate their maximum stresses in bending and shear. However, in practice the flanges carry some shear as well as bending stresses, and the webs not only carry bending stresses at parts distant from the neutral axis as well as shear stresses, but also, if they are thick enough, take over the role of the diaphragms.

Q.4.13
Determine all axial and shear stresses in the elementary beam shown in Fig. 4.35.

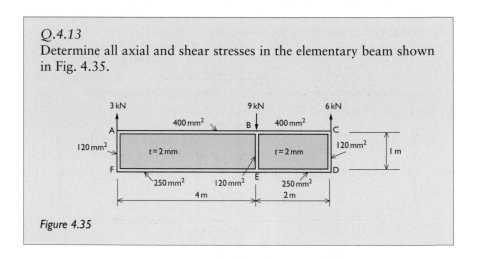

Figure 4.35

Q.4.14
Use the elementary beam concept to provide a rough check on the maximum bending stresses obtained in Ex. 4.4 for beams (i) and (ii) of Fig. 4.24. Explain why the simplified method underestimates the maximum bending stress in one case and overestimates it in the other.

4.6   Elastic shear in beams

General formula

Where the distribution of axial stress conforms with elastic bending theory as derived in Section 4.3, the distribution of shear stress must also be linked to it through equilibrium principles. Thus if two neighbouring cross-sections of a beam at coordinates x and $x + dx$ carry bending moments of M and $M + dM$, respectively, consider the longitudinal equilibrium of the material between them and lying above an imaginary cut PQRS as seen in Fig. 4.36. Using Eq. (4.5) for bending stress, the total force acting on the left-hand face is

$$\int_{\tilde{A}} \sigma \, dA = -\frac{M}{I} \int_{\tilde{A}} y \, dA$$

where integration is over the area $\tilde{A}$ of the cross-section which lies above RS. Similarly the total force acting on the right-hand face above PQ is

$$\int_{\tilde{A}} (\sigma + d\sigma) \, dA = -\frac{M + dM}{I} \int_{\tilde{A}} y \, dA$$

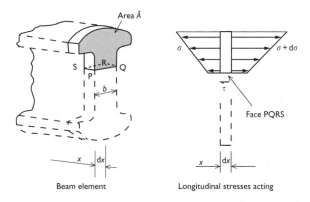

Figure 4.36 Equilibrium due to shear and bending.

If face PQRS carries a uniform shear stress τ, the shear force acting on it must be $\tau b\, dx$. Longitudinal equilibrium of the beam element therefore gives

$$\tau b\, dx = -\frac{dM}{I}\int_{\tilde{A}} y\, dA$$

However, using the differential relationship between shear and bending moment given by Eq. (4.2), it follows that

$$\tau = \frac{S}{Ib}\int_{\tilde{A}} y\, dA \tag{4.7}$$

Because of the complementary nature of the shear stresses identified in Section 3.2, Eq. (4.7) also defines shear forces acting on the cross-section at right angles to PQRS.

Ex. 4.5 Shear stresses in a rectangular beam
Determine a formula for the distribution of shear stress in a rectangular beam of cross-section $b \times d$ due to a shear force S, on the assumption that elastic bending theory holds.

Answer
Consider the area $\tilde{A}$ in Eq. (4.7) to be the area above the line $y = \tilde{y}$ shown in Fig. 4.37(a). The required integral is the first moment of area $\tilde{A}$ about the centroidal axis and can be determined from area × arm, where the area is $b(\frac{1}{2}d - \tilde{y})$ and the arm is the distance between the centroid of $\tilde{A}$ and that of the whole cross-section which is equal to $\frac{1}{2}(\frac{1}{2}d + \tilde{y})$. Alternatively strip integration could be used to give

$$\int_{\tilde{A}} y\, dA = \int_{\tilde{y}}^{d/2} by\, dy = \frac{b}{2}\left(\frac{d^2}{4} - \tilde{y}^2\right)$$

Hence from Eq. (4.7) with $I = bd^3/12$, the shear stress varies parabolically with $\tilde{y}$ according to

$$\tau = \frac{6S}{bd^3}\left(\frac{d^2}{4} - \tilde{y}^2\right) \tag{4.8}$$

as illustrated in Fig. 4.37(b) and (c).

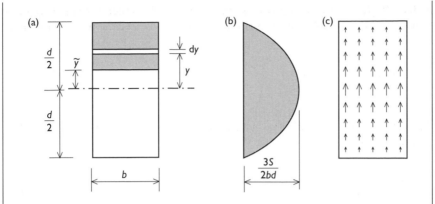

Figure 4.37 Shear stresses in a rectangular cross-section: (a) cross-section; (b) graph for τ; (c) vertical shear stresses.

Stress field for a rectangular beam

In order to obtain a more complete picture of what happens to the internal forces in a beam, it is possible to combine stresses due to bending and shear from formulae already derived and then predict principal stresses and their directions of action throughout a beam. Figure 4.38 shows such a stress field derived for half of a simply-supported beam with a central load. Whereas some of the stress lines link the loaded cross-sections at the centre and ends,

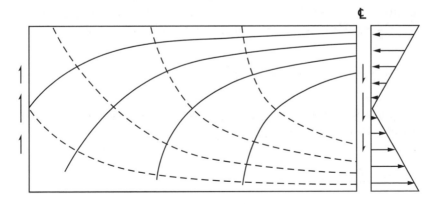

Figure 4.38 Basic elastic theory stress field for the left-hand side of a simply-supported rectangular beam with central concentrated load.

others start from the top or bottom of the beam, cross the neutral (centroidal) axis at 45° and amass at the central cross-section where the bending moment is largest. In this analysis, the role of the posts as described in Section 4.5 has not been considered. Therefore, if the external loads are not acting across the web, the stress field in a real beam will differ from that given in Fig. 4.38 in regions close to load application points.

Ex. 4.6 Shear stresses in a flanged beam

For the I section shown in Fig. 4.39(a), estimate using elastic bending theory, the shear stresses arising in the flanges and web due to an applied shearing force of 25 kN.

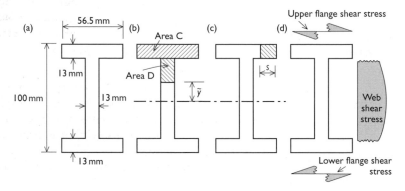

Figure 4.39 Elastic shear stress distribution in an I section subject to shear loading: (a) cross-section; (b) cut in web; (c) cut in flange; (d) shear stress distributions.

Answer

Consider $\tilde{A}$ to be the cross-sectional area above a horizontal line in the web at $\tilde{y}$ above the neutral axis. The integral $\int y \, dA$ for this area may be evaluated by summing the contributions from the two areas C and D shown in Fig. 4.39(b). Thus in Newton and mm units

$$\int_{\tilde{A}} y \, dA = 56.5 \times 13 \times 43.5 + 13(37 - \tilde{y})(37 + \tilde{y})/2$$

$$= 40{,}850 - 6.5\tilde{y}^2$$

Hence

$$\tau_{web} = 25,000(40,850 - 6.5\tilde{y}^2)/(3,239,000 \times 13)$$

$$= 24.25 - 0.00386\tilde{y}^2$$

It is not very meaningful to obtain shear stress estimates in the flanges by using a horizontal line to define $\tilde{A}$. This is because the distribution of shear force across the line will be far from uniform. In this regard, it is important to note that the section PQRS used to derive the shear formula 4.7 does not need to be horizontal or parallel to the neutral axis. Since the shear forces in the flange are mainly horizontal, it is more appropriate to use a vertical line at (say) distance s from the tip with $\tilde{A}$ being the area outboard of this (see Fig. 4.39(c)). Thus $\int y \, dA$ for this area is $13s \times 43.5 \, \text{mm}^2$ giving

$$\tau_{flange} = 0.3358s \, \text{N/mm}^2$$

The stresses in the left-hand side of the flange will be the same as for the right-hand side except for a change of sign. Hence the stress distribution computed is as shown in Fig. 4.39(d). (No attempt has been made to compute shear stresses acting in the region of the web/flange connection.)

Shear stresses in hollow sections

For a hollow beam, an area $\tilde{A}$ will require a boundary line to cut the cross-section twice. The equivalent formula to (4.7) would therefore be

$$\tau_1 b_1 + \tau_2 b_2 = \frac{S}{Ib} \int_{\tilde{A}} y \, dA$$

where b_1 and b_2 are the breadths at the two cuts and τ_1 and τ_2 are the corresponding average shear stresses. This would appear to provide insufficient information to determine shear stresses. However, if a hollow section is symmetrical about a vertical centreline, it is possible to employ an area $\tilde{A}$ which has $\tau_1 = \tau_2$ because of symmetry. Figure 4.40 shows the results of a shear stress analysis on a rectangular hollow section.

Shear flow due to applied shear

The concept of shear flow introduced in Section 3.6 for members subject to torque is also useful here when considering applied shear loading. The elastic beam theory shear flow is the integral of the shear stress across the

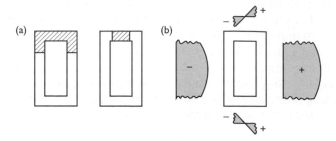

Figure 4.40 Elastic shear stress distribution in a hollow section subject to shear loading:
(a) cross-section showing cuts in webs or flange; (b) shear stress distribution
(positive clockwise for downwards shear).

cross-section. Thus Eq. (4.7) could be replaced by one for shear flow, $q = b\tau$,
thus

$$q = \frac{S}{I} \int_{\tilde{A}} y \, dA \qquad (4.9)$$

The magnitude of the shear flow is a more reliable quantity than an estimated
shear stress since it allows the user to decide whether to assume that the shear
stress is constant across the width of the imaginary cut or not.

Shear flow may be visualised by considering a real beam to be a conjunction of elementary beams. Each component elementary beam has flanges on
either side of the neutral axis representing longitudinal fibres of the real beam
and a web entirely within the cross-section. Figure 4.41 shows three such
representations (note that it does not violate any equilibrium conditions for
the webs of the elementary beams to be bent in cross-section). High shear
stresses occur where there is a thin web through which a large number of

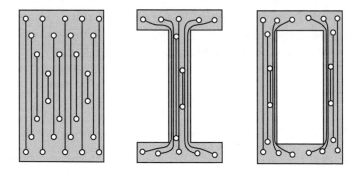

Figure 4.41 Different beam cross-sections visualised as combinations of elementary beams.

elementary beam webs need to squeeze. Shear flow lines will try to adopt the shortest path necessary to join the relevant flange fibres, thus crowding together at sharp concave corners. High local shear stress concentrations are avoided by rounding such corners. This is normal practice with rolled steel sections (see Fig. 4.28) and also extruded aluminium sections.

Shear flow is important when beams are fabricated using longitudinal joints (e.g. by welding flanges to webs or bolting flange plates onto an existing beam). These joints do not usually carry bending stresses, but they need to transfer the shear flow between the two parts.

Ex. 4.7
A simply-supported beam of span 7 m having the cross-section shown in Fig. 4.24 (iii) is to be fabricated by welding the flanges to the web. If the maximum loading for the beam is a distributed load of 65 kN/m, determine the maximum shear force/unit span that needs to be transferred across a weld securing one of the flanges to the web.

Answer
The maximum shear force will occur next to one of the supports and will be of magnitude $w\ell/2 = 227.5\,\text{kN}$. In Ex. 4.4, I was computed as $397.8 \times 10^6\,\text{mm}^4$. Assuming that $\tilde{A}$ represents one flange, $\tilde{A} = 3600\,\text{mm}^2$ and the arm (the distance between its centroid and the overall centroid) is 210 mm. Hence formula (4.9) gives

$$q = \frac{227.5 \times 10^3 \times 3600 \times 210}{397.8 \times 10^6} = 432\,\text{N/mm}$$

This is the maximum force/unit span requiring to be transferred across a weld.

Q.4.15
For the rectangular beam of Ex. 4.5, check that the resultant of shear stress distribution given by formula 4.8 is indeed S.

Q.4.16
A uniform beam with the cross-section shown in Fig. 4.42 has a span of 5 m. Determine the maximum concentrated load which may be placed anywhere on the beam if the direct stress is not to exceed $150\,\text{N/mm}^2$ and the shear stress is not to exceed $30\,\text{N/mm}^2$ (stresses due to self-weight may be ignored).

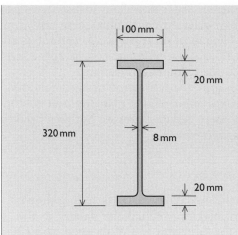

Figure 4.42

Q.4.17
If a shear force of 52 kN is applied to the T section in Fig. 4.43, determine the maximum shear stress according to elastic theory.

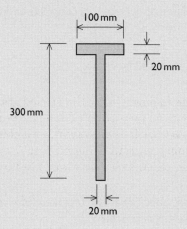

Figure 4.43

Q.4.18
The simply-supported beam of span 7 m having the cross-section shown in Fig. 4.24 (iii) and used for Ex. 4.7 is to be strengthened by bolting plates of dimensions 180 mm × 20 mm on to each flange as shown in Fig. 4.44. For a distributed load of 65 kN/m, determine the

maximum force/unit span required to be carried in shear by the bolts. Hence, determine the bolt spacing if each flange has two rows with the shear force/bolt not required to exceed 30 kN (for the purpose of this example, but not in practice, ignore any effect the bolt holes have on the properties of the cross-section).

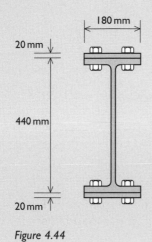

Figure 4.44

4.7 Tapered beams

It may often be advantageous to taper beams so that the I value is largest where the bending moment is greatest. Construction of tapered beams is easily accomplished when a beam is cast (such as with reinforced concrete) or when it is to be fabricated by joining separate components. A tapered beam may also be constructed from a rolled steel I section by cutting and rejoining in the manner shown in Fig. 4.45.

The effect of tapering a beam may be investigated most easily by adapting the concept of the elementary beam. Where an elementary beam has its upper flange horizontal and its lower flange sloping at an angle α, the flange and web forces at any chosen cross-section of depth d must be in equilibrium with the shear force and bending moment. With forces as designated in Fig. 4.46, taking moments about A and B in turn gives

$$P_U = M/d = P_L \cos \alpha$$

Thus, if the angle of taper is shallow, $\cos \alpha \cong 1$ and the flange forces are similar to what they would be in a uniform beam of depth d. However,

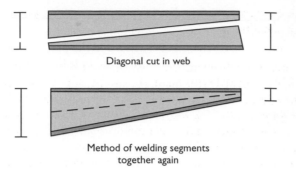

Diagonal cut in web

Method of welding segments
together again

Figure 4.45 Fabrication of a tapered member from a rolled steel I section.

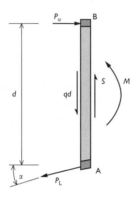

Figure 4.46 Forces acting at a cross-section of a tapered elementary beam.

vertical equilibrium gives

$$qd = S - P_L \sin \alpha$$

Here the last term is not negligible, and hence the vertical component of the force in the sloping flange needs to be taken into account when calculating web shear flow.

Ex. 4.8
A cantilever beam of span 2 m has a depth of section of 240 mm, flange areas of 800 mm^2 each and a web thickness of 6 mm. For the case of a uniformly distributed load of 50 kN/m, use elementary beam theory

to investigate to what extent tapering the beam in the manner shown in Fig. 4.45 has any benefits.

Answer
The shear force varies linearly from 0 at the tip to 100 kN at the root and the bending moment varies quadratically from 0 at the tip to -100 kNm at the root. Hence a uniform beam has maximum web shear stress at the root equal to $100{,}000/(6 \times 240) = 70 \, \text{N/mm}^2$ and the direct stresses in the flanges have a maximum value at the root of $\pm 100 \times 10^6/(240 \times 800) = 521 \, \text{N/mm}^2$. If, however, the web depth varies from 120 mm at the tip to 360 mm at the root, the angle of taper, α, is 6.84°. The forces in each flange at the root become $\pm 100 \times 10^6/360 = \pm 277{,}800 \, \text{N}$ giving direct stresses in the flanges of $\pm 347 \, \text{N/mm}^2$. Vertical equilibrium at the root, allowing for the lower flange to carry some of the shear force, gives

$$360q = 100{,}000 - 277{,}800 \times 0.1191$$

in Newton and mm units. Hence $q = 186 \, \text{N/mm}$ and the web shear stress is $31 \, \text{N/mm}^2$. Similar calculations at other spanwise positions and also considering the case of a beam tapered from 0 mm depth at the tip to 480 mm at the root gives the results shown in Fig. 4.47.

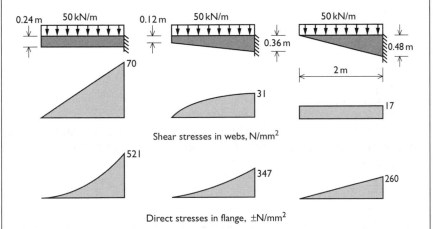

Shear stresses in webs, N/mm²

Direct stresses in flange, ±N/mm²

Figure 4.47 Shear and direct stress diagrams for a uniformly loaded elementary cantilever beam.

Ex. 4.9
Repeat the above example, but with a concentrated load of 50 kN placed at the tip.

Answer
The shear force is constant at 50 kN and the bending moment varies linearly from 0 at the tip to −100 kNm at the root. Figure 4.48 shows the computed values for web shear stress and flange bending stress.

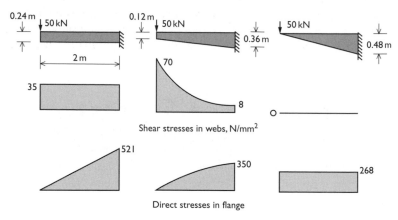

Shear stresses in webs, N/mm²

Direct stresses in flange

Figure 4.48 Shear and direct stress diagrams for a tip loaded elementary cantilever beam.

Comment on shear stress in tapered beams

In both examples, the maximum direct stresses in the flanges have been reduced by tapering the beam in the manner described. The effect on web shear stresses is not, however, always beneficial. In particular, where the load is concentrated, reducing the web depth near to the load application point leads to increased web shear stresses. The exception is when the web depth is reduced to zero (i.e. the beam tapers to a point), in which case a concentrated load placed there is carried by the flanges without any assistance from the web. These remarks also apply to support points. In particular, if a beam tapers virtually to a point near to a support and flange centre-lines are not concurrent with the line of the reaction (as in Fig. 4.49) large local stresses (particularly shear in the web) are likely to be generated.

Figure 4.49 Misaligned end support for a tapered beam.

Q.4.19

a A proposal for a simply-supported beam of span 9 m is to use tapered end sections as shown in Fig. 4.50. The most severe loading case is a uniformly distributed load of 22 kN/m. Using the elementary beam concept, estimate how the flange forces and web shear flows vary with span and compare with the corresponding values for a uniform beam of depth 700 mm (i.e. having the same total web area).

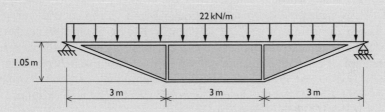

22 kN/m

1.05 m

3 m   3 m   3 m

Figure 4.50

b If the direct stresses in the flanges and the shear stresses in the web are to be limited to 400 N/mm^2 and 20 N/mm^2 respectively, suggest a suitable flange cross-sectional area and web thickness for both proposed designs.

c How could the configuration be altered to further reduce the required flange cross-sectional area without increasing the total area of the web?

4.8 Diaphragms

Diaphragms may be used in a beam to perform five often interrelated roles as follows:

 (i) To transfer applied loads to the web of the beam;
 (ii) To transfer internal forces at discontinuities;
(iii) To maintain the shape of the cross-section;
(iv) To prevent buckling of webs or compression flanges;

(v) To provide post-buckling strength in situation where webs may be allowed to buckle.

The possible need for diaphragms can be recognised from the banknote trick in Section 4.4 where, even for the best designs, collapse is caused by distortion of the cross-section in the vicinity of loading or support points rather than by tearing of the material.

The load transfer role of diaphragms can be illustrated as follows. Consider a simply-supported beam with a large central load. Stiffeners may be included at the ends and centre as shown in Fig. 4.51 with each one subject to the external load and the reverse of the shear input to the beam. Thus, if elastic beam theory is applicable, the internal forces acting on the diaphragms in Fig. 4.51 could be calculated using the shear flow formula (4.9).

Where beams have discontinuities, diaphragms may be needed to transfer forces. For instance, consider the partly tapered beam in Fig. 4.52 to be an elementary beam whose depth is d over the central part. If the total applied load, W, is symmetrically placed between B and C, the central web panel will have a shear flow of $-W/2d$ adjacent to B. This shear flow needs to be transferred by the diaphragm at B to react against the out of balance vertical component of the flange force as shown in the figure inset. Between A and B, there is no shear flow in the triangular web panel. (A similar situation occurs in Q.4.19 where the shear flow computed for the web exhibits a step change at the 1/3 span positions.)

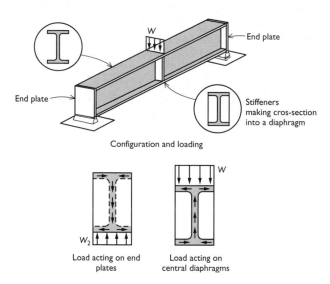

Figure 4.51 A simply-supported I beam with diaphragms.

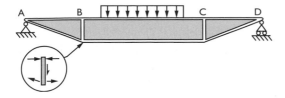

Figure 4.52 Forces acting at a discontinuity in a beam cross-section.

There are three possible design situations as follows:

a   Beams without a thin web (e.g. rectangular timber beams). These will maintain shape and hence do not need diaphragm support.

b   Flanged beams whose webs are sufficiently thick for most types of loading. Diaphragms may or may not be needed at positions of high loading or at discontinuities.

c   Thin-walled beams where diaphragms are needed at regular intervals (to be discussed further in Section 4.11).

Q.4.20
A weight is to be supported from a thin-walled circular cylinder which acts as a horizontal beam. At the cross-section where the load is to be applied, the cylinder is to be internally stiffened by a ring frame. The three alternative configurations for the support shown in Fig. 4.53 have been proposed. Indicate which you think would be the most acceptable solution and why?

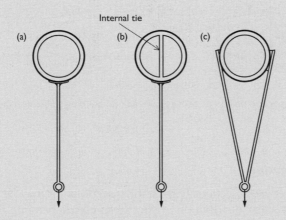

Figure 4.53

4.9 The shear centre

Definition

The shear centre is the point in the cross-section of a member through which forces must be applied to avoid any twisting. If a cross-section is doubly symmetric (Type I in Fig. 4.19), the shear centre (like the centroid) will coincide with the intersection of the planes of symmetry. However, in all other cases, the shear centre may not coincide with the centroid and problems could arise if this difference is ignored.

Two simple examples

Consider a T section turned on its side and used to support vertical loading. Because the web now lies along the neutral axis, it will carry negligible bending stresses. The bending characteristics will therefore be governed entirely by the flange with shear stresses solely located there (Fig. 4.54(a)). For the beam to bend without twisting, vertical loads must therefore align with the flange rather than the centroid (Fig. 4.54(b)). Alternatively, if a horizontal load is applied, this needs to be aligned with the horizontal plane of symmetry to avoid twist. Hence the shear centre will be located at the intersection of the flange and web.

Another simple example is an angle section. Because the only routes for shear stresses to take are along the two legs, the shear centre must be located where they intersect.

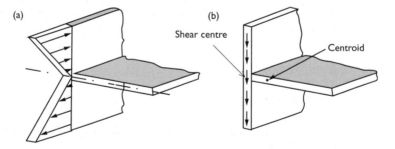

Figure 4.54 A T section on its side acting as a beam: (a) bending stress distribution; (b) shear flow.

A channel section

For some other cross-sections (e.g. channel sections), it may be necessary to determine the shear distribution arising from elastic bending theory if it is required to locate the shear centre.

Ex. 4.10
A thin-walled channel section has the dimensions shown in Fig. 4.55(a). Determine the position of the shear centre.

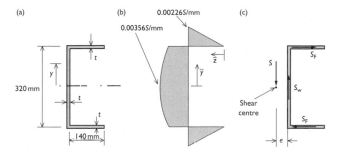

Figure 4.55 Determination of shear centre for a channel: (a) dimensions; (b) shear flow distribution; (c) shear force equilibrium.

Answer
Because the shear flow formula (4.9) is required, it is first necessary to determine the second moment of area. Ignoring the term in t^3 and using mm units

$$I = 2 \times 140t \times 160^2 + 320^3t/12 = 9.899t \times 10^6$$

For the top flange, at distance $\bar{z}$ from the tip, $\int y\,dA = 160t\bar{z}$. Hence $q_F = 160t\bar{z}S/I$, showing a linear increase from 0 at the tip to $0.00226S$ at the root. For the web, at distance $\bar{y}$ from the neutral axis,

$$\int y\,dA = 160 \times 140t + (160 - \bar{y})t \times (160 + \bar{y})/2$$

$$= (35{,}200 - \bar{y}^2/2)t$$

Hence $q_w = (0.00356 - 0.0505\bar{y}^2 \times 10^{-6})S$, the distribution being parabolic decreasing from $0.00356S$ at the neutral axis to $0.00226S$ at the top and bottom as shown in Fig. 4.55(b).

Integrating q_F for one flange gives $S_F = 0.158S$ and integrating q_w for the web gives $S_w = S$. The second result is expected, otherwise vertical equilibrium would be violated. However, the first result is required to determine the eccentricity, e, of the shear centre necessary to maintain torsional equilibrium (see Fig. 4.55(c)). Thus $Se = 320S_F$ giving $e = 50.7$ mm.

Open section behaviour

The surprising result that the shear centre of a channel section lies outside rather than within the cross-section can be verified by a simple experiment (see Q.4.22). Because an open section (particularly one which is thin-walled) is weak in torsion, the position of its shear centre is very important when there is no restraint against twisting. For instance, if a channel section were to be used as a beam to support a sliding door as shown in Fig. 4.56(a), because the loading is not aligned with the shear centre, there would be a tendency for the beam to twist. From elastic deflection theory, it can be shown that twisting would be about the shear centre. Hence the channel, apart from twisting, would have an increased sag on account of the twisting action. This could easily cause the door to bind or even come off its rail. These problems would be avoided most effectively by the use of a closed section instead of a channel (e.g. as shown in Fig. 4.56(b)).

Open sections are, however, suitable where there is restraint against torsion. If, for instance, two channel sections are used as beams to support a deck of planks as shown in Fig. 4.57(a), the planks, if stiff in bending and properly secured to the channels, will prevent them from twisting. Load transfer resultants will be directed through the shear centres of the channels as shown in Fig. 4.57(b). In practice, however, the connections between the planks and the channels are required to transmit both tensile and compressive forces as shown in Fig. 4.57(c).

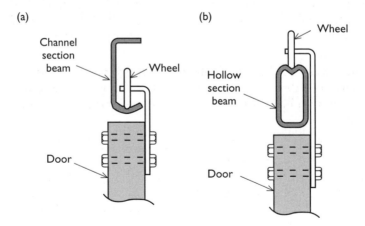

Figure 4.56 Different beam supports for a sliding door: (a) using a channel; (b) using a closed section.

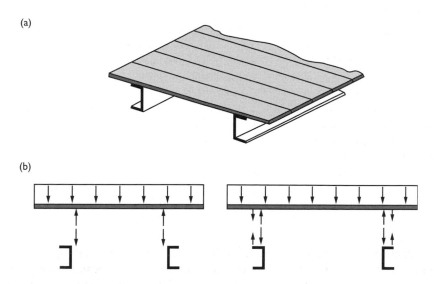

Figure 4.57 Two channels supporting plank decking: (a) configuration; (b) resultant internal reactions; (c) possible connection forces.

Q.4.21
Either
Choose values of web depth, a, and flange width, b, for a thin-walled channel section such that $a/b \neq 320/140$. Determine the position of its shear centre if the web and flanges all have the same thickness.

Or
Determine algebraically the position of the shear centre of a channel section with web depth a and flange width b for which the thickness t is uniform and small in comparison with the overall cross-section dimensions.

Q.4.22   Observation experiment
Using card or drawing paper, construct a channel section with end flaps as shown in Fig. 4.58. The ratio of flange width to web depth should be the same as for Ex. 4.10 or for the chosen values for Q.4.21. Secure one end to a firm support thus making a cantilever. Secure the other end to an end plate with holes for loading. Observe the angle of twist

when loads are applied at different positions and thus determine the shear centre experimentally. Compare this with the theoretical value already derived.

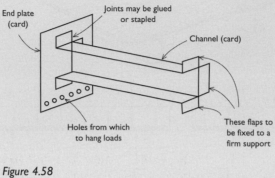

Figure 4.58

4.10   Unsymmetric bending

Warning recalled

In Section 4.3 the assumption was made that stress is proportional to the distance above the centroid for all but the type IV beams of Fig. 4.19 when subject to bending moment in a vertical plane. Type IV beams have no plane of symmetry and require special care. Not only will the shear centre not tend to coincide with the centroid, but also bending stresses calculated from the standard elastic bending formula (4.5) will be incorrect.

Ex. 4.11
Using elastic bending theory, estimate the thickness t required for the Z section, shown in Fig. 4.59, when it is subject to a bending moment of 2.6 kNm about the z axis and the bending stress is to be limited to 65 N/mm^2.

Incorrect solution
Ignoring the t^3 term gives $I_z = 816,700t\,\text{mm}^4$
Hence from Eq. (4.5), $\sigma = 222.9/t\,\text{N/mm}^2$
With $\sigma \leq 65\,\text{N/mm}^2$, this gives $t \geq 3.5\,\text{mm}$

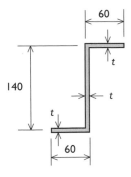

Figure 4.59 A Z section (dimensions in mm).

Reason for error
Figure 4.60 shows the stress distribution derived above viewed in elevation and plan. Whereas this corresponds to the correct applied bending moment about the z axis, the plan view shows that it requires also a bending moment to be applied about the y axis. If this bending moment is not present, the stress distribution cannot be correct.

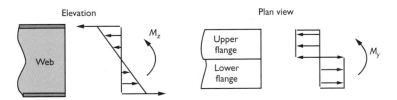

Figure 4.60 Incorrect stress distribution.

A correct solution
In practice, bending about the y and z axes will occur simultaneously. If σ_1 is the flange stress for bending about the z axis, each flange will develop a resultant force of $60\sigma_1 t$ acting at a distance of 30 mm from the web. This gives a bending moment about the y axis of $M_{y1} = 3600\sigma_1 t$ in mm units (Fig. 4.61(a)). However, for bending about the y axis, the flange stresses will vary linearly from 0 at the web to (say) σ_2 at the tip, giving a resultant for each flange of $30\sigma_2 t$ acting at a distance of 40 mm from the web. This gives a bending moment about the y axis of $M_{y2} = 2400\sigma_2 t$ (Fig. 4.61(b)). In order for there to be no resultant bending moment about the y axis, it is necessary that

$$M_{y1} + M_{y2} = 0$$

giving $\sigma_2 = -1.5\sigma_1$. Hence the stress distribution in the flanges must be as shown in Fig. 4.61(c). As a result, each flange will contribute a force of $15\sigma_1 t$ instead of the original incorrect value of $60\sigma_1 t$.

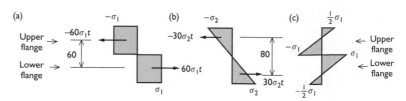

Figure 4.61 Plan view of Z section stress distribution: (a) due to vertical bending; (b) due to horizontal bending; (c) combined stresses with no resultant M_Y.

Because the web stresses are unaffected by bending about the y axis, they will remain at their previous values. For a bending moment applied about the z axis, the stresses will be as shown in Fig. 4.62 giving

$$M_z = (15 \times 140 + 35 \times 93.33)\sigma_1 t = 5367\sigma_1 t$$

[instead of $(60 \times 140 + 35 \times 93.33)\sigma_1 t = 11{,}667\sigma_1 t$]

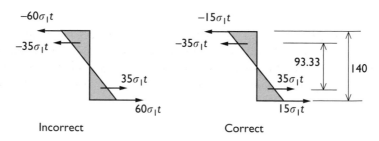

Figure 4.62 Elastic internal forces for the Z section due to M_Z (mm units, elevation shown).

For $M_z = 2.6\,\text{kNm}$ and $\sigma_1 \le 65\,\text{N/mm}^2$, this equation gives $t \ge 7.5\,\text{mm}$ (as opposed to the previously incorrect value of 3.5 mm).

The neutral axis position

For the Z section, the line in the cross-section having zero stress (called the neutral axis) passes through the 2/3 span positions of the flanges as shown in Fig. 4.63. Bending deflection will occur in a direction perpendicular to the neutral axis and, for this example, the horizontal component of bending deflection will be greater than the vertical component.

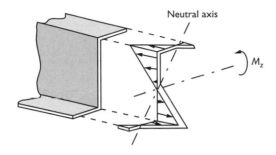

Neutral axis

M_z

Figure 4.63 Neutral axis position for the Z section due to M_z.

Elastic analysis of unsymmetric bending

Elastic bending analysis of any arbitrary shape of unsymmetric beam can be carried out using the three second moments of area:

$$I_z = \int y^2 \, dA, \quad I_y = \int z^2 \, dA \text{ and } I_{yz} = \int yz \, dA$$

Unlike types I, II and III of Fig. 4.19, unsymmetric sections tend to have non-zero values for the product second moment of area I_{yz}. Assuming that an elastic (linear) stress distribution occurs involving both horizontal and vertical bending, the stress at any point (y, z) of the cross-section can be expressed as

$$\sigma = c_1 y + c_2 z \tag{4.10}$$

where c_1 and c_2 are constants. Hence taking moments about the z and y axes (as shown in Fig. 4.64) gives

$$M_z = -\int \sigma y \, dA = -c_1 I_z - c_2 I_{yz} \text{ and } M_y = \int \sigma z \, dA = c_1 I_{yz} + c_2 I_y$$

$$\tag{4.11}$$

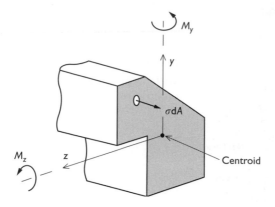

Figure 4.64 An unsymmetric section.

If only vertical loads are applied to the beam, $M_y = 0$. Hence solving for c_1 and c_2 and substituting into Eq. (4.10) gives

$$\sigma = \frac{\left(-I_{yy} + I_{yz}z\right) M_z}{I_y I_z - I_{yz}^2} \tag{4.12}$$

The neutral axis is the line corresponding to $\sigma = 0$, that is $y/z = I_{yz}/I_y$, and the fibres which contain the largest bending stresses are those which are furthest from the neutral axis.

Alternative solution for Ex. 4.11

Because the web lies on the y axis, it does not contribute to I_{yz}. The contribution of each flange is its area multiplied by $\bar{y}\bar{z}$ where $\bar{y}$ and $\bar{z}$ are the coordinates of its centroid relative to the centroid of the cross-section (see Q.4.23). Hence in mm units,

$$I_z = 816{,}700t, \quad I_y = 144{,}000t, \quad I_{yz} = -252{,}000t$$

Hence Eq. (4.12) gives in Newton and mm units,

$$\sigma t = -6.92y + 12.11z$$

From this formula, it can be deduced that the highest stresses occur when $y = \pm 70\,\text{mm}$ and $z = 0$. Because here $\sigma t = 484$, the maximum stress is limited to $65\,\text{N/mm}^2$ if $t \geq 7.5\,\text{mm}$.

Principal second moments of area

If a bending moment is applied at an angle to an unsymmetric section, it is possible to identify two angles for which elastic deflections will be in the

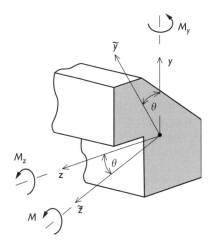

Figure 4.65 Rotation of axes.

same direction as the loading. The associated I values are called principal second moments of area, the determination of which can be expressed as a 2×2 eigenvalue problem as follows.

Consider loading to be applied at an angle θ to the vertical so producing a bending moment M about the $\tilde{z}$ axis as shown in Fig. 4.65. This bending moment will have components about the original axes of

$$M_z = M \cos\theta \quad \text{and} \quad M_y = -M \sin\theta$$

However, for this to be a principal direction of loading, the bending characteristics must conform to

$$\frac{M}{I} = -\frac{\sigma}{\tilde{y}}$$

where $\tilde{y}$ is the direction orthogonal to $\tilde{z}$ and I is the second moment of area about the $\tilde{z}$ axis. Because

$$\tilde{y} = y \cos\theta + z \sin\theta$$

it is necessary that

$$\sigma = -\frac{M(y \cos\theta + z \sin\theta)}{I}$$

Thus, from Eq. (4.10), $c_1 = -(M \cos\theta)/I$ and $c_2 = -(M \sin\theta)/I$. Substituting these values for c_1 and c_2 and also the values for M_z and M_y above into

Eq. (4.11) gives

$$\begin{bmatrix} I_z & I_{yz} \\ I_{yz} & I_y \end{bmatrix} \begin{bmatrix} \cos\theta \\ \sin\theta \end{bmatrix} = I \begin{bmatrix} \cos\theta \\ \sin\theta \end{bmatrix}$$

This is a 2×2 matrix eigenvalue problem of similar nature to the one obtained from the principal stress analysis of Section 3.2. Because the matrix is symmetrical, it can be shown that the two solutions to this equation are for axes at right angles (i.e. orthogonal) to each other and the principal second moments of area are the maximum and minimum I values that can be obtained for any angle θ.

Ex. 4.12
Determine the principal second moments of area and their axes for the Z section of Ex. 4.11 in which $t = 8$ mm. Also determine the maximum bending stress occurring in the cross-section when the direction of loading is aligned to either of these axes if $M = 2.6$ kNm. Is the maximum stress derived from this analysis bound to be the maximum stress occurring for any direction of loading?

Answer
From the second moments of area derived from the alternative solution to Ex. 4.11 and using $t = 8$ mm, the matrix requiring eigensolution is

$$\begin{bmatrix} 6.533 & -2.016 \\ -2.016 & 1.152 \end{bmatrix} \times 10^6 \text{ mm}^4$$

This gives eigenvectors of {0.949 −0.316} and {0.316 0.949} with corresponding eigenvalues 7.204×10^6 mm^4 and 0.481×10^6 mm^4. The eigenvalues are the principal second moments of area and their directions, defined by the eigenvectors, are $\theta = -18.41°$ and $71.59°$ respectively (see Fig. 4.66).

a   When the bending moment is applied about the $\tilde{z}$ axis in Fig. 4.66, the maximum $|\tilde{y}|$ occurs at the outside tip of a flange with a value $74 \cos 18.41° + 60 \sin 18.41° = 89.16$ mm. Hence from the simple bending theory,

$$\sigma_{max} = \frac{M|\tilde{y}|_{max}}{I} = \frac{2.6 \times 89.16}{7.204} = 32.2 \text{ N/mm}^2$$

b   When the bending moment is applied about the $\tilde{y}$ axis in Fig. 4.66, the maximum $|\tilde{z}|$ occurs at the inside, tip of a flange with a value of $60 \cos 18.41° - 66 \sin 18.41° = 36.09$ mm. Hence from the simple theory,

$$\sigma_{max} = \frac{2.6 \times 36.09}{0.481} = 195.1 \, \text{N/mm}^2$$

Because, for different angles θ, $\tilde{y}_{max}$ varies as well as I, the highest stress is not necessarily obtained when I has its lowest value.

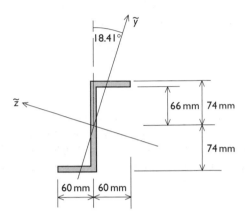

Figure 4.66 Principal axes.

Z sections as purlins

A purlin is a horizontal roof support secured at regular intervals to inclined rafter members, which are normally part of the roof trusses. The two main loadings are the weight of the roof acting vertically downwards and wind loading acting approximately perpendicular to the roof structure. These must be carried by beam action. A Z section is commonly used being more effective than an I section or channel at carrying vertical loads because of the more favourable alignment of the largest principal second moment of area (see Fig. 4.67(a) and (b)). Also the flanges are more easily accessed for the purpose of bolting or welding.

Under wind loading conditions roofs are usually subject to outward rather than inward forces. If wind loads can be approximately equal to or even greater than the gravity forces in magnitude, the resultant forces acting on the purlin, although smaller than when the wind is not acting, could be aligned in the direction of the weak principal second moment of area (see Fig. 4.67(c)). Hence care may need to be taken in assessing which are the worst loading cases.

232 Beams

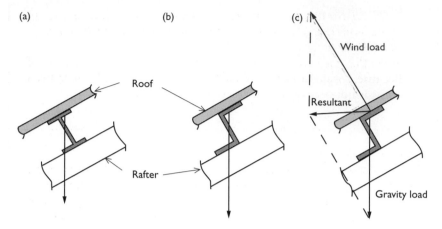

Figure 4.67 Loading directions of purlins: (a) I section subject to vertical loading; (b) Z section subject to vertical loading; (c) Z section subject to combined vertical and wind loading.

Q.4.23
Extend the parallel axis theorem of Section 4.3 to determine the contribution of any area of a cross-section to I_{yz}.

Q.4.24
An unequal angle section of the dimensions shown in Fig. 4.68 is to be used as a simply-supported beam of span 4 m with a maximum (vertical) loading of 5 kN/m.

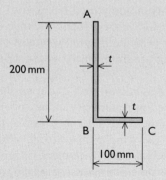

Figure 4.68

a   Working on the basis that the thickness, t, is small, obtain an expression for the maximum bending stress in the beam as a function of t.

b   Show that the neutral axis passes through the 2/3 span position of the lower leg and explain theoretically why this could have been anticipated.

Q.4.25

If the solution to Q.4.24 has produced stress levels which are too high for a thin angle section to be used, a thicker section such as that shown in Fig. 4.69 would be required. Determine, using classical unsymmetric bending theory, the maximum bending stress when this cross-section is used for the beam.

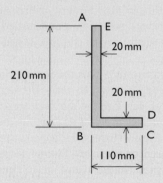

Figure 4.69

Q.4.26 Experiment

Devise a simple experiment to investigate the response of an angle section to vertical bending. The apparatus could be a cantilever fairly similar to that shown in Fig. 4.58.

Points of note are as follows:

- Twisting should be avoided if the load is hung in the plane of the vertical leg.
- The cantilever will not buckle so easily if the horizontal leg is at the bottom rather than at the top. (This is because the ends of the legs will be in tension rather than compression.)

- A method should be devised of measuring the deflection of the end of the cantilever.
- Because the neutral axis should pass through the centroid and the 2/3 span position of the horizontal leg, a considerable amount of horizontal bending should take place. Aim to identify whether the cantilever end deflection is at right angles to the expected neutral axis position.

4.11 Thin-walled beams

Developments in ship hull design

Iron started to be used in the hulls of ships in the nineteenth century with I.K. Brunel's 87 m long S.S. Great Britain (1843) as one of the major early examples, Fig. 4.70. Despite changes of use, two groundings and abandonment for 33 years, she has been restored and can now be seen at Bristol dock (Beckett, 1980). The fact that she has survived where wooden hulled ships would have been lost many times over, is testimony to the robust nature of her hull construction.

Closely associated with the early use of iron hulls was the realisation that a ship hull is a beam and should be designed as such to carry the bending, shearing and twisting loads to which it is subject. Even when anchored in still water, the longitudinal weight distribution of a ship and its cargo is unlikely to match that of the supporting water buoyancy forces. However, the most severe loading cases will occur in rough sea conditions. Two critical cases considered in design are when waves support the bow and stern, thus producing a sagging bending moment, or when a wave supports the centre of the ship, thus producing a hogging bending moment (Fig. 4.71). In both cases, the ship may be considered to be in equilibrium. Dynamic situations will also occur, but many of these will not produce any larger bending moments than those which arise for the static cases. They can, however, produce high local water pressures in situations where the hull 'slams' against a water surface.

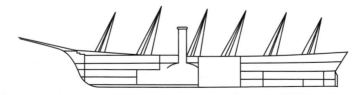

Figure 4.70 Hull construction of the S.S. Great Britain.

(a)

(b)

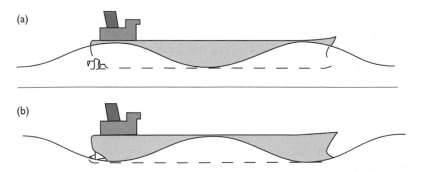

Figure 4.71 Critical sea conditions for hull bending: (a) producing sagging bending moment; (b) producing hogging bending moment.

The *Titanic*, launched in 1911, like all ships of that era, had riveted steel plating. However, by 1950 riveting had been largely replaced by welding. Welding saves on time and cost of manufacture. It also avoids any problems over watertightness, eliminates the weight of plate overlaps and avoids protruding rivet heads. Therefore, during the Second World War, when ship production needed to be rapidly increased to make up for losses, a standardised form of construction was adapted in America using all welded hulls. The first 'Liberty ship' took 245 days to construct and fit out. However, using mass production techniques, this was reduced to, at fastest, eight days with three ships a day being produced.

During this period, a lot of experience was gained about their performance. In particular, Liberty ships were found to be very susceptible to fracture. Some would break up in heavy seas, one even broke up in harbour in calm water and with the temperature mild (Schlager, 1994). Because many fractures started at hatch corners, these were soon reinforced and other remedial measures were taken. Investigations into welded ship design, both in the USA and the UK, approved welding for ship construction, but confirmed the need for sound design, good workmanship and the use of ductile steel.

Ship hulls

Oil tankers are constructed with transverse bulkheads which divide the hull into separate tanks. They are also likely to have one or two longitudinal bulkheads which either create further subdivision of the storage space or else may contain holes to act as dampers against sloshing of the oil cargo in rough conditions. When a hull is subject to bending in a vertical plane, the deck and bottom plating act as flanges and the side skins and longitudinal bulkheads act as webs of a beam. Because of the box nature of the hulls they have good torsional characteristics. Figure 4.72(a) shows a possible structure for a tanker with a double hull (now preferred to reduce the risk of oil pollution in the event of groundings or collisions).

In order for the various plates to be structurally effective, they need to be stiffened. Longitudinal stiffeners have three functions:

(i) To resist buckling of the plates when they are subject to compressive and shear stresses.
(ii) To support the plates when they are subject to pressure from one side.
(iii) To carry some of the longitudinal bending stresses themselves.

Because transverse bulkheads tend to be too far apart for stiffened plates to easily span between, transverse members are also included which link up to form frames as shown in Fig. 4.72.

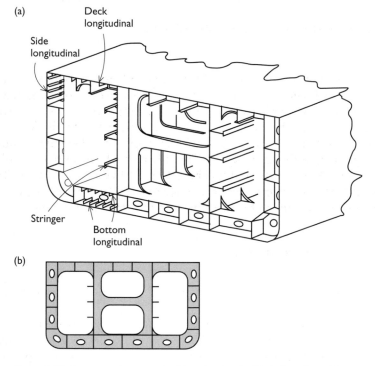

Figure 4.72 Structure between bulkheads of a double hull tanker: (a) cutaway view at midships (only a few longitudinals shown); (b) a transverse frame.

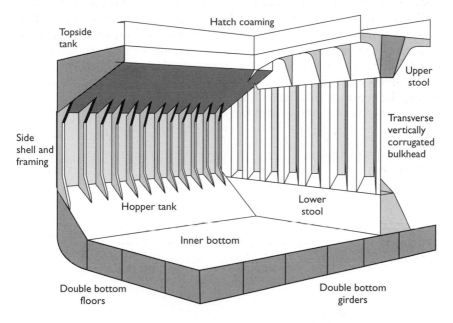

Figure 4.73 Part of a hold for a single side skin bulk carrier.

Bulk carriers and container ships differ from tankers on account of their lack of internal longitudinal bulkheads and the presence of large hatch openings. Hence each upper flange needs to be separated into two parts running either side of the hatch openings. The structure of the bulk carrier shown in Fig. 4.73 has been designed in such a way that there are no horizontal ledges within the cargo space. The side plates have stiffeners which span vertically. Also the internal bulkhead (which does not need to be flat) has been stiffened by giving it vertical corrugations. However, the single side skins are now considered to be a weakness of this particular design.

The Britannia Tubular Bridge

Sutherland (1983) described the Britannia Tubular Bridge, built in 1850, as 'perhaps the greatest step forward in structural understanding and practice in the last 200 years'. A railway crossing of the Menai Straits required two spans of 140 m. To use cast iron arches might have been feasible despite requiring twice the span of previous such bridges. However, this was ruled out because the arches would interfere with shipping. Although suspension bridges of that span would have been feasible, they had already been found to be too flexible for railway traffic.

To overcome this virtual impasse, Robert Stephenson approached William Fairbairn, who had experience with iron ships, with a view to constructing a girder bridge out of wrought iron. Thus the first tubular metal bridge was developed over 100 years before their benefits started to be appreciated more widely (Fig. 4.74). Separate tubes were used for each of the two railway lines and their bottom and top flanges were both of cellular construction. One of the important findings of load tests on models was that more material was required in the compression flange than the tension flange. This was a surprising result at the time because previous experience was with cast iron, which could be safely loaded to higher stress in compression than in tension. However, with wrought iron, tensile strength was much more reliable, but the thin plates tended to buckle easily in compression (hence the difference in flange details in Fig. 4.74). Another feature visible in Fig. 4.74 is the

Figure 4.74 Off-site construction of a tube of the Britannia bridge.

vertical web stiffeners and corner cleats. These were sufficient to maintain the cross-sectional shape in the absence of full diaphragms or cross-bracing members, neither of which would have been feasible because they would have prevented trains from travelling through the boxes.

Although the individual spans were constructed off site and raised onto piers by hydraulic jacks so that they were simply-supported at their ends, adjacent tubes were later joined together over the piers to produce two parallel continuous beams. Furthermore, by jacking individual tubes up to predetermined slopes before securing the joints, hogging bending moments were induced in the tubes over the supports so relieving, to some extent, the maximum sagging moments due to self-weight (see Fig. 4.75). Thus some control over bending moment distributions was achieved, but by a different method to that described for multispan bridges in Section 4.2. Understanding of the deflection characteristics of beams as well as their static equilibrium was required to achieve the desired effect.

The Britannia bridge performed satisfactorily for 120 years, but was unfortunately damaged beyond repair in 1970 when a lightweight timber roof (which had been put in to preserve it by preventing rusting) caught fire. The only similar bridge still standing is a smaller span bridge on the same railway line at Conway.

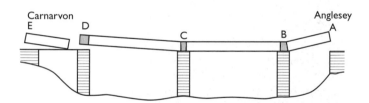

Figure 4.75 On-site fabrication of the Britannia tubes (one end of a tube was raised off its support and then, after the joint at the other end was made, it was lowered again. Joints were made in the order B, C, D).

Modern box girder bridges

The Britannia and Conway bridges remained curiosities in bridge development for over 100 years until bridges started to be built with steel box girder decks in the 1960s. Since then, steel box girders have been a favoured form of deck construction, often as free standing beams, but with longer spans being supported from suspension cables by means of hangers or directly from towers by means of cable stays. For highway bridges, the traffic normally travels on top of the boxes, the material is steel instead of wrought iron and the sides are often angled to reduce wind loads and associated dynamic effects. However, some of the features of the construction technique for

Figure 4.76 Construction of the Foyle Bridge, Londonderry. Once the central part of each of the boxes was lifted into position and joined to the corresponding outer lengths, the ends were pulled down onto their supports from their raised positions in order to reduce the central sagging bending moment.

the Britannia bridge can still be recognised in modern bridges. Figure 4.76 shows the Foyle bridge, Northern Ireland, which is a steel box girder bridge continuous over three spans. Like at Menai, two parallel boxes were used (one for each carriageway) and the boxes were fabricated in sections off site, floated there on rafts and jacked into position on the main piers. Also like at Menai, the sagging bending moments due to self-weight at centre span were reduced. This was done by pulling the ends of the boxes down onto their abutments after joining the sections together.

Aircraft structures

When aircraft speeds increased in the 1930s, lightweight canvas 'skins' became unsuitable and 'monocoque' construction was introduced in which a sheet aluminium skin took over from a space frame as the main structure.

Fuselages act as beams carrying the payload and are supported mainly via the wings both when in the air and when on the ground (the main undercarriage usually being extended from the wings). Hence the fuselage is normally subject to a hogging bending moment and the wings a sagging bending moment. However, critical situations for aircraft structures tend to be in manoeuvring flight or when strong gusts are encountered, with design requirements expressed in terms of g, the acceleration due to gravity. Gusts normally provide the most severe loading cases for commercial aircraft. Typically they need to be designed for $+3\,g$ to $-1.5\,g$ with the addition of an appropriate load factor. In contrast military aircraft need to be highly manoeuvrable and so their worst loading cases tend to be at least twice as severe. Negative g loading cases reverse the normal situation subjecting the fuselage to sagging and the wings to hogging bending moment.

The need to pressurise cabins for high altitude flight has led to circular cross-sections becoming standard (as discussed in Section 3.7). Longitudinal stiffeners called 'stringers' and transverse frames are required to make the skin effective as a beam. However, the requirement to keep the structural weight as low as possible, means that fuselage frames are not made stiff enough to prevent any distortion of the fuselage cross-section. As a result, bending theory, as described in this chapter, may not predict very well the internal forces, particularly those in the frames themselves. The largest loads are applied to the fuselage at the wing connections and there much stronger fuselage frames are required.

Wings also act as beams, but their shapes are untypical of other types of beam, being governed by aerodynamic rather than structural considerations. Bending action is achieved mainly by the inclusion of one or more 'spars' acting as webs linking the top and bottom skins. The skins are stiffened by stringers so that, in the neighbourhood of the spars, they make effective flanges (Fig. 4.77). Also the resulting multicell box construction, by supporting Bredt–Batho shear flow, provides wings with good torsional characteristics (important in preventing dynamic instability, to be discussed in Section 7.4). Diaphragms called 'ribs' are also included at regular spacing to maintain the shape of the cross-section.

Traditionally aircraft structures have been riveted, thus avoiding the fatigue risks of welding. However, fatigue is a significant risk, even with riveted structures, because stresses are higher around rivet holes, thus providing a place for cracks to start. One method of combating this risk has been to divide the skin into longitudinal panels so that a crack will stop when it reaches the edge of the panel it is in, thus disabling one panel rather than

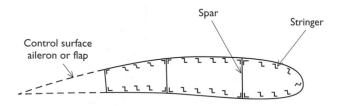

Figure 4.77 Typical cross-section of an aircraft wing.

a whole wing. To work effectively, this 'fail-safe' technique also requires the following:

- A structure designed to perform satisfactorily with any one panel unable to carry load.
- Regular inspections to identify any cracks.
- Ability to replace defective panels easily.

Most of the gravity and inertia loading of an aircraft comes from the fuselage. Hence the bending moment will tend to peak at the roots of the wings where they intersect with the fuselage. Normally wings are designed to link at either the bottom or top of the fuselage so that the high bending moments in the wings can react against each other rather than against opposite sides of fuselage frames.

Because reductions in the weights of airframes lead to increases in payload capacity, strength/weight is a more important ratio than strength/cost. On account of this, aircraft designers are particularly keen to use improved materials and construction techniques. Some examples of modern practice are as follows:

- Use of well-monitored welding in place of riveting in compression areas.
- Machining skin and stiffeners from solid metal to avoid the need for riveting or welding.
- Use of sandwich honeycomb construction for skins (i.e. two skins separated by a light honeycomb core).
- Use of titanium instead of steel for highly loaded components.
- Use of carbon fibre reinforced plastics.
- Employment of non-linear techniques to model the structural performance as accurately as possible.

4.12   Problems at the Yarra (Barber et al., 1971; Bignell et al., 1978)

The project

In the 1960s, rapid growth in the use of vehicular traffic led to the need to improve road systems around large cities often involving crossings of major rivers by means of tunnels or bridges. The West Gate bridge at Melbourne was part of one such road project. To clear shipping, the roadway needed to be 58 m high where it crossed the Yarra river. A large number of approach spans were in pre-stressed concrete with the central five spans to be of steel box girder construction, the longest span over the river being supported by cable stays (Fig. 4.78). The Australian consultants contracted to design

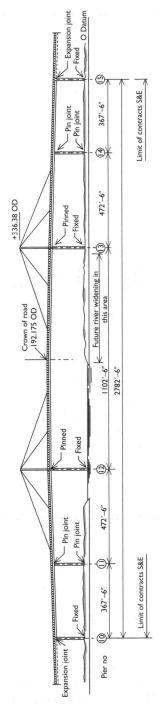

Figure 4.78 Elevation of the steel spans for the proposed West Gate Bridge.

the bridge engaged Freeman Fox and Partners, pioneers of steel box girders construction, to design the steel spans. The bridge was particularly wide for its time, having eight traffic lanes, with the steel spans being designed as a three cell trapezoidal box as shown in Fig. 4.79.

Construction technique

The outer steel spans 10–11 and 14–15 were the first to be erected as these were intended to act as platforms from which the central three cable supported spans could be cantilevered out. These outer steel spans were to be simply-supported beams between piers spaced 112 m apart, until such time that the neighbouring spans were connected to them. They could have been constructed in place using temporary supports or constructed at ground level and raised into place by jacking. However, the steel contractors drew up a scheme, not previously tried, in which each span was to be fabricated on trestles at ground level in two halves, split down the longitudinal centre-line (Fig. 4.80). Each half-span was then to be raised onto the piers and joined to its partner at high altitude. This halved the weight to be lifted at any one time.

Although the contractors successfully erected span 14–15 this way, they encountered particular difficulties aligning the two parts of the upper flange where they needed to be joined. The main reasons for these difficulties were:

a   The two outstanding parts of the upper flange were under compressive bending stress and had a tendency to buckle in waves (this is despite the addition of temporary stiffening members and diagonal bracing wires).
b   The precise profile of each half-span depended on the way seven 16 m length boxes were bolted together on trestles at ground level. Temperature variations during construction, for instance, would affect the finished profile. As a result, a vertical misalignment of 89 mm at centre span was discovered once the half-spans had been lifted into place.
c   Due to self-weight, each half-span bowed outwards creating a horizontal gap of approximately 63 mm at centre span.
d   The job of correcting these gaps was made difficult because of their altitude and also the stresses present in the half-spans due to their self-weight.

These problems were overcome by a process of jacking and temporarily removing and relocating some bolts to get rid of the buckles (although significant residual stresses were left in the material due to this work).

Countdown to disaster

Because the steelwork erection was very much behind schedule, the work for spans 10–11 was handed over to the contractors who had originally

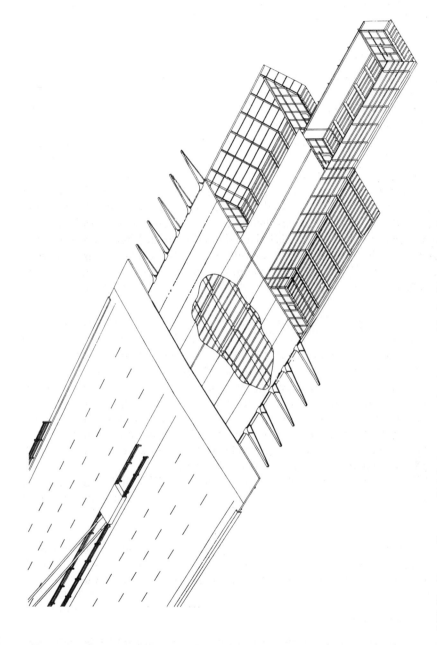

Figure 4.79 The three cell trapezoidal deck structure showing assembly details.

Figure 4.80 Two matching half cross-sections as raised.

been engaged just for the concrete parts of the construction and who had no previous experience of steel erection. They encountered similar problems to the previous contractors when trying to connect the two half-spans at altitude, although this time the vertical misalignment had increased to 114 mm. Ten 800 kN concrete blocks were lifted on to the deck over a pier to use as 'kentledge'. When seven of the blocks had been moved to the centre of the upper half-span, the vertical gap at centre span was reduced to 25 mm. At this stage, a large buckle appeared near to centre span so no more concrete blocks were moved there. Instead jacks were used to try to close the various remaining gaps and diaphragms were connected on the north end of the span thus linking the two half-spans, but it was not possible to link the diaphragm affected by the buckle. At this stage, the kentledge was taken off and some bolts were removed to try to eliminate the large buckle. However, on 15 October 1970, with the beam in a weakened state precisely where its bending moment was largest, the span collapsed killing 35 workmen and seriously injuring others (some of whom were having a lunch break in huts situated directly beneath the span).

Contributory causes

There were many contributing causes to this disaster:

- Poor labour relations leading to strikes and delays.
- Use of contractors not familiar with steel erection.
- Choice of a difficult and untried erection technique.
- Failure to learn from problems during the erection of span 14–15 by improving the procedure for span 10–11.
- Failure to review procedures when signs of excessive stresses were observed (i.e. yielding of material and severe buckles).
- Poor communications between the steel consultants and contractors.

However, the reason for including this item of practical experience in a chapter on beams is because of the effect that unsymmetric bending of the split cross-section had on the events. In considering erection procedures and also estimating stresses during erection, it may have been assumed that the split cross-sections would each bend about their horizontal axis. The asymmetry of the cross-sections contributed in two ways to the problems of erection:

a   Each split cross-section deflected outwards as well as downwards due to self-weight, so creating the horizontal gap where they needed to be joined.
b   The bending stresses in the outstanding parts of the upper flange which were to be joined were greater than anticipated because of their increased distances from the neutral axes of the half-spans (Fig. 4.81).

Outstanding upper flange

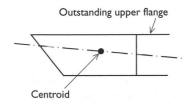

Centroid

Figure 4.81 Approximate position of the neutral axis due to gravity loading.

Not only were stresses higher in the critical region because of unsymmetric bending, but also, while the kentledge was present, the critical bending moment would have been nearly 19% greater. If the bridge had been designed as two separate boxes, one for each carriageway, instead of one box to carry the whole eight lanes, there would have been no need to join two split cross-sections at altitude. The precedent was already set for this at Menai where the two rail tracks ran through separate tubes. Some later road bridges have been constructed with separate carriageways, the Foyle bridge (Fig. 4.76) being an example. The failure on the Yarra emphasizes an advantage of design–build contracts, where the method of construction is thoroughly investigated at the time of the design rather than at a later date.

Aftermath

A court of inquiry into the collapse (Barber *et al.*, 1971) identified various contributory causes finding many people in part responsible. However, the West Gate bridge was not the only steel box girder collapse at about this time. Other collapses occurred at Vienna in 1969, at Milford Haven, Wales in June 1970 and at Koblenz, Germany in 1971. As a result, the principles and procedures adopted for the design and construction of box girder bridges were critically reviewed (Merrison, 1973). More care was called for in design in order to achieve continuity of load paths and avoid buckling, and in construction in order to maintain accurate geometry. Since then many steel box girder bridges have been built successfully, proving that, when well designed and built, they can fulfil a very useful role in infrastructure development.

4.13　Gondolas of the Falkirk Wheel

Marking the millennium

Funding to stimulate new ventures celebrating the start of the third millennium AD has encouraged designers to produce adventurous and innovative new structures. Notable constructions in Britain include the London Eye (a giant Ferris wheel), the Millennium Dome (which produced the world's largest covered area), the Eden Project (whose biomes are described in Chapter 5), the Gateshead Millennium Bridge (discussed in Chapter 6), London's Millennium Bridge (mentioned in Chapter 7) and the Falkirk Wheel, the subject of this section.

A 'revolutionary' concept (Ballinger, 2003)

Originally the Union canal linked Edinburgh to the Forth and Clyde canal at Falkirk by a cascade of 11 locks. However, in the 1960s, the canals were closed and the locks were dismantled. The Falkirk Wheel, completed in 2002, has been a key part of the reinstatement of this canal system so that it can be used for pleasure craft. It is a rotating boat lift which lifts and lowers boats to a height of 23.7 m in seven minutes, a small fraction of the time that it used to take boats to pass through the eight locks it replaces. It is the first boat lift to be built in Britain in over 100 years, and the first in the world to use a rotating mechanism. The 'wheel' operates by interchanging two lengths of canal between the lower and upper levels. The movable lengths of canal called 'gondolas' or 'caissons' are swung about an axle on two massive arms (Fig. 4.82). The arms, the axle and the gondolas all posed interesting challenges with regard to structural design. The beam action of the gondolas is discussed below.

The gondolas

The gondolas are supported on circular rails within the arms, being kept horizontal by means of a system of gears. They each hold a 21.45 m length of water and are simply-supported at their ends. The cross-section shown in Fig. 4.83 consists of the base plate acting as the lower flange of the beam, plates acting as walkways constituting the upper flange, and four webs, two of which provide the water containment.

Water pressure on the base plate is distributed to the webs by means of cross-beams at closely spaced (0.625 m) centres (Fig. 4.84(a)). The inner webs are strengthened against lateral water pressure by vertical stiffeners

Figure 4.82 The Falkirk Wheel with the near gondola being raised to link with the Union Canal.

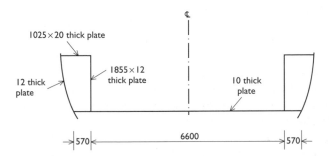

Figure 4.83 Gondola cross-section (dimensions in mm).

which cantilever from the stiffening beams of the base plate (Fig. 4.84(b)). Whereas most of the loading on each gondola is symmetrical about its centre-line, unsymmetrical loading does arise, for instance, on account of the rolling forces. The gondolas would, like the tray of a matchbox, be torsionally very weak except for the resistance of the two closed boxes shown in Fig. 4.84(c), which contain internal diaphragms.

Another major design consideration has been the effect of wind loading, not only on the stresses in the gondolas, but also on their deflections and the way these might affect the operation of the gearings, rolling mechanisms and the canal gates.

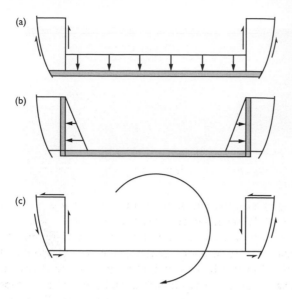

Figure 4.84 Action and reaction within a cross-section of a gondola: (a) water pressure on base plate reacted through shear webs; (b) water pressure on inner webs (self-equilibrating); (c) torsion (e.g. due to rolling forces) reacted by torsion boxes.

Q.4.27
By considering a gondola of the Falkirk Wheel to be a beam of span 21.45 m, estimate the maximum stresses in the top and bottom flanges due to water being present to a depth of 1.85 m.

NB   To obtain a quick estimate simplify the analysis by ignoring the bending stresses in the webs. Flange dimensions are as shown in Fig. 4.83.

Q.4.28
Discuss how horizontal wind loads acting on a gondola may be reacted.

Q.4.29
Determine the greatest possible increase in the maximum bending moment in a gondola when a boat of overall weight 60 kN and length 18 m is moored in it, assuming that the water level is unchanged by the presence of the boat.

Chapter 5

Trusses

Trusses are used for many types of structure. Some of these are hidden in such places as the roof spaces of domestic houses whilst others form some of the greatest examples of the structural engineer's art. In this book trusses are introduced after beams so that aspects of their behaviour, design and use may be discussed more effectively. Whereas basic analytical techniques for trusses are easily understood, care has often to be taken with classification of truss types, implementing and interpreting computer solutions and coping with 3D aspects.

On completion of this chapter you should be able to do the following:

- Understand the logic behind the geometrical form of some common types of truss both in 2D and 3D.
- Identify whether a framework can be analysed as a truss and whether it is statically determinate or not.
- Use joint equilibrium or the method of sections to obtain member forces of statically determinate trusses.
- Quickly identify under a specific loading which members of a simple truss take tension, which take compression and which are not loaded.
- Appreciate the usefulness of trusses and some aspects of their historical development.

5.1 An appreciation of trusses

Trusses are structures comprised of bar members which are designed to carry load by means of axial forces alone. They can be recognised through the frequent use of triangular arrangements of members.

Timber trusses have undoubtedly been used from ancient times, but because of the susceptibility of timber to various forms of deterioration and fire, it is difficult to know how extensive their use has been. Many fine timber trusses can be found supporting roofs of churches and public halls (Figs 5.1 and 5.2). Timber members are good at carrying both axial tension and compression forces if the grain direction is parallel to the axis of the member

Figure 5.1 A hammer-beam roof at Queen's University, Belfast, built in 1932. Oak pins, rather than iron bolts or screws were employed.

Figure 5.2 Internal structure of the main hall of Old Faithful Inn, Yellowstone National Park, Wyoming, the largest timber structure in the world.

(as would normally be the case) and if there are no knots present. Further-more, compared with masonry, timber is a light material which is easy to lift into place and does not impose much load on the foundations. One of the main difficulties with timber trusses is the need to develop joints of sufficient strength. It is tension joints which have given most difficulty. Because of this, older timber structures rarely had spans greater than the available length of pieces of timber.

With the industrial revolution more opportunities opened up for timber trusses. There was a need for economic forms of roof structure to cover larger spans than had been previously common. In shipbuilding, for instance, the lofting halls required large open plan covered areas and the Belfast truss which supported a barrel shaped felt roof on timber 'sarking' was widely used (see Fig. 5.3) (Gould *et al.*, 1992). The contrast between these soft-wood, efficient and economical trusses with low rise to span ratio and the ornate hardwood hammer-beam trusses with steeply pitched roofs seen in some churches could not be greater. Timber trestle bridges were used extensively in the early development of railways, particularly in North America where they played a major role in colonising the west (Fig. 5.4). However the railroad engines imposed heavy loads on these structures. There were many collapses

Figure 5.3 Mould loft at Harland and Wolff shipyard, Belfast, c.1910, used for laying out profiles, showing Belfast roof trusses supporting a shallow barrel shaped felt covered roof.

Figure 5.4 Gilahina Trestle, Alaska in use from 1911 to 1938.

and also some were lost because sparks from the engine smokestacks initiated fires.

As iron became available in larger quantities, engineers found this to be a better material for the construction of railway bridges because of its strength and permanence. The nineteenth and early twentieth centuries saw the development of large truss bridges to take railways across rivers and estuaries. There was much innovation in both materials and form. However, once steel became available, its ductility and greater strength gave it superiority over cast iron and wrought iron.

Iron and steel were not just used for railway bridges, however. They were put to many other uses often in the form of truss structures. Although Gustav Eiffel constructed railway bridges, he is most well known for the trussed tower he designed for the 1889 Paris exposition (Fig. 5.5) which, over 100 years later, still dominates the Paris skyline (Harris, 1975). Also very well known is the Statue of Liberty at New York. But in contrast to the Eiffel tower, those who admire it may not realise that it is supported by yet another of Eiffel's trusses (Fig. 5.6).

Trusses are used frequently in modern structures, but technology has advanced and priorities have changed. Welding has replaced riveting as the main method of jointing in steel. Also increased labour costs relative to material costs have been the catalyst for several changes. For instance fewer members tend to be used and the surface areas of exposed steel are kept low to facilitate painting. Also large components of major trusses are often prefabricated and bolted to their neighbouring components on site.

Figure 5.5 Part of the Eiffel Tower.

Q.5.1 Observation question
a Write a list of modern day uses of trusses.
b From memory sketch the shape of a truss you are familiar with. Then, if you can inspect the truss itself or a photograph of it, check your sketch for accuracy.
c Suggest what may be the worst loading cases for this truss and indicate where and in what direction these loads will be applied to it.

Figure 5.6 Truss supporting the Statue of Liberty.

5.2   Classification of frames and trusses

Class types

A 'frame' or 'framework' is the general term for structures which comprise a set of interconnected bar members. Frames can be classified into

- 'Sway frames' which can be distorted by applying loads to the joints;
- 'Trusses' which cannot be so distorted.

Figure 5.7(a) shows a simple sway frame where application of loading to opposite joints will produce a sway 'mechanism'. The structural performance of a sway frame depends on the ability of the members to resist bending and the joints to maintain the angles between attached members (Fig. 5.7(b) and (c)). Their structural performance is not to be discussed in this chapter. On the other hand, a truss will maintain the relative position of its joints

Figure 5.7 A simple sway frame: (a) original shape; (b) with members weak in bending; (c) with weak joints.

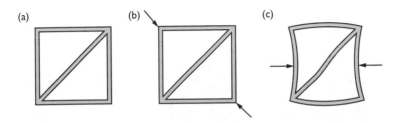

Figure 5.8 A simple truss: (a) original shape; (b) with load applied at joints; (c) with load applied to members.

when loads are applied to its joints. The only deflections that will occur arise through axial extension or shortening of the members due to axial loading, providing that the members are not so heavily loaded that any of them fracture, yield or buckle. It will normally tend to be stiffer and stronger than a similar sway frame provided that loading is applied only at the joints (see Fig. 5.8).

Another important classification depends on whether or not there are extra members included which do not help to prevent sway and are therefore classified as 'redundant'. A truss which has just sufficient members to prevent sway is statically determinate, because equilibrium considerations alone are sufficient to determine the member axial forces for any case involving joint loads only. A truss which has redundant members is described as statically indeterminate. Although equilibrium conditions still need to be satisfied for these trusses, other information is required or assumptions have to be made before the axial forces can be predicted. Hence it is possible to identify four classes of frame as shown in Fig. 5.9.

Where axial forces in members are compressive, there is a need for them to have good bending characteristics in order to prevent buckling (discussed further in Chapter 7). Some bending strength will also be required to resist loads not applied at joints or to allow for possible 'eccentricities' where the centroids of the cross-sections are off-set (either deliberately or accidentally)

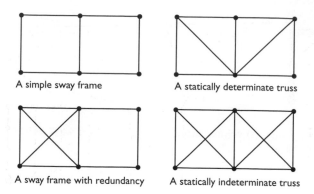

Figure 5.9 Classification of different frame types.

from the straight lines connecting the joints. However, this chapter is concerned with the axial forces in trusses rather than these other secondary effects. For this purpose it is convenient to consider the joints as pinned, whether or not pin-joints are actually present. This does not change the classification shown in Fig. 5.9, but makes identification easier. Furthermore, for the case of statically determinate trusses, it provides a means of estimating axial forces in the members.

Recognising class types

This chapter is mainly concerned with statically determinate trusses. Figure 5.10 shows a number of plane statically determinate trusses which may be recognised as such by noting that there are just sufficient members to prevent relative movement of the joints as follows:

Frame a is the basic triangle.

Frame b is frame *a* with one more joint. The new joint only requires to be connected to the basic triangle by two members, thus restraining its two possible directions of movement and forming an additional triangle.

Frame c can be obtained by starting with any triangle as the basic one and adding extra joints in the same way as for *b*.

Frame d is similar to frame *c* but with the three possible body movements prevented by three external restraints (two at A and one at B).

Frame e is similar to frame *d* in so far as its triangular form can be built up in the same way and it also has three external restraints to prevent body movement.

Frame f has four external restraints which not only prevent body movements but also fix the distance between joints C and D. Thus the extra external restraint acts in lieu of a bar member CD in making a basic unit.

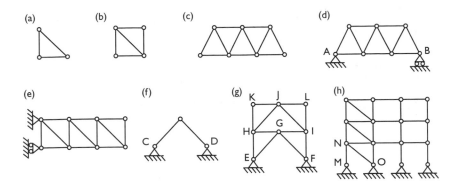

Figure 5.10 Some plane statically determinate trusses.

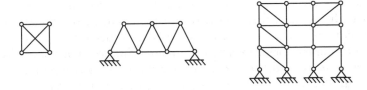

Figure 5.11 Some plane statically indeterminate trusses.

Frame g has EFG as its basic unit similar to frame *f* with extra joints included in the order H, I, J, K, L each restrained by two new members.

Frame h can be obtained by starting with triangle MNO as the basic one and adding further joints each connected by two members to previously named joints or to the ground.

Whereas triangular shaped panels are a feature of these statically determinate trusses, not all panels are necessarily triangular as illustrated by frame *h*. The frames shown in Fig. 5.11 have been obtained from frames *b*, *d* and *h* by adding members or external restraints without altering the number of joints, thus creating redundancies. The frames shown in Fig. 5.12 have been obtained by subtracting members, thus creating sway mechanisms.

Use of formulae

Consider plane frames which are restrained against body movements (e.g. frames *d–h* of Fig. 5.10). Let the total numbers of joints, members and restraints be j, m and r, respectively. There are $2j$ possible joint movements with $m + r$ possible constraints (each member providing just one constraint that its length should be correct). For the frame to be a statically determinate

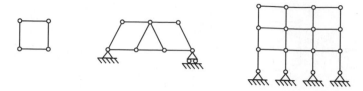

Figure 5.12 Some plane sway frames shown with joints pinned.

truss, there must be just enough constraints for joint movements all to be prevented. Hence a necessary condition for this to happen is $m + r = 2j$. This formula is well known and can also be used for plane frames in which body freedoms are allowed (e.g. frames a, b and c of Fig. 5.10) provided that r is assumed to be three. Furthermore simple sway frames must have $m + r < 2j$ and statically indeterminate trusses must have $m + r > 2j$. You are invited to check that all the fourteen frames shown in Figs 5.10–5.12 are correctly classified.

However, these conditions do not take account of the possibility that frames may have both sway mechanisms and redundant members. If the correct number of members are present, but they have been put into the wrong places to restrain all relative joint movements, frames such as those in Fig. 5.13 are obtained. These have both redundancies and sway mechanisms. Thus the formula $m + r = 2j$ is not a sufficient means of recognising whether a plane frame is statically determinate. Instead the following robust formula is suggested:

$$m + r - e = 2j - s \tag{5.1}$$

where e is the number of extra (or redundant) members and s is the number of independent sway mechanisms present. This formula has been obtained by equating the number of effective constraints to the number of possible movements successfully restrained. If, for a frame, $m + r = 2j$ this formula indicates that $e = s$. To verify that the frame is indeed statically determinate, it is necessary to determine if either $e = 0$ or $s = 0$. For instance, in the first

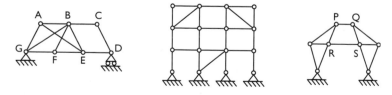

Figure 5.13 Some plane sway frames with redundant members.

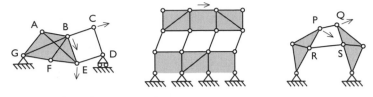

Figure 5.14 Sway mechanisms for the frames of Fig. 5.13.

frame of Fig. 5.13, the panels ABEFG can be recognised as not distorting even without member AE being present. Hence member AE is seen to be redundant, and as it is the only such redundancy, $e = 1$. Alternatively, it may be recognised that panel BCDE can distort. Because this is the only possible sway mechanism, $s = 1$. Figure 5.14 indicates the sway mechanisms for the frames of Fig. 5.13 whilst showing shaded those panels which cannot distort. In the third of these frames, the sway mechanism is only a mechanism for incremental movements (i.e. whilst the lengths of members PQ and RS are not affected by the vertical components of their end displacements). Nevertheless the frame cannot be classified as statically determinate.

Q.5.2
a   Figure 5.15(a) shows a frame with internal members crossing but unconnected. Show that this is a statically determinate truss.
b   Figure 5.15(b) shows the same frame, but with the internal members connected at the intersection point which is therefore classified as a joint. Does this change the classification of the frame?

(a)   (b)

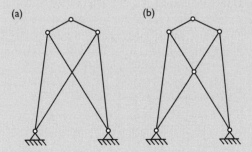

Figure 5.15

Q.5.3
Classify all the six plane frames shown in Fig. 5.16 and show diagrammatically how any possible sway mechanisms may displace.

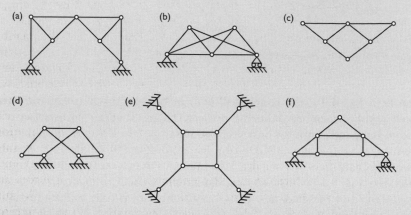

Figure 5.16

Q.5.4
Figure 5.17 shows projected views of a statically determinate space truss. With fully pinned joints, there are just enough members to prevent any sway mechanisms occurring. By examining this and other simple space trusses, devise a robust formula which could be used to classify space frames. Also identify how it should be used when there are no external restraints.

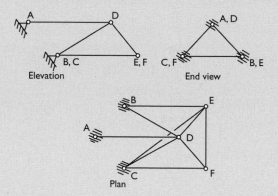

Figure 5.17

5.3 Equilibrium of plane trusses

Force variables

When a plane truss is assumed to have pin-joints and is only loaded at the joints, each member is subject to just the two forces exerted by the hinges. However, these forces must both act along the line joining the hinges, be equal in magnitude and opposite in direction. Provided that the member is straight with its centroidal axis passing through the hinges, its loading will be purely axial. With axial force as the only unknown for each member and, using the notation of Section 5.2, there will be $m + r$ unknown forces in total. If each member of the frame is sectioned into two parts, each component of the dissected truss will comprise one joint together with parts of the attached members. Since there are two equilibrium equations which can be specified for each component (moment equilibrium being automatically satisfied), there will be $2j$ equilibrium equations in total. Hence, when $m + r = 2j$, there are as many equations as unknown forces and it should normally be possible to solve them for the unknowns. The statically determinate case when $e = s = 0$ corresponds to the equations having a viable solution. If $e = s > 0$ however (as for the frames of Fig. 5.13), the coefficient matrix of the equations will be 'singular' in which case its determinant is zero and a unique solution of the equations cannot be found.

Ex. 5.1
Determine the member axial forces and reactions for the truss shown in Fig. 5.18(a).

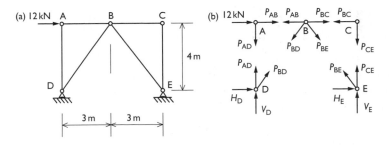

Figure 5.18 Specification of forces for a plane truss: (a) geometry and loading; (b) exploded view of the free-body diagrams for every joint.

First answer
Since $m + r = 2j$ and inspection reveals no mechanism present, this is a statically determinate truss. Figure 5.18(b) shows an exploded view of the free-body diagrams for the various joints. The member forces have all been designated as tension positive. There are ten unknown forces for which the ten joint equilibrium equations are as follows (in kN units):

$$
\begin{aligned}
A \rightarrow && P_{AB} + 12 &= 0 \\
A \uparrow && -P_{AD} &= 0 \\
B \rightarrow && -P_{AB} + P_{BC} - 0.6P_{BD} + 0.6P_{BE} &= 0 \\
B \uparrow && -0.8P_{BD} - 0.8P_{BE} &= 0 \\
C \rightarrow && -P_{BC} &= 0 \\
C \uparrow && -P_{CE} &= 0 \\
D \rightarrow && 0.6P_{BD} + H_D &= 0 \\
D \uparrow && P_{AD} + 0.8P_{BD} + V_D &= 0 \\
E \rightarrow && -0.6P_{BE} + H_E &= 0 \\
E \uparrow && 0.8P_{BE} + P_{CE} + V_E &= 0
\end{aligned}
$$

which can be specified in matrix form as

	Horizontals		Verticals		Diagonals		Reactions						
A	1									P_{AB}		-12
			-1							P_{BC}		0
B	-1	1			-0.6	0.6				P_{AD}		0
					-0.8	-0.8				P_{CE}		0
C		-1								P_{BD}	$=$	0
				-1						P_{BE}		0
D					0.6		1			H_D		0
			1		0.8			1		V_D		0
E						-0.6			1	H_E		0
				1		0.8				1	V_E		0

(Note that the zeros have been omitted from the coefficient matrix.)

Equations of this size and larger can be solved rapidly on a computer. That is one way of obtaining the answer which is given below. Hand solutions, however, can be cumbersome unless advantage is taken of the sparsity pattern of the matrix.

Second answer
It is noticeable that the equations for joints A and C can be immediately solved giving

$$P_{AB} = -12\,\text{kN}, \quad P_{AD} = P_{BC} = P_{CE} = 0$$

The other equations can then be solved rapidly by examining joints in the order B, D, E giving

$$P_{BD} = 10\,\text{kN}, \quad P_{BE} = -10\,\text{kN}, \quad H_D = -6\,\text{kN}, \quad V_D = -8\,\text{kN},$$
$$H_E = -6\,\text{kN}, \quad V_E = 8\,\text{kN}$$

Rapid hand solution

The rapid solution technique shown in the second answer above can be applied to frames e, f, g and h of Fig. 5.10 by starting first with an unrestrained joint which only has two attached members. In these cases a suitable order for solving the joint equations is the reverse of an ordering required to show that no mechanisms were present (e.g. L, K, J, I, H, G, F, E for frame g). It cannot, however, be universally applied to all frames which are externally restrained. For frame d, for instance, there is no suitable joint to start the process. In such cases, one or more of the overall equilibrium equations may be used to determine reactions, so that a restrained joint may be used to start the solution process.

Use of the overall equilibrium equations

The overall equilibrium equations can be derived from the joint equilibrium equations. For if rows 1, 3, 5, 7 and 9 of the matrix equation (given before) are added together, the horizontal equilibrium equation for the whole structure,

$$H_D + H_E = -12\,\text{kN}$$

is obtained. Although the overall equilibrium equations are not adding any new information, they are valuable as a partial check on results computed by means of the joint equilibrium equations. When one or more of these equations have been used to start off the rapid solution technique, a corresponding number of joint equilibrium equations will be superfluous except as a check.

Where a truss has body freedoms (e.g. frames a, b or c of Fig. 5.10), either the overall equilibrium conditions will have needed to be satisfied in advance, or extra variables (such as accelerations) will need to be introduced so that all the joint equilibrium equations may be satisfied simultaneously.

Computer implementation

Whereas it is possible to develop computer programs to automatically construct and solve joint equilibrium equations, this is not the preferred computational technique. Instead it is usual for computer programs to employ the stiffness method, which can be applied more universally to different types of structure and which can be used to predict deflections as well as member forces (as discussed in Chapter 6).

The effect of a mechanism on the solution

If an attempt is made to solve a frame problem using the assumption that the joints are pinned, either by forming the joint equilibrium equations or by using the stiffness method, when a mechanism is present in the frame, this will cause the coefficient matrix to be singular. If an attempt is made to solve these equations by elimination (the usual equation solving procedure used by computer programs), a divisor should become zero making it impossible to continue. Whereas this would always happen if exact arithmetic were to be used, rounding errors may frequently occur making the divisor close to, but not precisely zero. In such circumstances a computer solution may fail completely presenting the user with a form of error message or it may produce completely erroneous results. If, therefore, unanticipated problems of this sort arise in truss analysis, the original problem should be carefully checked for the presence of mechanisms and the geometrical input data should be checked for errors.

Q.5.5
Determine the member forces in the two trusses shown in Fig. 5.19 under the given loadings.

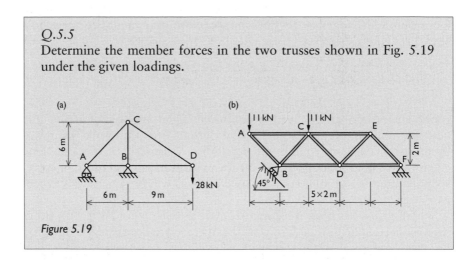

Figure 5.19

Q.5.6

Figure 5.20 shows the geometry of a truss and loads arising through wind action. A computer analysis has given the following results:

$P_{AB} = 89.2\,\text{kN}, \; P_{AC} = -93.3\,\text{kN}, \; P_{BC} = -38.3\,\text{kN}, \; P_{BD} = 76.9\,\text{kN},$

$P_{CD} = 27.6\,\text{kN}, \; P_{CE} = -121.2\,\text{kN}, \; P_{DE} = -20.2\,\text{kN}, \; P_{DF} = 124.1\,\text{kN},$

$P_{EG} = 1.1\,\text{kN}, \; P_{EH} = -129.7\,\text{kN}, \; P_{FG} = -71.5\,\text{kN}, \; P_{GH} = 58.0\,\text{kN},$

$V_F = -121.0\,\text{kN}, \; H_G = -114.0\,\text{kN}, \; V_G = -1.0\,\text{kN}, \; V_H = 116.0\,\text{kN}$

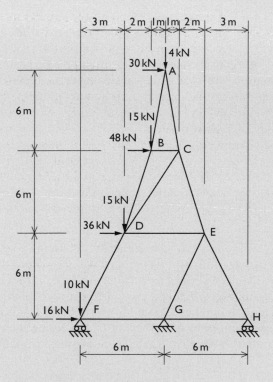

Figure 5.20

Check whether any mistakes have been made in the analysis or input data.

Q.5.7

A software company has received complaints about a frame analysis program because clients have not obtained satisfactory results for the

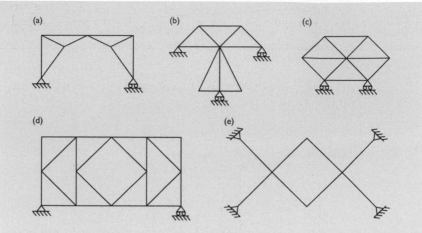

Figure 5.21

truss problems given in Fig. 5.21 which have been treated as if they were pin-jointed. The reasons for the complaints could be

(i)   The structure contains a sway mode and hence cannot be classified as a truss. The program cannot be expected to analyse it satisfactorily.
(ii) The input data have been wrongly specified.
(iii) The computer program gives erroneous results.

The program has already analysed both statically determinate and indeterminate trusses satisfactorily so the company will only check the coding of the program if nothing can be found amiss with (i) or (ii). For each of the structures, indicate whether a sway mechanism is present or whether the input data should be checked for possibility (ii). Where a mechanism is present suggest a modification to the frame which would convert it into a truss and allow results to be obtained from the program.

5.4   Direct analysis for specific member forces

The method of sections proves to be very effective and versatile for direct evaluation of forces in individual members of plane trusses. Not only can it be used to bypass the tedious solution of all the joint equilibrium equations, but it also gives an insight into the factors with regard to geometry and loading which affect individual member forces. Thus it is particularly useful in preliminary design work, assessment of critical components of existing structures and carrying out spot checks on results computed by other methods.

Ex. 5.2
Determine the forces in members DE and DF of the loaded truss shown in Fig. 5.22(a).

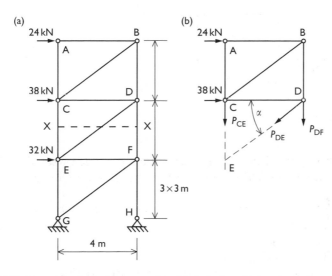

Figure 5.22 Use of method of sections: (a) a loaded truss; (b) free-body diagram of the part above section X–X.

Answer
Consider equilibrium of the part of the truss above the section X–X whose free-body diagram is shown in Fig. 5.22(b). A direct equation for P_{DE} may be obtained from the horizontal equilibrium of the upper part of the truss, thus

$$P_{DE} \cos \alpha = (38 + 24) \, \text{kN}$$

where $\tan \alpha = 3/4$. Hence $P_{DE} = 77.5 \, \text{kN}$.

A direct equation for P_{DF} is obtained by taking moments about E for forces acting on the upper part of the structure, thus in kNm units:

$$4P_{DF} + 6 \times 24 + 3 \times 38 = 0$$

which gives $P_{DF} = -64.5 \, \text{kN}$, the negative sign implying compression.

Ex. 5.3
Determine the force in the member PR of the loaded truss shown in
Fig. 5.23(a).

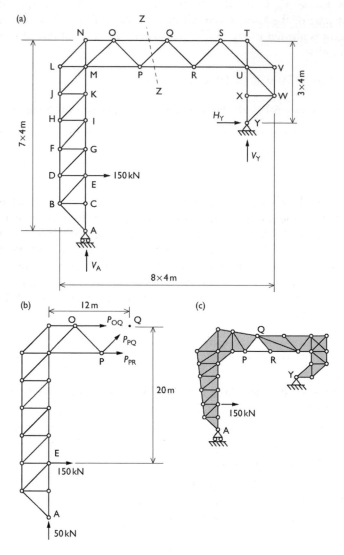

Figure 5.23 A trussed gantry: (a) geometry, loading and reactions; (b) free-
body diagram for part to left of section Z–Z; (c) a modified truss
with the same force P_{PR} (provided that the labelled joints are in
the same position).

Answer

To use the method of sections for this truss it is first necessary to obtain the support reactions either at A or Y. Taking moments about Y for the whole frame, V_A is found to be 50 kN. Then consider the equilibrium of the part of the truss to the left of the section Z–Z whose free-body diagram is shown in Fig. 5.23(b). Taking moments about Q gives in kNm units

$$4P_{PR} + 150 \times 20 = 50 \times 12$$

Hence $P_{PR} = -600$ kN, the negative sign implying compression.

Comment

Not only has the above member force been easily determined, but it is also apparent from the analysis that its magnitude is independent of the positions of most of the members and joints of the truss. For instance the truss shown in Fig. 5.23(c) will have the same force for member PR despite geometric changes within the shaded areas of the truss even to the extent of including two redundant members. The analysis also highlights the influence of the local depth of the truss (i.e. the height of Q above PR) on the magnitude of PR.

Q.5.8

For the loaded truss shown in Fig. 5.24, determine the force in member EG.

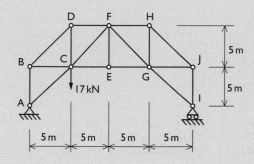

Figure 5.24

Q.5.9
Consider the above one or two stage use of the method of sections to determine individual member forces. Specify

a   What are the requirements necessary for the method to be successful?
b   What steps are required to determine a particular member force?

Q.5.10
a   Use the method of sections to directly obtain the force P_{LO} in the truss whose geometry and loading is shown in Fig. 5.25.

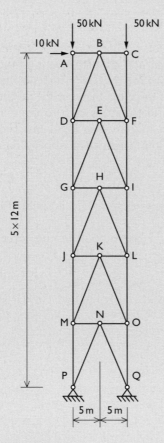

Figure 5.25

NB   There is a way of doing this despite the fact that any section cutting member LO must also cut at least three other members.

b   By using the method of sections and only three joint equilibrium equations, determine the magnitudes of forces in members KM, KO, MN and NO.

5.5   Shapes for plane trusses

Some design requirements affecting shape

Design requirements for any form of plane truss will include the following considerations:

a   There needs to be joints where required to carry the externally applied loads and reactions.
b   Members which are subject to axial compression should not be unduly long (the problems of designing compression members are discussed in Chapter 7).
c   Design and fabrication are simplified if members can be continuous through joints and if joint and member details are repeatable.

Roof trusses

The basic unit for a simply-supported roof truss to support a roof of constant pitch is the triangle shown in Fig. 5.26(a) consisting of two rafter members and a tie. The rafter members will always be loaded in compression and the tie in tension except in situations where pressure due, for instance to wind or an explosion, might create upward forces in excess of the self-weight

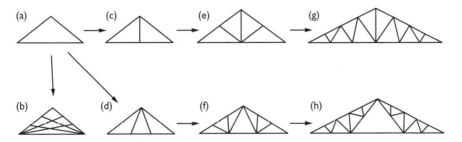

Figure 5.26 Derivation of some roof-truss profiles.

of the roof. This basic truss, however, only resists loading applied at the apex by means of truss action. To perform its job adequately there needs to be extra joints along the rafters, not only to act as loading points, but also to reduce the unsupported lengths of the compression members. One solution shown in Fig. 5.26(b) is not practical because of the accumulation of members meeting at the eaves joints and the long lengths required for bracing members. Most roof trusses establish strong points on the tie member by linking them to the apex, as shown in Fig. 5.26(c) and (d), from which more convenient arrangements of internal members can be derived such as the other trusses shown in Fig. 5.26. All of these trusses are statically determinate when pin-jointed and simply-supported at the eaves joints.

Trusses with parallel chords

Many trusses consist of two parallel 'chords' or 'booms' interpersed by bracing members which triangulate the intervening space. The structural action of such trusses is like that of beams with the chords taking on the role of flanges in resisting bending moment and the bracing members performing the functions of webs as far as shear transfer is concerned.

The Warren truss, having alternating diagonals is common. Sometimes vertical members are added to provide extra loading points at deck level or to shorten unsupported lengths of the compression chord as in Fig. 5.27. Figure 5.28(a) and (b) show two other classical forms of beam truss. The Pratt truss acting as a simply-supported beam has all but the end diagonal bracing members taking tension in the dead-load case, with the shorter vertical members taking compression. This configuration is generally preferred to the equivalent Howe truss where the longer bracing members are the ones which take compression. On the other hand, by adding some more bracing members to a Howe truss, it is possible to provide more loading points at deck level whilst shortening the lengths of diagonal compression members (Fig. 5.28(c)).

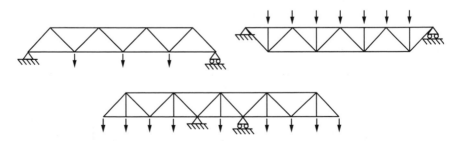

Figure 5.27 Warren trusses without and with additional vertical members.

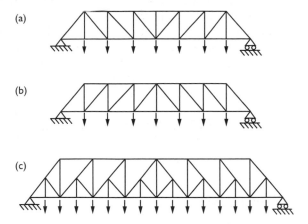

Figure 5.28 Some other types of trusses with parallel chords: (a) Pratt truss; (b) Howe truss; (c) Howe truss with additional bracing members.

Bowstring trusses

Bowstring trusses (Fig. 5.29) have the advantage that the depth is largest at the centre where the bending moment on the equivalent beam is generally largest. This results in both chord members being more uniformly loaded along their lengths. The inclined chords of bowstring trusses acts to a certain extent like arches or suspension cables depending on whether they carry compression or tension.

Figure 5.29 Bowstring trusses.

Towers and pylons

Normally, for towers and pylons, trusses are preferred which are symmetrical about their vertical axis. The K truss (see Fig. 5.25) and lattice trusses (Fig. 5.30(a)) are frequently used. In the case of electricity pylons a major consideration is to stabilise the vertical members against buckling. Figure 5.30(b) shows one way in which this can be achieved by the use of additional bracing members.

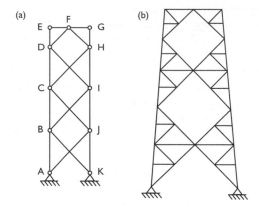

Figure 5.30 Lattice towers: (a) a statically determinate truss; (b) with secondary
bracing.

Identification of load paths

To gain a qualitative understanding of the structural action of trusses and, in
particular, to find where tensile or compressive forces predominate, it is often
useful to consider the application of a single concentrated load to various
joints in turn so that load paths through the structure can be identified. This
process is usually easy to perform when the truss is statically determinate by
inspecting the equilibrium of individual joints.

Ex. 5.4
For the roof truss shown in Fig. 5.31(a) having a single downwards
load applied at joint B, determine which members carry compression
and tension forces.

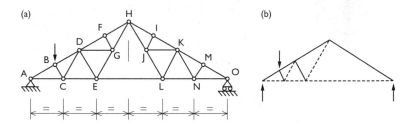

Figure 5.31 Load path identification for a roof truss: (a) geometry and loading;
(b) load path (compression: full line, tension: dotted line).

Answer
Resolving forces at joint B in the direction perpendicular to ABD indicates that member BC must be loaded in compression. Then, resolving perpendicular to ACE at C, it follows that member CD is loaded in tension. By similar reasoning, because there are no external loads acting on joints F, I and M, it can be shown that no forces act in members FG, IJ and MN and consequently also in members DG, JK, KN, KL and HJK. Examining forces acting on joints D, E and G in a similar way identifies how the load path connects to the apex joint H giving the result shown in Fig. 5.31(b).

Q.5.11
Some forms of roof truss, which do not fit in with the profiles derived in Fig. 5.26, are scissors trusses as shown in Fig. 5.32 and queen post trusses having the profile shown in Fig. 5.16(f). Discuss their classification (are they true trusses according to the definitions of Section 5.2?) Also, discuss their merits or demerits.

Figure 5.32 Scissor trusses: (a) without a central vertical post (this roof in Scotland is adorned with antlers); (b) with a central vertical post (this roof is over the dining hall of the Old Faithful Inn, Yellowstone National Park, USA); (c) diagrammatic representations.

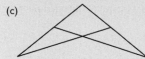 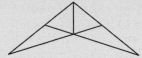

Figure 5.32 Continued.

Q.5.12
If the double cantilever truss shown third in Fig. 5.27 carries the same loading at each deck joint, identify which members will be loaded in tension and which in compression.

Q.5.13
Consider a single downwards load applied at the fourth deck joint from the left (counting the support joint as one) in the braced Howe truss of Fig. 5.28(c).

a   Which members can be quickly identified as being in tension, which in compression and which unloaded assuming that no other applied loads are acting on the truss?
b   Under the above single load application, can you deduce which are the most highly loaded segments of the upper and lower chord?

Q.5.14
For the lattice tower shown in Fig. 5.30(a), indicate which members
are loaded in tension and which in compression when:

a   a vertical downwards load is applied at E,
b   a vertical downwards load is applied at F,
c   a horizontal load directed to the left is applied at D.

5.6 Trusses with redundant members

Statical determinacy and indeterminacy

The main advantage of a statically determinate structure is that the forces
are governed by statics alone and are not, therefore, influenced by other
factors some of which the designer may have difficulty in quantifying. On
the other hand, there are no alternative load paths. If any one member fails,
the structure will collapse.

Conversely, a structure which is statically indeterminate (e.g. a truss with
redundant members) has alternative load paths and is not so reliant on every
member performing correctly. However, the various other factors which
influence behaviour should be appreciated and, where necessary, taken into
account. These factors include initial stresses in the members, lack of fit, temperature effects and load-deflection characteristics of the members. Further
discussion of this topic takes place in Sections 6.11 and 9.1.

Self-equilibrating systems

Figure 5.33 shows a truss which has one redundant member which could
be any one of the six comprising the central panel CDHI. If 40 kN loads
are applied to joints J and I and the force in member CH is designated as

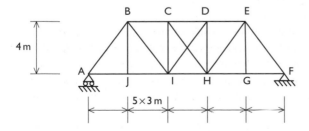

Figure 5.33 A truss with one redundancy.

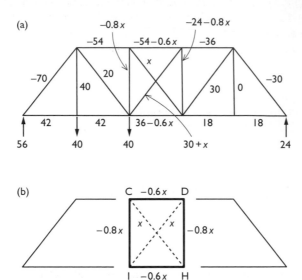

Figure 5.34 Equilibrium solutions for the truss of Fig. 5.33: (a) with external loading (kN units); (b) with no external loading.

the variable x, the joint equilibrium equations may be solved to give all the member forces in terms of x as shown in Fig. 5.34(a). From this solution it is apparent that there is scope for change in the load paths for the forces, but only within the panel CDHI. Thus if member CH could not carry load for some reason, the solution in which $x = 0$ corresponds to the case where the load path avoids this member. Similarly, if member DH could not carry load, the solution in which $x = 30$ would avoid it having to carry load. The possibility of alternative load paths is a feature of all redundant structures and usually has a beneficial effect on structural performance.

However the presence of the redundant member in the above truss also creates the possibility that additional forces occur over and above those necessary, from an equilibrium standpoint, to support the external loads. If, for instance, member CH was the last to be put in place by bolting and, because of misalignment of the bolt holes, it had been necessary to stretch the member to make it fit, there would be a tension force in the member which would pull inwards on the rest of the truss at joints C and H. The inward pull on these joints would put compression into members CD, CI, IH and DH and tension into member ID with the forces being in the ratios depicted in Fig. 5.34(b). Self-equilibrating systems may include balanced reaction forces. For instance, the third frame shown in Fig. 5.13 has already been identified as having one mechanism and one redundant member. If member RS is treated as the redundancy and a tension force is introduced into it, joint equilibrium considerations yield a force system which is represented diagrammatically in Fig. 5.35.

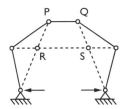

Figure 5.35 The self-equilibrating force system for the third frame of Fig. 5.13.

Tie bracing

When a rectangular panel, which is braced across its diagonals by two wires, is distorted, one wire will be stretched and take up load, but the other will go slack rather than accept compressive force. Hence, in the absence of self-equilibrating forces, the tower shown in Fig. 5.36(a) containing wire cross-bracing will resist lateral forces from either direction by means of statically determinate truss action as shown in Fig. 5.36(b) and (c). Cross-bracing using wires (or round bars, which have a very low buckling load) are sometimes used to keep structure weight low. Furthermore, if the ties are tightened up when being installed (for instance, by including turn-buckles), self-equilibrating forces will be developed in which the ties are

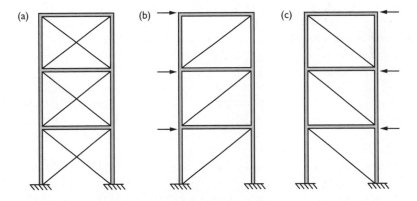

Figure 5.36 A tower with wire cross-bracing: (a) unloaded; (b) with loading from the left; (c) with loading from the right.

in tension and the columns and beams in compression. In that situation, the braces do not go slack immediately the truss is subject to side loading. Whereas this manoeuvre will not increase the total amount of lateral loading that the truss can carry, it will improve the deflection characteristics of the structure for loadings insufficient to cause the ties to become slack.

The use of tie members in this way illustrates just one way in which, despite the introduction of uncertainty, statically indeterminate forms of structure may have advantages over statically determinate ones.

A modern airship hangar (Janner et al., 2001)

Airships are being developed to lift heavy loads (up to 160 t) over large distances. Unlike the airships of the early twentieth century, they are helium filled and do not contain rigid frames, being called 'blimps'. Hence they are both lighter and safer than their earlier counterparts. The very large hangar constructed to house the first two blimps provides a maximum clearance height of 101.5 m and a floor area of 66,000 m², enough to accommodate eight football pitches (Fig. 5.37). The rounded form of the structure avoided unnecessary internal space and kept wind loading low. Apart from wind loading, the other main loading case is due to self-weight. One way in which the self-weight of the central part of the structure was kept as low as possible was by cross-bracing the panels between the five semi-circular arches and the horizontal bars supporting the textile membrane. Figure 5.38 shows this lightweight cross-bracing in place during construction.

Figure 5.37 Brand airship hangar.

Figure 5.38 The central structure of the Brand airship hangar before the fabric covering was in place.

Q.5.15
Identify diagrammatically any self-equilibrating systems arising from redundant members in the trusses of Figs 5.11 and 5.16.

Q.5.16
Figure 5.39 shows a simply-supported lattice truss. By assuming that joints around the edges are pinned and that bracing members are unconnected to each other where they cross, determine how many sway mechanisms and how many redundant members are present. Indicate diagrammatically the deflection shape of any mechanism and the loaded members of any self-equilibrating stress systems.

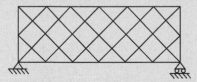

Figure 5.39

5.7 The Forth Rail Bridge

Prelude *(Koerte, 1992)*

The benefits to be gained in the nineteenth century by the development of railways were immense. They provided a rapid means of transporting both goods and people by land, unmatched by canal transport or horse drawn coaches and wagons. Route development was very expensive, but the rewards were high for enterprises which were successful in capturing trade. In this era of railway 'mania', Thomas Bouch, who was certainly ambitious and a visionary, but has also been unkindly described as a 'chancer', proposed to the North British Railway Co. (NBR) that the wide estuaries of the Forth and Tay should be bridged, in order to complete a fast east coast route in Scotland to rival the more successful route of the Caledonian Railway Co. After the NBR told Bouch that it was 'the most insane idea that had ever been propounded' he initiated steam ferries to carry the railway coaches and trucks over the estuaries. However, he did eventually gain support to bridge the Firth of Tay and a world record 3 km long bridge to his design was opened in 1878. It had 85 spans across this relatively shallow estuary. His main materials were masonry bases for the piers with cast iron columns supporting wrought iron trusses.

In December 1879, while Bouch was consolidating plans to bridge the deeper Firth of Forth, some of the central 'high girders' of the Tay Bridge collapsed taking with it a train and 75 people. Although, in the inquiry, much poor workmanship and cost cutting were queried, the accident was almost certainly due to underestimation of wind loads about which there was a large degree of uncertainty at the time (Martin and Macleod, 1995). As a result, too narrow a bridge had been built (Fig. 5.40). This disaster had very wide repercussions. Sir Thomas Bouch was no longer trusted and his novel proposal for a suspension bridge across the Firth of Forth was abandoned. In fact, suspension bridges had had a poor history in the nineteenth century and on two occasions, at Stockton on Tees in England and at the Niagara Gorge in America, they had performed poorly when subjected to the heavy loading of railway trains. Instead Sir Benjamin Baker, a particular advocate of cantilever bridges was commissioned to design a bridge with Sir John Fowler, one of the most experienced British engineers, being the engineer-in-chief. The bridge, which was the largest structure of its age was completed in 1898 (Fig. 5.41).

The design concept *(Westhofen, 1890)*

The small island of Inchgarvie made a convenient place for a central pier. However, there was a need for spans on either side which were much larger than any existing truss bridge. Previous cantilever bridges were by Gerber

Figure 5.40 Pier bases of Bouch's Tay Bridge exposed at low tide. Note the broader supports for the replacement bridge built alongside.

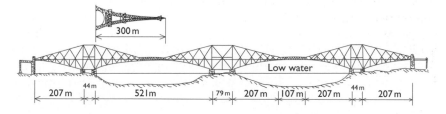

Figure 5.41 Elevation of the Forth Rail Bridge showing the Eiffel Tower to the same scale.

over the Maine at Hassfort of span 130 m and by Charles Shaler Smith over the Kentucky river at Cincinnati with three 114 m spans. Baker and Fowler's proposal was for cantilevers to be built out from the central pier simultaneously in both directions so that their self-weights were balanced at all times during construction. Three such pairs of balanced cantilevers provided the crossing of the main part of the channel. In order to allow some independent movement of the cantilevers (e.g. due to train loads or temperature strains), two 32 m span trusses connected the central double cantilevers to the neighbouring ones. These trusses were simply-supported on the ends of the cantilevers, being hinged at their shoreward end and resting on rocking posts (in order to allow horizontal relative movement as well as rotation) where they were attached to the Inchgarvie double cantilevers. The two main spans of 521 m were four times any previous cantilever span. Including approach viaducts, the overall length was 2528 m.

Design features

In America, steel was first used in cast form for Eade's Mississippi bridge at St Louis completed in 1874. However, steel was not able to be used in Britain in large quantities until a process had been developed for reducing the sulphur content. In using steel for the first time, Baker acknowledged its importance in enabling him to adopt higher working stresses than with iron, thus making the large spans possible.

The shoreward ends of the two outer balanced cantilevers were loaded down by a weight greater than half the weight of a beam truss, sufficient to prevent the weight of trains lifting the cantilevers off their piers (see Fig. 5.42). However, because the central double cantilever was not supported at either end, its central tower was given a wider base so that the weight of

Figure 5.42 A demonstration of cantilever action organised by Sir Benjamin Baker.

two trains on the end of one of the cantilevers would not cause it to lift off any of its piers. Of the twelve main cantilever piers, only three (one for each double cantilever) were completely fixed. The others were allowed to slide both laterally and longitudinally in a bed of oil to allow for movements due, for instance, to wind or temperature expansion. Great care was thus taken with support conditions to ensure that, as near as possible, the reactions would act in the way predicted during design. The cantilevers were close to being statically determinate when viewed as 2D trusses in elevation (Fig. 5.43).

Apart from gaining as much evidence as possible about wind forces, Baker made his cantilever tower bases particularly broad with the main columns inward sloping (Fig. 5.44). This not only gave the bridge adequate resistance to wind forces, but also gave it the appearance of stability which would assure travellers that the Tay Bridge disaster would not be repeated at the Forth. Despite the 3D nature of the whole bridge, design against applied loads was able to be broken down into many individual 2D analyses of trusses such as those in Figs 5.43 and 5.44. Figure 5.44 also shows that the two railway tracks needed to be supported on a separate viaduct within the much larger cantilever trusses.

One of the distinctive design features of the Forth Rail Bridge was the use of hollow circular steel cylinders for the main compression members. These were fabricated from steel plates bent to the correct curvature and drilled on shore before being carried to site. Internal frameworks of longitudinal I sections and rings to maintain shape were first erected and then the curved plates were riveted round them. This process was labour intensive and required holes to be accurately drilled to avoid alignment problems during erection. Cylindrical compression members were chosen because

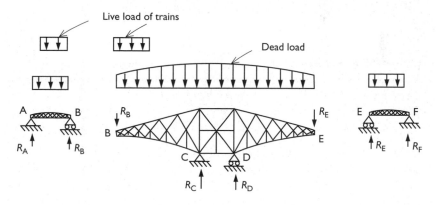

Figure 5.43 Exploded view of Inchgarvie balanced cantilevers and linking suspended spans simulating a severe loading case.

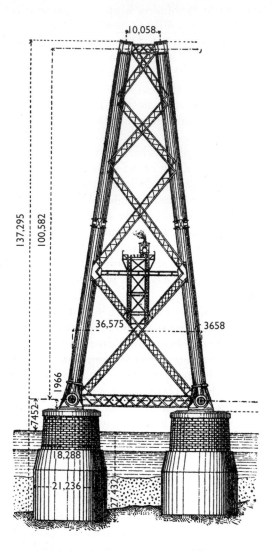

Figure 5.44 Cross-section at a tower (dimensions in mm).

of their efficiency in carrying large compression forces. Whereas cylinders strengthened by longitudinal members and ring stiffeners have been used extensively in the twentieth century (e.g. for aircraft fuselages and submarine construction), experience in the nineteenth century was more limited. Two previous bridges had used cylinders for their compression chords. These were both by Isambard Kingdom Brunel at Chepstow in 1852 (since dismantled) and at Saltash in 1859. Also boilers were used extensively for steam engines. However, all of these uses were on a much smaller scale. The satisfactory design and fabrication of the various quite complicated joints was clearly a major achievement. Lattice girders were used for other members because they offered low wind resistance and could be designed to be stiff enough to carry compressive as well as tensile forces.

Aftermath

Despite the magnitude of the construction undertaken by Sir William Arrol, the bridge, after being opened in 1889, recovered its construction costs from the approximately 20,000 trains per year that crossed it in the first 20 years. It was not without its critics. Theodore Cooper, one of the most senior bridge builders in America described it as 'the clumsiest structure ever designed by man'. However, he no doubt regretted saying that after the collapse of the Quebec bridge under construction in 1907 (see Section 7.3). The Forth Rail Bridge on the other hand, has been reassessed and found to be safe for rail traffic after more than 100 years of service.

It was also attacked from an aesthetic standpoint. William Morris, who criticised all iron architecture, is recorded as calling it 'the supreme specimen of all ugliness'. Baker's reply to this slur was as follows: 'Probably Mr. Morris would judge the beauty of a design from the same standpoint, whether it was a bridge a mile long or for a silver chimney ornament. It is impossible for anyone to pronounce authoritatively on the beauty of an object without knowing its function. The marble columns of the Parthenon are beautiful where they stand, but if we took one and bored a hole through

Figure 5.45 The Forth Rail Bridge still in use after more than 100 years.

its axis and used it as a funnel for an Atlantic liner, it would, in my mind, cease to be beautiful, but of course Mr. Morris might think differently'.

Not all were critical though. Alfred Waterhouse, an architect wrote to Fowler saying: 'the simple directness of purpose with which it does its work is splendid and invests your vast monument with a kind of beauty of its own'. This comment rather than Morris's is the one which would be considered appropriate today. Of all the large truss bridges throughout the world, the Forth Rail Bridge must surely rank as the most elegant (Fig. 5.45).

Q.5.17   Observation question
Can you identify a metal truss bridge near to you? If so, can you answer the following?

- What type of metal was it made from?
- When was it constructed (if there is no date on the bridge, make a guess)?
- What type of truss is it? (e.g. parallel chord, cantilever, arch)
- How many plane trusses can you identify within its structure?
- Are these trusses statically determinate or indeterminate?
- Are the principal loads applied through the joints of the trusses or not?
- How are the members jointed?

Lastly can you find a method of checking your answers (by talking to a structural engineer or finding a publication about it?)

5.8   Towers as 3D trusses

Taking account of the third dimension

It is essential that designers take into account the 3D nature of all structures. Many structures, like the Forth Rail Bridge, have 2D trusses as their primary structure. Even then there will be a need to carry load which is applied transversely to these main trusses and to ensure sufficient lateral restraint is provided to prevent buckling of their compression members. In the Forth Rail Bridge bracing between corresponding compression members on either side of the main cantilevers performs most of these functions by creating secondary 2D trusses in lateral planes (see Fig. 5.44). The remaining sections in this chapter consider trusses from the outset as three-dimensional.

Analysis of 3D trusses by hand using joint equilibrium requires much care. Furthermore, it is often difficult to find cross-sections which enable direct methods to be determined for specific member forces. Hence reliance tends to be placed on computer programs to provide formal analyses for specific loading cases. Indeed much of the development of 3D or 'space trusses' has been since digital computers have been available to facilitate the analysis process.

Classification

Classification of 3D trusses is important both to be sure that sway mechanisms are not present and also to identify whether or not there are redundancies and hence alternative load paths. The main difference from plane frames is that each joint of a 3D frame has three instead of two possible directions of movement. Hence, the robust classification formula for 3D frames using the notation developed in Section 5.2 is

$$m + r - e = 3j - s$$

It is also important to note that in order to be fully restrained against all possible body movements, it is necessary for $r \geq 6$. This formula may be checked for the statically determinate truss shown in Fig. 5.17. Also, it should be noted that the basic 3D truss unit is a tetrahedron, which conforms to this formula with $m = 6$, $j = 4$ and $e = s = 0$ provided that $r = 6$ to allow for the fact that body freedoms are unrestrained.

Ex. 5.5
How can the space frames shown in Fig. 5.46 be classified?

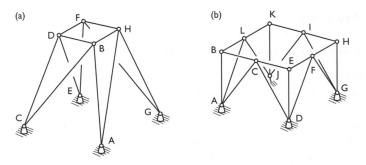

Figure 5.46 Two space frames: (a) a statically determinate truss; (b) a frame with four sway mechanisms.

Answer

a   Since, for the first frame, joints at A, C, E and G are restrained in all three directions, $r = 12$. With $m = 12$ and $j = 8$, the robust formula gives $e = s$. However, joint B cannot sway in the plane ABC because of members AB and BC. It is also prevented from swaying out of this plane by the planar panel, BHGA. Therefore, it cannot be involved in a sway mechanism and by similar arguments, neither can joints D, F and H. Thus no sway mechanism is present giving $s = 0$ and, for this problem, $e = 0$. The frame is classified as a statically determinate truss.

b   For the second frame, $m = 20$, $r = 12$ and $j = 12$. Substituting in the robust formula gives $32 - e = 36 - s$ implying that there must be at least four sway mechanisms. In fact, there are four sway mechanisms involving lateral displacements of the joints C, F, I and L, respectively.

Coping with 2D joints in 3D truss analysis

Joints supported in one plane only are a common feature of many structures which are generally considered to be trusses, but which do not meet the criterion given in this chapter. In practice, if these joints are part of a bracing system, they are not intended to be subjected to large external forces in the lateral direction. If a member connecting principal joints of the frame (i.e. ones not involved in sway movements) is continuous through one of these joints, small amounts of lateral load which might be applied to the joint (e.g. because of wind) will be able to be supported by bending action.

The main problem arises if an attempt is made to analyse such a structure by computer without the user appreciating the presence of sway mechanisms. To perform a computer analysis successfully, it will be necessary either to include bending stiffness for sufficient members to restrain the possible sway mechanisms (Fig. 5.47) or, if no lateral loads are to be applied, artificial external restraints may be included which prevent just these lateral movements (Fig. 5.48). For the case of lattice bracing, it is possible to avoid specifying joints at the cross-over points of the bracing members (Fig. 5.49).

Towers and pylons

Consider the one storey structure shown in Fig. 5.50 in which the joints are all vertices of a polygon of arbitrary planform. Each side panel is plane

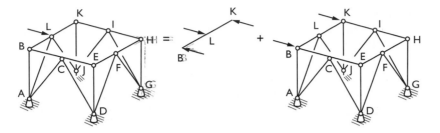

Figure 5.47 A lateral force at joint L resisted by bending action of member BLK together with truss action.

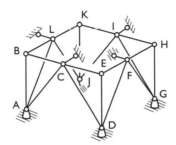

Figure 5.48 Use of artificial restraints to enable a space truss to be analysed as pin-jointed.

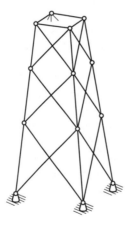

Figure 5.49 A lattice tower.

(a)   (b)

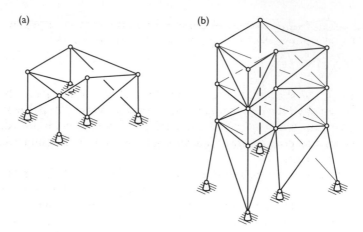

Figure 5.50 A tower with polygonal planform: (a) one storey; (b) multistorey.

and contains one diagonal bracing member. The argument used to classify the frame of Fig. 5.46(a) can be used to prove that this frame is a statically determinate truss. Furthermore, by adding additional storeys which have the same characteristics, tower structures can be formed which also classify as statically determinate trusses (e.g. Fig. 5.50(b)). Additional members, either forming cross-bracing or placed across the central void will amount to redundancies providing additional load paths. However, symmetric bracing within panels is usually employed for towers and pylons, lattice and K bracing or derivatives of these being common (Figs 5.49 and 5.51).

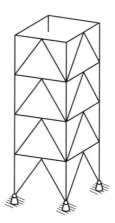

Figure 5.51 A K-truss tower of square planform.

Q.5.18
If the results from a space truss analysis by computer are suspect, what are the possibilities for checking the validity of the results?

Q.5.19
a   The frame shown in Fig. 5.52(a) has all the lower joints fully restrained whilst all the upper joints form a square. All of the supporting members are of the same length. Classify this frame identifying any possible sway mechanisms and any self-equilibrating force systems.
b   Also classify the frame shown in Fig. 5.52(b) which is similar in specification except that its upper joints form a regular pentagon.
c   Can you extend the results above to classify frames which have *m* equally spaced joints round the circumference of a circle and *n* storeys with a general arrangement typified by Fig. 5.52(c) (with $m = 8$ and $n = 2$).

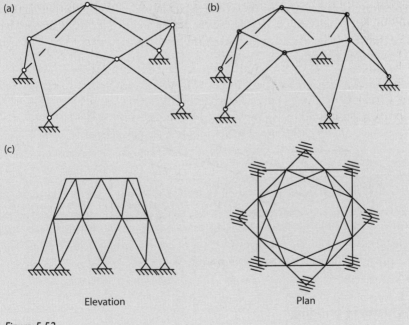

(a)

(b)

(c)

Elevation

Plan

Figure 5.52

5.9 3D truss action in buildings and bridges

Roof structures

To form a roof structure using roof trusses of the kind described in Section 5.5, it is necessary to place them on supporting walls or columns at a regular spacing. Frequently the roofing material is attached to purlins which run longitudinally across the trusses. Bending moments are minimised in the trusses if the purlins connect to the joints in the rafter members. However, such a configuration does not provide any means of resisting forces applied to the ends of the building (normally due to wind) which could push all the trusses (connected via the purlins) over sideways. A common method of stabilising the roof structure against this possibility is to include 'wind bracing' as shown in Fig. 5.53. Theoretically only one bay is required to be braced, but it is generally preferred to brace both end bays so that the structure resisting the wind forces is close to their points of application.

Another question which needs to be considered is whether any of the lower joints of the roof trusses need to be braced laterally. If the dead load of the roof definitely exceeds any possible upload due, for instance, to wind or explosion, tension forces in the principal members will ensure their stability without lateral support. Lateral bracing of lower chord joints of the end bays may, however, be useful in assisting the end walls to carry wind or other forms of side loading.

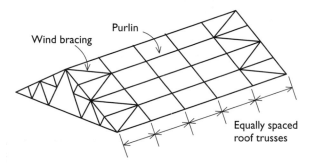

Figure 5.53 Structure to support a pitched roof.

Multistorey buildings

The floors in a multistorey building are usually provided with in-plane stiffness so that, as far as horizontal movements are concerned, their distortion can normally be neglected. Each floor can, however, move horizontally in any of three body freedoms and to prevent this sway bracing is sometimes employed. For a building with a few storeys, it is sufficient to include brac-ing between columns in three parts of the building. The bracing arrangement shown in Figs 5.54 and 5.55(a) would be suitable because sway movements in the *x* direction are prevented by the panel DE, whilst sway movements in the *y* direction and also rotations in the horizontal plane are prevented by panels AB and FG.

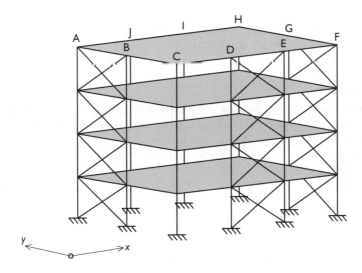

Figure 5.54 A multistorey building with braced panels to resist sway.

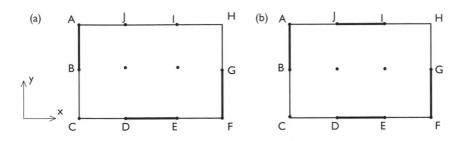

Figure 5.55 Plan of a multistorey building showing possible positions for braced panels to prevent horizontal sway movements: (a) with three braced panels; (b) with four braced panels.

Some general points regarding sway bracing in buildings:

- Walls which are designed to carry shear loading (called shear walls) are often used in the place of shear bracing.
- Use of four braced panels (e.g. one on each outer face as shown in Fig. 5.55(b)) will provide less tendency for the building to twist under side loading.
- For buildings in earthquake zones, it is good practice for them to be designed to be doubly symmetric in both inertia and structure, so that twisting motions are not provoked by horizontal earth movements (discussed in Section 9.4).
- The forces requiring to be carried by bracing systems will increase as the height of the building increases.

Truss bridges

Common forms of truss bridges comprise two parallel primary trusses which span between end supports which are linked together at the chord levels. Secondary structure is needed to

- resist side loading (e.g. due to wind);
- transfer deck loading to joints of the primary trusses;
- stabilise the compression chords of the primary trusses.

One way of creating a space truss is to link both bottom and top chords. If these links have sufficient bracing members to form two secondary trusses in horizontal planes, a tubular structure of rectangular cross-section will be formed. The only possible remaining sway mechanisms will be a distortion of the cross-section which will be prevented by the inclusion of one or more internal members (Fig. 5.56). Figure 5.57 shows two possible arrangements for a complete bridge, in which the only requirements for members to be stiff or strong in bending, is to resist compression buckling and to transfer applied loads (particularly those on the deck) to the nearest joints. Figure 5.58 shows a possible load path for a lateral force applied to an upper chord joint of the through bridge of Fig. 5.57(b).

In many cases, however, it may not be convenient or practical to develop a complete space truss in this way. For instance, if trusses are to be placed above the deck of a bridge, the height of traffic may preclude the upper chords of the primary trusses being linked. In such cases, bending stiffness is required in members to resist sway movements, particularly when they can initiate buckling (see Fig. 5.59).

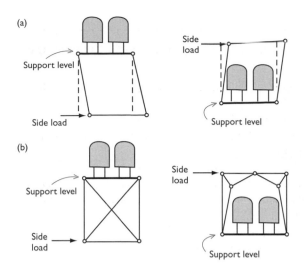

Figure 5.56 Maintaining the cross-sectional shape of truss bridges (cases are shown where the deck is level with the upper or lower chords): (a) distortion of the rectangular cross-section; (b) bracing to maintain the shape of the cross-section.

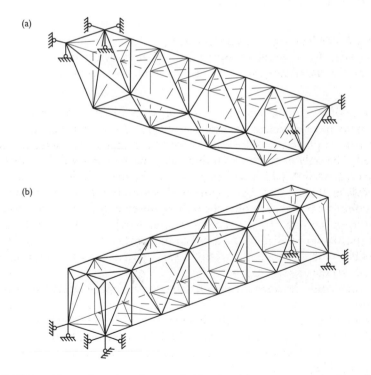

Figure 5.57 Possible configurations for truss bridges involving no sway mechanisms: (a) with the truss below deck level; (b) with the truss above deck level (a through bridge).

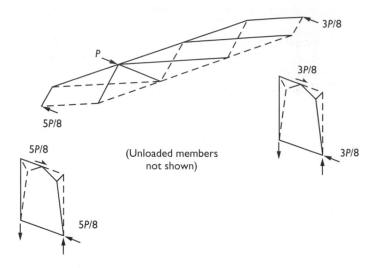

P

3P/8

3P/8

5P/8

3P/8

5P/8

(Unloaded members
not shown)

3P/8

5P/8

5P/8

Figure 5.58 Exploded view showing a possible load path for a lateral force applied to the upper chord joint of the truss shown in Fig. 5.57(b).

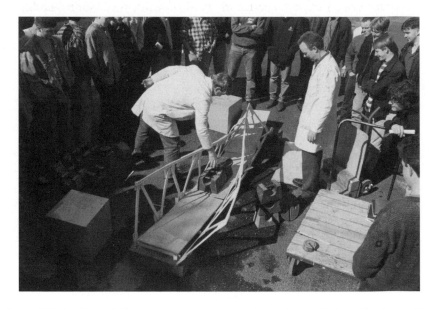

Figure 5.59 Buckling of the top chords of a small timber bridge during a load test. The bridge was designed and built by stage 1 students as an exercise at Queen's University, Belfast.

Q.5.20

One of the simplest forms of 3D truss, which can be used as a beam or column, has a triangular cross-section with each face braced (e.g. the series of linked pyramids shown in Fig. 5.60). Do you know of, or can you think of, any possible uses for such trusses?

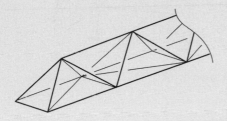

Figure 5.60

Q.5.21   Observation question

In order not to impede spectators' lines of sight, most sports stadia feature large cantilevered roof structures. If you have the opportunity to inspect any such roof, sketch the visible structure. Does it include plane, cantilevered trusses as the principle component? Is there bracing present in order to resist forces lateral to the cantilevers or stabilise compression members against buckling? Describe the primary and secondary structural components within the roof and suggest their functions? Can you tell how the forces exerted on the roof are carried down to the foundations?

Q.5.22

Specify factors influencing whether a truss bridge should be designed with the primary trusses situated below or above the deck. Indicate, for each factor, which is the most suitable type of bridge.

5.10   Truss domes

Dome geometries

The dome truss shown in Fig. 5.61 is a variation of the polygonal tower shown in Fig. 5.50 which can be shown to be statically determinate when pin-jointed. A problem with this form of construction, however, is how to

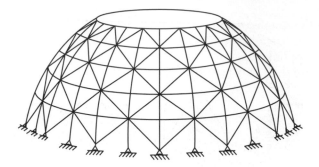

Figure 5.61 A statically determinate trussed dome.

Figure 5.62 The Buckminster Fuller Dome at Expo'67, Montreal.

avoid the clutter of members that occurs if the ribs are continued through to the apex.

The geodesic dome provides an interesting means of creating spherically shaped trusses invented by Buckminster Fuller. The Buckminster Fuller dome designed for the American Pavilion at the 1967 Montreal world exhibition (later dismantled after a fire) did a lot to popularise the concept (Fig. 5.62). Geometries for geodesic domes can be generated from the regular icosahedron which has identical equilateral triangles forming its twenty faces. Figure 5.63 shows a regular icosahedron, a template for making a model

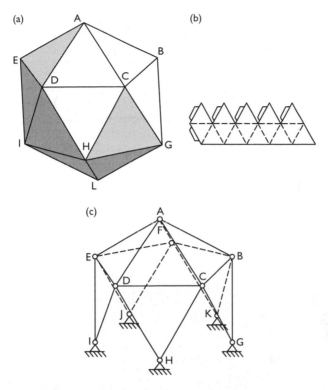

Figure 5.63 The icosahedron basis for geodesic domes: (a) a regular icosahedron; (b) a template to make a model of part of an icosahedron; (c) a space frame based on the icosahedron.

of part of it and a simple truss derived from it. Its particular merit is that all members have the same length and all joints above ground level have the same geometry. Geodesic domes having more members than this can be generated from it by triangulating each face in such a way that the additional joints are projected onto the surface of the circumscribing sphere. If an edge AB is divided into four equal lengths by intermediate joints M, N and O, these joints will lie on the great circle passing through A and B. Triangulation for the whole of the segment ABC is as shown in Fig. 5.64(a). By placing joints Q, R and U in suitable positions and dividing other segments in an identical way, the following results are obtained:

- All joints will have six impinging members except for the icosahedron joints which will have five.
- There will only be four different member lengths (see Fig. 5.64(b)).

- Excepting those at ground level, there will only be four types of joint required (see Fig. 5.64(c)).
- Furthermore, the differences in length of members and angles of joints may be small enough to be accommodated within tolerance limits.

Single layer trusses of this form are suitable where each icosahedron triangle is not too extensively subdivided (see Fig. 5.65). However, if there is a large degree of subdivision, the members impinging on one joint lie almost in one plane. This results in a danger of buckling (including snap through buckling discussed in Section 7.4). To guard against this possibility, double layer domes are sometimes used. The Montreal Expo'67 dome had inward facing pyramids constructed on each triangle of the outer structure with the apexes joined to form an inner layer. The inner layer members were arranged in

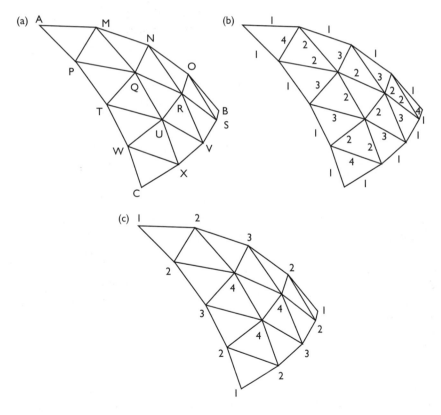

Figure 5.64 A typical geodesic dome segment using four joints on each edge: (a) triangulation; (b) member types; (c) joint types.

Figure 5.65 A geodesic dome with three joints on each edge.

hexagons, except adjacent to each icosahedron joint, where there needed to be a pentagon. The dome had a diameter of 76 m and a height of 61 m. It contained 24,000 tubular steel members and 6000 welded joints and was covered with plastic sheeting.

In most cases, the important requirement of a dome is to cover a large floor area without necessarily being very high. Many different geometrical forms have been used for shallow domes.

Biomes for the Eden Project *(Jones et al., 2001)*

The Eden Project has been created to tell the fascinating story of man's relationship with plants. For the first stage of the development, two large greenhouses were required to contain different climatic conditions. The site chosen was the Bodelva pit in Cornwall which, at the time, was soon to become vacant.

The use of geodesic domes enabled the structural design to be undertaken even though the ground profile would not be known until china clay workings had been finished and the site handed over (one of the objectives being to alter the dramatic landscape as little as possible). Figure 5.66 shows how one of the two greenhouses (dubbed 'biomes') was constructed by linking four geodesic dome segments (or 'bubbles') together using trussed arches to support them where they intersect. The principal members of the biomes form a honeycomb of mainly hexagonal panels, rather than the triangular

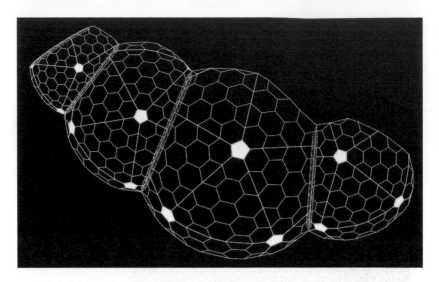

Figure 5.66 Plan of the Warm Temperate Biome for the Eden Project.

pattern discussed earlier for other geodesic domes. Its geometry is similar
to the inner layer of the Expo'67 dome. With an average of 1.5 members
per joint this structure has many sway modes and hence requires bending
strength in the members and joints to make it into a viable structure.

This pattern of members maximises transparency for light transmission.
It also provides a convenient shape for the Ethyltetraflouroethylene (ETFE)
foil cushions used as cladding. These cushions are held within aluminium
frames which are bolted to the outside of the biome members. They are
light-weight, very transparent and also self-cleaning. However, there is an
operational need to maintain pressure in them in order to prevent them from
flapping about.

The main structure of the biomes is supplemented by an inner layer of
members. This consists of parallel lines of members which intersect to form
triangles and either hexagons or, under icosahedron points, pentagons.
Each triangle forms the base of a tetrahedron with one of the joints of the
main layer as the apex. This lightweight secondary structure gives increased
strength but does not eliminate the need for bending strength in members.
Thus the biomes are described as semi-braced. Inner bracing members can
be seen behind the main structure in Fig. 5.67. All the steel of the biomes,
being inside the cladding, has been galvanised for corrosion protection from
the aggressive hot and humid climate. In total the two biomes carry 625
hexagonal cushions with side lengths varying from 2.725 to 5.488 m as well
as other pentagonal and triangular cushions.

Figure 5.67 Placing of an ETFE cushion (not yet inflated) on a Biome.

> Q.5.23
> What is the relationship between an icosahedron and the way in which footballs are often constructed?

5.11   The Alexander L. Kielland (Naesheim, 1981; Lancaster, 2000)

Prelude

During the twentieth century, society developed a tremendous thirst for oil for transport, power generation and a multitude of uses in the manufacturing industry, so making it financially viable to tap undersea sources. Provided that water depths were not too large, extraction tended to be from a production platform on which the important operating functions were carried out above the water level. Platforms were normally supported by 3D steel trusses which were founded on the seabed straddling the oil wells. However, for drilling purposes it was advantageous to use rigs which could be moved easily from place to place, and semi-submersible drilling platforms were developed in the 1960s in the USA for this purpose. In some respects, they were an ingenious compromise between a ship and a production platform. Buoyancy was obtained from pontoons which, when the platforms were moored, were situated well below the water surface away from the buffeting of waves. Like production platforms, their decks were supported by means of 3D steel trusses, but like ships, they could be moved.

The rig

The Alexander L. Kielland was one of a series of platforms each of which had five 'flattened doughnut' pontoons situated at the vertices of a regular pentagon (hence they were called pentagone rigs). Each pontoon and connecting column was divided into 11 separate tanks allowing water ballast to be pumped in or out (see Fig. 5.68). Adjustment of the amount of water ballast allowed the rig to be lowered or raised and allowed any tendency to tilt (due to shifts in the platform's centre of gravity for instance) to be corrected. When the rig needed to be moved ballast was pumped out, thus allowing the pontoons to rise to the surface thereby providing less drag resistance.

Apart from the need to ride the most severe storms, a major design consideration was for the rig to survive an impact with a ship. The main danger came from supply vessels which needed to dock against the rig. As a result, it was necessary to cater for incorrect ballast in one column or pontoon due to one or two tanks becoming breached. Also, the possibility of pump malfunction needed to be considered. With the pentagone rigs, a ballast misalignment produced less of a tilt to the platform than would have been the case if only four pontoons were to have been used. However, the main selling point of this configuration was that it reduced the vertical motions caused by wave action.

The main structure of the rig consisted of four components as follows:

a   Five columns, A–E, provided access to the pontoons and also additional buoyancy.
b   Horizontal bracing members connected the columns immediately above the pontoons.
c   Deck structure was attached to the top of column C and was linked by two braces with a walkway at deck level to the top of each of the other columns.
d   Diagonal bracing members provided connections between nodes 5 and 6 and the columns, all at the level of the lower bracing members, with nodes 1, 2, 3, 4 and 7 as part of the deck structure.

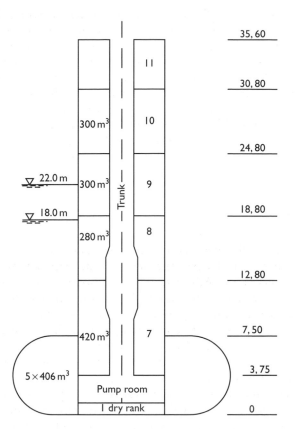

Figure 5.68 One pontoon and column showing capacities of watertight tanks and heights in metres.

Figure 5.69 shows an elevation and two sections of the rig and Fig. 5.70 shows all the bracing in plan.

Assuming that the deck structure provided in-plane, but not out-of-plane stiffness, it may be replaced, for analysis purposes, by an equivalent truss connecting the top of column C with all the nodes at the deck level. The equivalent deck truss requires 7 additional members to give in-plane rigidity (see Fig. 5.71). Hence the equivalent frame has 17 joints in all, 10 at the deck level and 7 at the lower level. The number of members is 17 at the deck level, 10 at the lower level, 15 diagonals and 5 columns giving 47 in total. Because an unrestrained 3D frame has six body freedoms, $r = 6$ should be used in the robust formula from Section 5.8 to give

$$53 - e = 51 - s$$

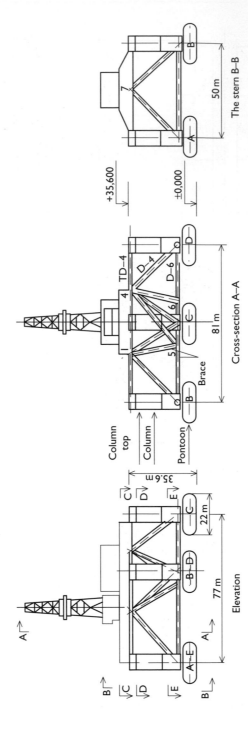

Figure 5.69 Elevation and vertical sections.

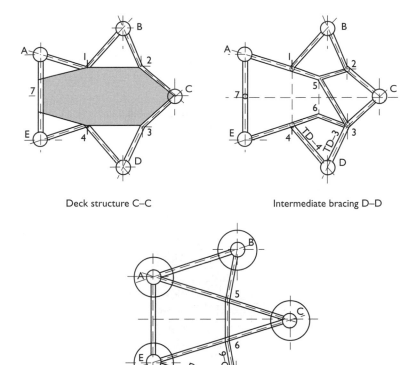

Deck structure C–C   Intermediate bracing D–D

Lower level bracing E–E

Figure 5.70 Horizontal sections.

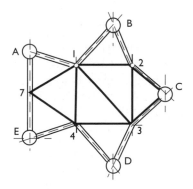

Figure 5.71 Equivalent truss to represent deck structure.

Since no mechanisms are present, there must be two redundant members. (The first rig in this series did not have a lower horizontal brace connecting columns A and E so that must have only had one redundant member.)

The structure was designed in this way because of the following reasons:

- There was a need to keep self-weight as low as possible not only to keep down costs, but also to facilitate moving the rig.
- A truss with many redundancies could have had self-equilibrating forces present whose magnitudes depended critically on the sequence in which joints were fabricated.
- Lower level bracing members were omitted between columns C and its neighbours to allow access for docking supply vessels.

The accident

Four years after its construction, the Alexander L. Kielland had not been used for its design purpose of drilling. It was moored next to a production platform and linked to it by a movable walkway. Its purpose was to provide accommodation for workers from this and neighbouring platforms. When, on 27 March 1980, winds arose to 16–20 m/s with wave heights of 6–8 m, the gangway was removed and the Alexander L. Kielland was moved further away from the other platform. Some 40 minutes later, loud bangs were heard and it suddenly developed a list of 30–35 degrees. Subsequently, it gradually heeled over further until the one remaining anchor wire broke and the rig overturned. Four pontoons were visible above the water, but the other, still attached to its column, was floating independently. Of the 218 workers on board at the time, 123 lost their lives. It was a long time before a method was found to right the upturned rig.

The primary cause

Examination of the detached leg soon revealed that lower brace D–6 had developed a brittle fatigue fracture which started from a deficient weld design at a hydrophone attachment. Once this brace fractured completely, all the other bracing members linking column D to the rest of the rig fractured in quick succession causing the initial sudden list. The list increased further due to water entering ventilators of the trunk of column E and the buildings on deck (see Fig. 5.72).

Fatigue

The Comet disasters (Section 3.7) highlighted the danger posed by fatigue in aluminium. Fatigue is also a possibility in steel and it is well known that

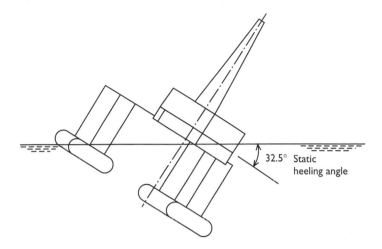

Figure 5.72 Estimation of probable stable floating position in calm water for 'Alexander L. Kielland' after the loss of column D.

welding has to be carried out carefully to avoid flaws occurring which will be crack starters in situations where high repeated loading could take place. There were two factors of relevance to the Alexander L. Kielland:

a   The main structural members of the rig were fabricated from a good quality steel permitting higher tension stresses to be used than for mild steel. Unfortunately the fracture toughness characteristics after welding were not good.
b   Because the hydrophone (a device to help to position the rig over oil wells, which had not actually been used) was deemed to be non-structural, the degree of weld penetration was considered to be unimportant. The welding was not carefully designed, controlled or inspected. Indeed paint traces revealed that a large crack must have been present before the rig left the construction yard.

The behaviour of the damaged structure

Once brace D–6 had completely fractured, the buoyancy force acting on the pontoon and column D, together with any other forces arising through the motions of the sea, air and rig itself, were applied to the remaining five braces supporting this column. Although the severance of this one member did not

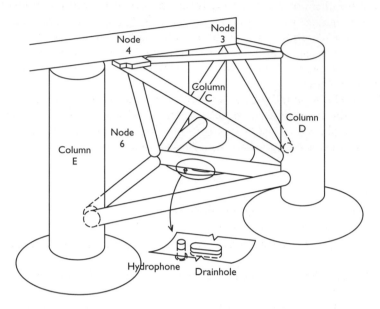

Figure 5.73 The brace D–6 and surrounding structure showing an enlargement of the hydrophone attachment.

result in a sway mechanism being formed in the equivalent pin-jointed frame, the only member restraining the leg against rotation about an axis through nodes 3 and 4 was the lower brace D–E. However, this brace had its own axis not far from parallel to the axis 3–4 (Fig. 5.73). Hence, because the buoyancy force, in particular, would have a large moment about the axis 3–4, extremely large unanticipated forces would have been developed in this member and through equilibrium of joints, in other members also. When any one of these members failed, a sway mechanism would have been created which the remaining structure would have been unable to restrain through bending action.

Therefore once one member was severed, the break away of the leg arose through the lack of much redundancy in the structure. Alternative load paths that were available were not sufficiently robust to cope with that situation. Inclusion of lower bracing members between columns C and its neighbours might have provided extra load paths sufficient to have enabled the damaged structure to stay in one piece.

Lessons

One of the clear faults was the failure to appreciate the structural significance of a weld carried out for a non-structural purpose. Also it would have been useful if the chain of events arising from a failure of any one component had been investigated. In the aircraft industry, design against fatigue has been a major consideration because aluminium, like steel in a seawater environment, has no stress limit below which fatigue does not occur. To cope with fatigue and other types of failures, fail-safe philosophy has been developed (see Section 4.11).

The off-shore oil industry has taken on board lessons from this tragedy. Even so, the operations required to extract oil and gas off-shore remain hazardous. Ships may head for port or heave to when the worst storms rage, but off-shore installations have to ride them out.

Q.5.24

Find a way of checking that there is not a mechanism in the equivalent pin-jointed frame indicated by Figs 5.69, 5.70 and 5.71.

In case you are not sure how to do this, here are three suggestions:

a Make a 3D model using, for instance, straws for members and plasticine for joints.

NB   Do not make the lower braces A–5–C and C–6–E and the upper braces A–7–E continuous through their middle nodes. Inspect for mechanisms by testing the rigidity of the model.

b Find four nodes interconnected by six members in the form of a tetrahedron. With this as the basic unit, add any node which is linked to it by three members not in one plane. See if by continuing in this way, you can include all the joints. Any members left over will be redundant.

c Input data to a 3D truss analysis program to simulate this frame and see if it gives sensible answers to a variety of loading cases (in which case it can be assumed that there are no mechanisms).

Chapter 6

Linear elastic behaviour

Because structures must be able to maintain their shape sufficiently well under load, being able to predict deflections is a vital part of assessing their performance. One of the objectives of this chapter is to examine the load–deflection characteristics of statically determinate structures which are loaded within their linear elastic range. However, being able to calculate deflections also makes available analysis methods for statically indeterminate structures. Hence another objective is to gain an appreciation of their more complicated characteristics.

On completion of this chapter, you should be able to do the following:

- Estimate elastic deflections for statically determinate trusses and both statically determinate and indeterminate beams.
- Calculate elastic stresses and deflections for beams made of composite materials.
- Understand the principles of the direct stiffness method for computer structural analysis.
- Appreciate how redundancy affects the performance of structures.
- Make use of symmetric properties in the analysis of structures or in checking computed results.

6.1 Strain and stiffness

The linear elastic range

All materials change shape when subject to load. At microscopic level, the stresses acting on an incremental element will cause changes in the dimensions of the element. These changes called 'strains' cause the structure to deflect. Most structures obey Hooke's Law, which states that stress is proportional to strain, over a significant part of their loading range. For loading within this range, the material is said to be linear elastic. The term 'elastic' implies that there is no loss of energy and so the material will revert to its original state when unloaded.

If all the material within a structure is loaded within the elastic range, the structure itself often also exhibits linear elastic properties. It is structures with this type of characteristic which will be examined in this chapter (structures which behave in a non-linear fashion, despite their material all being loaded within the elastic range will be discussed in Chapter 7). The slope of the curve of load vs deflection is called 'stiffness' (a stiff structure is one which requires a large load to produce a small deflection and therefore has a large stiffness). If a structure behaves linearly, as in Fig. 6.1, there is only one value of stiffness for the whole of the loading range and the theory of superposition holds. Many structures are designed with the intention that all the material remains within its linear elastic range when subject to normal working loads. Often deflections need to be limited in order to prevent adverse effects. Hence this chapter is most relevant for investigating the behaviour (and hence the serviceability) of structures subject to working loads. It is also relevant for investigating failure when brittle materials are used which are linear elastic up to failure. It is not applicable for determining collapse loads if the materials used have non-linear load–deflection characteristics.

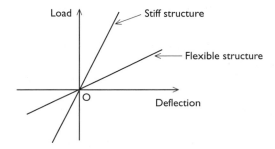

Figure 6.1 Linear load–deflection characteristics.

Direct strain

Direct strain is defined as extension/length and is thus non-dimensional. If an element of material is subject to unidirectional stress σ_x, the corresponding direct strain is designated ε_x. The stiffness of the material is described by the ratio $E = \sigma_x/\varepsilon_x$ called the 'modulus of elasticity' or 'Young's modulus' and has the units of stress. Approximate values for different materials are shown in Table 6.1. Note that the stresses these values represent are extremely large. They are the stress levels at which, if the linear elastic behaviour were to continue far beyond the actual elastic limit, a specimen loaded in tension would double in length.

Table 6.1 Approximate values of Young's modulus

Material	$E(kN/mm^2)$
Steel	200
Aluminium	70
Timber[a]	7–11
Stone	20–100
Brick	14
Concrete	20–25

Note
a In the direction of the grain.

Ex. 6.1
a   A member of a truss of length ℓ, cross-sectional area A, and modulus of elasticity E carries an axial-load P. Obtain a formula for the extension e of the member.
b   For a steel member of length 4 m and cross-sectional area 880 mm^2, estimate the extension resulting from a tensile load of 160 kN.

Answer
a   The axial stress in the member (assumed to be constant across the cross-section) is $\sigma = P/A$. Hence the axial strain is $\varepsilon = P/EA$ giving an extension of

$$e = P\ell/EA \tag{6.1}$$

b   Stress $= 160 \times 10^3/880 = 182\,\text{N/mm}^2$, strain $= 182/(200 \times 10^3) = 0.91 \times 10^{-3}$, extension $= 4000 \times 0.91 \times 10^{-3} = 3.6\,\text{mm}$.

Poisson's ratio effect

When a stress σ_x is applied to an element of material in the x direction and no other stresses are present, strains are found to occur not only in the x direction, but also in the y and z directions. These are almost invariably opposite in sign to ε_x. For an isotropic material $\varepsilon_y = \varepsilon_z = -v\varepsilon_x$ where the constant v is known as 'Poisson's ratio'. This generally has a value close to 0.3 for metals.

With bar members Poisson's ratio effects are of little significance. However, where structural elements have 2D or 3D stress systems (e.g. slabs, panels, domes, cylinders, etc.) or where restraints against movement are present in other than the loaded direction, Poisson's ratio should be taken into account in determining deflections and stiffnesses and in diagnosing the presence of secondary stresses.

Ex. 6.2
The material of a bearing block of dimension 100 mm × 150 mm in plan and 60 mm deep has $E = 8\,\text{kN/mm}^2$ and $v = 0.28$. Determine the stiffness of the block in kN/mm when subject to vertical compression loading if

a   the sides are unrestrained;
b   the block is encased in a box with effectively rigid walls.

Answer
(Note: compressive stresses and strains are treated as positive)

a   If the deflection is e mm, the strain is $e/60$. Hence the stress is $8e/60 = 0.1333e\,\text{kN/mm}^2$. The total force on the block is therefore $0.1333e \times 15,000 = 2000e\,\text{kN}$ and the stiffness equals force/displacement$= 2000\,\text{kN/mm}$.

b   With the other two directions restrained, the vertical stress σ_1 will be augmented by restraining forces σ_2 and σ_3. The strains ε_2 and ε_3 will be zero. Therefore, the stress–strain relationships for the block in the three orthogonal directions will be

$$E\varepsilon_1 = \sigma_1 - v(\sigma_2 + \sigma_3) \qquad E\varepsilon_2 = \sigma_2 - v(\sigma_1 + \sigma_3) = 0$$
and
$$E\varepsilon_3 = \sigma_3 - v(\sigma_1 + \sigma_2) = 0$$

Using the last two equations to eliminate σ_2 and σ_3 gives

$$E\varepsilon_1 = \sigma_1\left(1 - \frac{2v^2}{1 - v}\right)$$

Hence the effective stiffness of the material is multiplied by $(1 - v)/(1 - v - 2v^2) = 1.28$. The block stiffness will be factored by the same amount giving a value of $2560\,\text{kN/mm}$.

Ex. 6.3

A steel pressure vessel comprises a circular cylinder of radius r closed by hemispherical ends. The thickness t_1, of the cylinder, and t_2, of the hemispheres, may be considered to be negligible in comparison with r. If $\upsilon = 0.3$, determine the ratio t_2/t_1 which is required to ensure that the hoop strains in the cylinder and hemispheres match where they meet.

Answer

Using formula derived for Ex. 3.9, the hoop and longitudinal stresses in the cylinder are

$$\sigma_h = \frac{pr}{t_1} \quad \text{and} \quad \sigma_\ell = \frac{pr^2}{t_1(2r + t_1)} \simeq \frac{pr}{2t_1}$$

where p is the internal pressure. (There are also radial stresses, but these are relatively small and will be ignored.) Allowing for the Poisson's ratio effect from the longitudinal stress, the hoop strain in the cylinder will be

$$\varepsilon_h = \frac{\sigma_h}{E} - \upsilon \frac{\sigma_\ell}{E} = \frac{pr}{2Et_1}(2 - \upsilon)$$

The internal forces in the hemispheres will be the same as in a sphere of the same dimensions and loading, a diametral section of which gives the situation shown in Fig. 3.53(a). Hence $\sigma_1 = \sigma_2 = pr^2/t_2(2r + t_2) \simeq pr/2t_2$ are the circumferential stresses acting on planes at right angles to each other. The circumferential strain will therefore be approximately

$$\varepsilon_c = \frac{\sigma_1}{E} - \upsilon \frac{\sigma_2}{E} = \frac{pr}{2Et_2}(1 - \upsilon)$$

For the circumferential strains to match at the interfaces $\varepsilon_h = \varepsilon_c$ giving $t_2/t_1 = (1 - \upsilon)/(2 - \upsilon) = 0.412$.

Comment

Pressure cylinders with a thickness ratio t_2/t_1 greater than 0.4 will expand more than their hemispherical ends, resulting in some local bending necessary to produce compatibility at the joints. Local self-equilibrating stress systems which prevent a mismatch in deflections between adjoining parts of a structure in equilibrium are called 'secondary stresses'.

Shear strain

It is also possible to identify shear strain, designated γ through the application of shear stress τ. A shear stress τ_{xy} distorts an originally square element of material into a rhombus (alternatively called a diamond or lozenge shape) in which edges, originally parallel to the x and y axes, deflect into lines at $90° \pm \gamma_{xy}$ to each other. The shear modulus, G, relates shear stress and strain according to

$$\tau_{xy} = G\gamma_{xy}$$

Shear stress on its own has no tendency to alter the volume of isotropic material. Thus there is no Poisson ratio effect involving shear stress.

Direct strains on a surface

The three strains ε_x, ε_y and γ_{xy} entirely describe the in-plane deformation at a surface. The direct strain at an angle θ to the x axis may be evaluated by examining a rectangle having a diagonal of length ℓ in the required direction. Figure 6.2 shows the deformed rectangle AB′C′D′ placed over the undeformed one ABCD in such a way that the position of the corner A and the direction A $\rightarrow$ D coincide for both (because strains are expected to be small, deflections need to be exaggerated whenever they are shown on a diagram). The projection of CC′ in the y direction will be the same as that of BB′, namely $\varepsilon_y \ell \sin \theta$. However, the projection of CC′ in the x direction will be $\varepsilon_x \ell \cos \theta$ due to ε_x plus $\gamma_{xy} \ell \sin \theta$ due to γ_{xy}. Projecting these movements on to the diagonal gives the extension FG + GC′ shown in the inset to Fig. 6.2. Because the displacements are small, the angle EC′F can be assumed

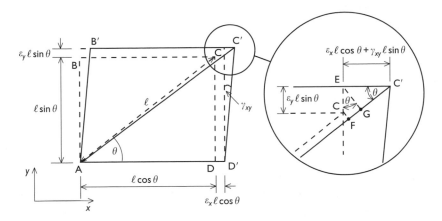

Figure 6.2 2D strain of a rectangular element.

to be θ, thus giving

$$FC' = \varepsilon_x \ell \, \cos^2\theta + \varepsilon_y \ell \, \sin^2\theta + \gamma_{xy} \ell \, \sin\theta \, \cos\theta$$

and hence

$$\varepsilon = \varepsilon_x \cos^2\theta + \varepsilon_y \sin^2\theta + \gamma_{xy} \sin\theta \, \cos\theta$$

is the direct strain on the diagonal AC'.

   This formulae and a corresponding one for shear strain at an angle θ bear striking similarities with the corresponding formulae (3.1) for stress. As a result, the following deductions may be made:

- It is possible to draw a Mohr's circle of strain with the horizontal axis representing ε and the vertical axis representing $\frac{1}{2}\gamma$ (note the $\frac{1}{2}$!) as shown in Fig. 6.3.
- Principal strains, being the maximum and minimum direct strains, occur in directions at right angles to each other (their directions coincide with those of the principal stresses).
- The shear stress is zero when measured between the principal directions and maximum when measured between directions at 45° to them.
- It can be shown that, for an isotropic material,

$$G = \tfrac{1}{2}(1 + v)E$$

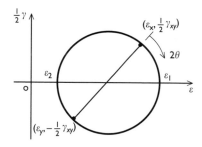

Figure 6.3 Mohr's circle of strain.

Strain measurement

The usual method of measuring strain is by means of electrical resistance strain gauges. In the common form of gauge (Fig. 6.4), a printed circuit in the shape of a concertina, sandwiched in between two insulating layers, is stuck to the surface of the material to be tested. Because changes in

Connections

Figure 6.4 The printed circuit of an electrical resistance strain gauge.

resistance caused by strain are small and also because gauges can be temperature sensitive, a Wheatstone bridge is used to compare the gauge resistance with that of an unloaded dummy gauge (Fig. 6.5). By only passing a very small pulse of current through the gauges in order to take readings, their temperature is not affected by the reading process. Hence only one dummy gauge needs to be used to compare against many active gauges. It is important that temperature differences between the gauges are minimised (e.g. by keeping them out of direct sunlight).

Where one-directional stress is expected (such as in a member of a truss or at the outer fibres of a beam flange), one strain gauge will suffice. In other cases, however, it is necessary to use three gauges in order to determine the three possible independent surface strains ε_x, ε_y and γ_{xy}. In that case, it is normal to set the three gauges out in a 45° or 60° 'rosette' (e.g. Fig. 6.6).

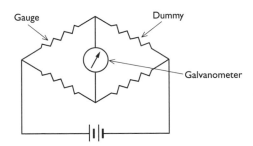

Gauge   Dummy

Galvanometer

Figure 6.5 A Wheatstone bridge circuit.

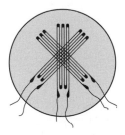

Figure 6.6 A 45° strain gauge rosette.

Stress measurement

Stress is surprisingly difficult to measure directly (an exception being when the material is photoelastic). Instead the normal procedure is to first determine the stress–strain characteristics of the material through strain measurements for which there are controlled stress regimes. Then the stresses required may be deduced from strain measurements by referring to the controlled tests.

Material elastic properties

Whereas an isotropic material has only two independent elastic constants (E and v), materials which are not isotropic (such as fabric or timber) have much more complicated properties. Franz Newmann determined that up to 21 elastic constants are possible for anisotropic crystalline material (Timoshenko, 1983). Other problems arise when the material is not homogeneous (e.g. concrete). If used at all on concrete (as opposed to reinforcement within the concrete), the results may be meaningless unless the gauges are significantly longer than the maximum dimension of the aggregate.

Q.6.1
Specify some reasons why structural deflections may cause problems.

Q.6.2
If the spherical housing of Q.3.14 is to be constructed of steel with $v = 0.3$ and a shell thickness of 4 mm, determine what cross-sectional area the reinforcing ring will need to have if the strains in the shell and ring are to be compatible.

Q.6.3
a   A uniform solid circular bar of diameter 20 mm is loaded in tension to determine the stress–strain properties of the material. Two strain gauges are placed on the surface at the centre of the bar, the first being parallel to the bar axis and the second being wrapped around the bar at right angles to the first. If the strains are found to vary linearly with the tension force, T, with a typical set of readings being

$\varepsilon_1 = 0.00103, \quad \varepsilon_2 = -0.00026 \quad$ when $T = 8.8\,\text{kN}$

predict values of E and v for the material.

b   A circular cylinder made from this material has two strain gauges attached to its outer surface. The first is set with its axis longitudinally and the second with its axis circumferentially. If, when the cylinder is pressurised, the readings are $\varepsilon_1 = 0.00024$ and $\varepsilon_2 = 0.00067$, estimate the longitudinal and circumferential stresses at the surface of the cylinder.

Q.6.4
a   Why should strain effect the electrical resistance of strain gauges?
b   Why is it advantageous to broaden the hairpin bends of foil gauges as shown in Fig. 6.4?

6.2   Beam bending

Radius of curvature

In Section 4.3, the assumption was made that the elastic bending stress in a beam is proportional to the distance from the centroid. This assumption is consistent with a beam, subject to a uniform bending moment, deflecting in the arc of a circle with plane cross-sections remaining plane. Consider a segment of beam of original length ℓ bent such that the centroidal axis has a radius of curvature R as shown in Fig. 6.7. The centroidal axis AB will

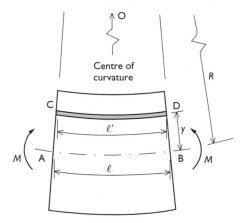

Figure 6.7 A beam segment subject to uniform bending.

not show any change in length and so will be also the neutral axis. If the distorted length of a filament, CD, of material at distance y above the neutral axis is ℓ', from similarity of the two segments OAB and OCD,

$$\frac{\ell'}{R-y} = \frac{\ell}{R}$$

However, because the strain ε, on the filament CD is given by $\varepsilon = (\ell' - \ell)/\ell$, it follows that

$$\varepsilon = -\frac{y}{R} \tag{6.2}$$

Since, in a beam, no lateral direct stresses will be present to give a Poisson's ratio effect, the bending stress, σ, will be related to the radius of curvature through

$$\frac{\sigma}{y} = -\frac{E}{R} \tag{6.3}$$

Although few beams carry constant bending moment, this result will also apply to beams with a linearly varying bending moment (i.e. constant shear). In this case, the shear deformation distorting the cross-section will be similar at both ends of a short segment, so cancelling any effects these would have on longitudinal strains, as shown in Fig. 6.8. For sections of beam where the BMD is non-linear, bending theory stress distributions do not strictly apply. However, they are almost invariably used and will not normally be very much in error. Shear force or shear flow distributions derived through Eqs (4.7) or (4.9) are likely to be less accurate than bending stress distributions.

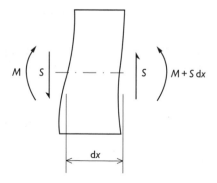

Figure 6.8 Elastic deflected form of a beam element subject to constant shear force.

Combining Eqs (6.3) and (4.5) gives

$$\frac{M}{I} = -\frac{\sigma}{y} = \frac{E}{R} \tag{6.4}$$

These formulae encapsulate elastic bending theory relating bending moments, through longitudinal stress and strain, to radius of curvature and are worth memorising.

Differential equation for bending

Consider a horizontal beam whose deflection at spanwise coordinate x is v. The curvature of the beam is the inverse of the radius of curvature R and can be specified accurately in terms of x and v by

$$\frac{1}{R} = \frac{\mathrm{d}^2 v}{\mathrm{d}x^2} \Bigg/ \left[1 + \left(\frac{\mathrm{d}v}{\mathrm{d}x} \right)^2 \right]^{3/2}$$

However, where beam slopes are small, the denominator has negligible effect, making

$$\frac{\mathrm{d}^2 v}{\mathrm{d}x^2} \simeq \frac{1}{R} = \frac{M}{EI} \tag{6.5}$$

The product EI relating bending moment to curvature is known as the 'bending stiffness'.

Use of integration

Deflections of statically determinate beams may be found by integrating the curve of M/EI twice and using the known deflections or slopes at support points to determine the constants of integration.

Ex. 6.4
A uniform cantilever beam, which is built-in at end A, has a bending stiffness EI and length ℓ, carries a concentrated load, P, at end C. Obtain an expression for the end deflection.

Answer
From equilibrium, the bending moment at any point distance x from A is

$$M = EI\frac{\mathrm{d}^2 v}{\mathrm{d}x^2} = -P(\ell - x)$$

Integrating with respect to x gives

$$EI\frac{dv}{dx} = -P\left(\ell x - \frac{1}{2}x^2\right) + c_1$$

However $dv/dx = 0$ at $x = 0$, hence $c_1 = 0$.
Integrating again gives

$$EIv = -P\left(\frac{\ell x^2}{2} - \frac{x^3}{6}\right) + c_2$$

But $y = 0$ at $x = 0$, hence $c_2 = 0$.

These expressions are plotted in Fig. 6.9 in which the maximum deflection, $v = -P\ell^3/3EI$, is seen to occur at the tip.

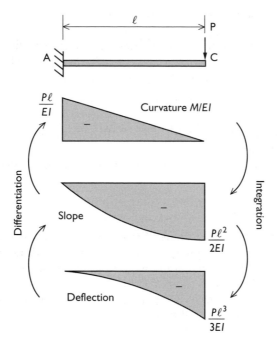

Figure 6.9 Variational relationships for elastic deflections of beams.

This result is included in the set of standard solutions shown in Fig. 6.10.

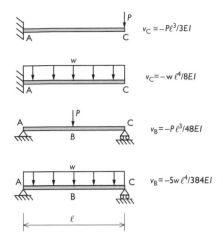

$v_C = -P\ell^3/3EI$

$v_C = -w\,\ell^4/8EI$

$v_B = -P\,\ell^3/48EI$

$v_B = -5w\,\ell^4/384EI$

Figure 6.10 Maximum deflections for some uniform beams.

Integration of the curve of M/EI to obtain deflections has parallels with the integration of the loading diagram to obtain the BMD (compare Figs 4.6 and 6.9). From the variational relationships, it follows that

(i)   Local maximum and minimum deflections occur where the slope is zero (or at the ends).
(ii) The area of the diagram of slope, measured between any two points that have zero deflection, must be zero.
(iii) Between any two points that have zero slope, the area of the diagram of M/EI must be zero.

The moment–area method

Mathematical integration was easy to use for Ex. 6.4 because it did not contain any discontinuity in the loading. The moment–area method, which uses the graph of M/EI as its starting point, is more suitable for general use and is attributed to Otto Mohr (Matheson, 1971).

A general formulae for integration of the curve of M/EI gives the change in the slope between two points A and B as

$$i_B - i_A = \int_A^B \frac{M}{EI}\,dx \qquad\qquad (6.6)$$

where i represents the slope dv/dx. If EI is constant,

$$EI(i_B - i_A) = A_m$$

where A_m is the area of the BMD between A and B.

However, the most useful formula for determining deflections comes from integration by parts and is

$$\delta_B^A = \int_A^B \frac{M|x - x_A|}{EI} \, dx$$

Here δ_B^A is the distance of A above the tangent at B and the right-hand side is the moment, measured about A, of the area between A and B of the diagram of M/EI. If EI is constant, this may be specified in the form

$$EI\delta_B^A = A_m \bar{x}^A \tag{6.7}$$

where $\bar{x}^A$ is the distance of the centroid of A_m from A (always treated as positive). The various quantities involved in the two moment–area formulae are illustrated in Fig. 6.11.

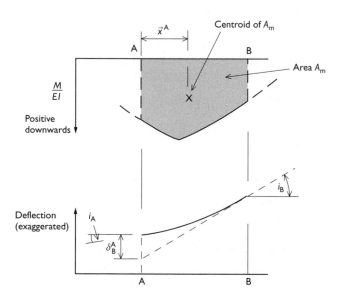

Figure 6.11 Moment–area quantities in Eqs (6.6) and (6.7).

Ex. 6.5
A uniform beam ABC of length ℓ and bending stiffness EI which is simply-supported at the ends carries a uniformly distributed load w over BC, its right-hand half. Determine an expression for the central deflection.

Answer
Figure 6.12 shows the end reactions and the BMD obtained from equilibrium. The BMD may be divided into a triangle A'DC' of height $w\ell^2/16$ and a parabola DEC' with a height at the 3/4 chord position E of $w\ell^2/32$. On the figure is also shown a sketch of the (exaggerated) deflection shape. The required deflection of B is $B''H = B''G - GH$. However, $B''G = \frac{1}{2} C''F = \frac{1}{2}\delta_A^C$ and $GH = \delta_A^B$. Therefore the answer can be deduced from δ_A^B and δ_A^C.

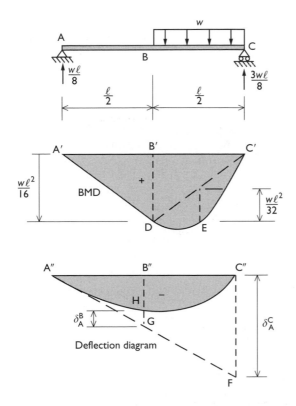

Figure 6.12 Use of the moment–area method for determining deflections.

To determine δ_A^B, A_m involves only the triangle A'DB'. Hence

$$EI\delta_A^B = \left(\frac{1}{2} \times \frac{w\ell^2}{16} \times \frac{\ell}{2}\right) \times \frac{\ell}{6} = \frac{w\ell^4}{384}$$

In order to determine δ_A^C, the area A_m is the whole of the BMD. Splitting this into the triangle A'DC' and the parabola DEC' gives

$$EI\delta_A^C = \left(\frac{1}{2} \times \frac{w\ell^2}{16} \times \ell\right) \times \frac{\ell}{2} + \left(\frac{2}{3} \times \frac{w\ell^2}{32} \times \frac{\ell}{2}\right) \times \frac{\ell}{4} = \frac{7w\ell^4}{384}$$

Thus the required deflection, v_B, is such that

$$-EIv_B = \frac{1}{2} \times \frac{7w\ell^4}{384} - \frac{w\ell^4}{384}$$

giving $-v_B = 5w\ell^4/768EI$.

Deflections due to shear

The previous analysis conforms to Euler's concept of elastic beam behaviour in which deflections are solely dependent on bending action. However, beam deflections are also affected by shearing action. To investigate the significance of shear deflections, consider the following example:

Ex. 6.6
A uniform beam of length 4 m is simply-supported at its ends and carries a central concentrated load of 410 kN. It has a symmetric cross-section with a total depth of 607 mm and a second moment of area of 863×10^6 mm^4. The web depth is 547 mm and its thickness is 11.2 mm. The elastic moduli are $E = 200$ kN/mm^2 and $G = 130$ kN/mm^2. Make a rough estimate of the deflection under the load due to shear and compare this with the deflection due to bending.

Answer
From Fig. 6.13, it follows that each of the two web panels are subject to uniform shear forces of 205 kN. If this force is taken uniformly over the web, as with an elementary beam, the shear stress is $\tau = 205,000/(547 \times 11.2) = 33.5$ N/mm^2 and the shear strain is $\gamma = 33.5/130,000 = 0.000257$.

For the deflection of the beam to be compatible throughout, the vertical cross-sections must remain vertical when only shear forces are present, as shown in Fig. 6.13. Hence the central deflection due to shear, δ_s, is equal to 2γ m, that is,

$$\delta_s = 0.000257 \times 2000 = 0.514 \text{ mm}$$

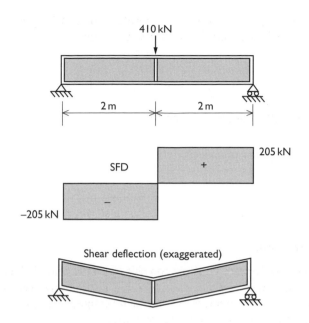

Figure 6.13 Shearing action of a simply-supported beam.

However, a bending analysis based on the formula given in Fig. 6.10 gives a deflection due to bending of

$$\delta_B = \frac{P\ell^3}{48EI} = 3.17 \, \text{mm}$$

(the maximum bending stress being $144.3 \, \text{N/mm}^2$). Hence in this case, the deflection due to shear is less than a sixth of the deflection due to bending.

Comment
With the same cross-section, increasing the span, ℓ, of a beam increases δ_s in proportion to ℓ and δ_B in proportion to ℓ^3. Furthermore, if the web thickness is increased, δ_s will be reduced whilst δ_B will not be affected very much. Hence, unless a beam has a low span/depth ratio, or its web is thin, deflections due to shear will be negligibly small in comparison with deflections due to bending. Shear deflections will normally be ignored in this book.

Q.6.5

A uniform beam of bending stiffness EI and length ℓ, simply-supported at its ends, carries a uniform loading w per unit length. Use mathematical integration to determine an expression for the deflection at distance x along the beam. Hence verify the fourth entry for maximum deflection given in Fig. 6.10.

Q.6.6

A uniform beam ABC is simply-supported at A and C and carries a concentrated load P at B, distance a and b from ends A and C respectively (where $a + b = \ell$).

a   Determine an expression for the deflection under the concentrated load.
b   Show that, when $a = b = \frac{1}{2}\ell$, the third result in Fig. 6.10 is obtained.

6.3   Composite beams

Why go composite?

A composite beam is one made up of two materials with different E values and properly bonded together. There are many different reasons why it may be advantageous to fabricate beams in this way. Some examples are

- *A lintel beam supporting brickwork*   Lower layers of brickwork may act in conjunction with a lintel to form a composite beam as shown in Fig. 6.14(a). (Since the brickwork needs to be present anyway, such composite action is worth investigating.)
- *A steel beam supporting a concrete slab*   Where a steel I beam supports a reinforced concrete slab, it is possible to utilise part of the slab as part or all of the upper flange of the beam (Fig. 6.14(b)). Since concrete is effective in resisting compression, but is weak in tension, the ideal situation is where the beam only needs to resist sagging bending moments.
- *A ship hull of steel and aluminium*   The use of aluminium alloy for the upper decks of ships permits higher superstructures to be used without impairing their hydraulic stability.

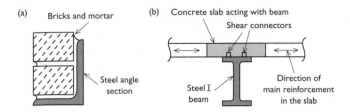

Figure 6.14 Examples of composite beams: (a) a lintel; (b) a beam/slab.

This section is primarily concerned with material combinations other than steel and concrete, the special considerations for which will be the subject of Section 6.4.

The need for strong and effective connections

It was seen in Section 4.6 that longitudinal joints in a beam need to transfer shear forces. Hence, in order to make different materials act together in the form of a composite beam, it is necessary to give special consideration to the longitudinal interfaces between them. For the lintel beam, composite action is only possible whilst the mortar has the requisite shear strength. In the beam/slab structure, shear connectors in the form of studs or inverted U shaped hooks are welded to the top flange of the steel. The slab is then cast with the shear connectors protruding into the mould so providing the shear resistance when the concrete has set.

It is also necessary for there to be no undesirable chemical or electro-chemical reactions between the materials. For instance, cathodic protection is required to prevent electro-chemical reactions between steel and aluminium in composite ship hulls. In other cases adverse reactions may be avoided by insulation.

The modular ratio method

Where all the material behaves elastically, it is possible to perform a structural analysis by the modular ratio method in which the section properties are obtained for an equivalent beam made of one material only.

Consider a beam made from steel and timber with moduli of elasticity E_s and E_t respectively. An element of cross-sectional area dA of steel subject to strain ε will develop an incremental force $E_s \varepsilon\, dA$ which is the same as if the area of dA is replaced by an area $m\, dA$ of timber, where $m = E_s/E_t$, the modular ratio. Hence the equivalent timber beam needs to have all the areas of steel replaced by m times as much timber, with the constraint that the substitute material must have the same strain as the original, that is, it must be at the same distance from the neutral axis.

Ex. 6.7
A simply-supported timber joist of span 3.6 m and cross-section
100 mm × 50 mm can safely carry a uniformly distributed load of
1.2 kN/m. However, it is required to be reinforced to carry extra
distributed loading.

a   Determine the total load that can be carried if the beam is reinforced
    with a steel plate of dimension 50 mm × 10 mm attached to the
    bottom surface as shown in Fig. 6.15 and the stress in the timber
    is to be no greater than in the original beam. It may be assumed
    that both materials behave elastically with $E_s = 200\,\text{kN/mm}^2$ and
    $E_t = 16\,\text{kN/mm}^2$.
b   For the new maximum load, determine the maximum stress in
    the steel plate and the maximum shear force to be resisted at the
    material interface.

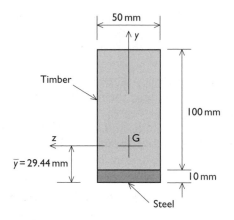

Figure 6.15 A composite beam cross-section.

Answer
a   The maximum bending moment in the original beam was (from
    the second case of Fig. 4.7): $M = w\ell^2/8 = 1.2 \times 3.6/8 =$
    1.944 kNm. For the original cross-section: $I = 50 \times 100^3/12 =$
    $4.167 \times 10^6\,\text{mm}^4$. With $y_{max} = 50\,\text{mm}$: $\sigma_{max} = 1.944 \times$
    $50/4.167 = 23.3\,\text{N/mm}^2$. When the steel plate is added, the posi-
    tion of the centroid changes and needs to be determined. Because
    $m = 12.5$, the cross-sectional area of the equivalent timber rep-
    resenting the steel is $12.5 \times 500 = 6250\,\text{mm}^2$. Hence, if $\bar{y}$ is
    the distance of the neutral axis above the bottom of the total

cross-section, equating moments of area gives

$$(5000 + 6250)\bar{y} = 5000 \times 60 + 6250 \times 5$$

$$\therefore \bar{y} = 29.44 \, \text{mm}$$

The revised second moment of area $\bar{I}$ is therefore

$$\bar{I} = \frac{50}{3}(80.56^3 + 19.44^3) + \frac{50 \times 12.5}{3}(29.44^3 - 19.44^3)$$

$$= 12.6 \times 10^6 \text{mm}^4$$

The maximum timber stress will occur in compression at the top fibres for which $y_{max} = 80.56$ mm. Hence the maximum permissible bending moment is given by

$$\bar{M}_{max} = \sigma_{max}\bar{I}/y_{max} = 23.3 \times 12.6 \times 10^6/80.56 \, \text{Nmm}$$

$$= 3.64 \, \text{kNm}$$

But $\bar{M}_{max} = \bar{w}_{max}\ell^2/8$ giving $\bar{w}_{max} = 2.24$ kN/m as the maximum allowable distributed loading on the composite beam (an improvement factor of 1.87).

b   The maximum stress in the equivalent timber replacing the steel occurs at the bottom fibres with $y_{max} = -29.44$ mm. Thus

$$\sigma = 3.64 \times 10^6 \times 29.44/(12.62 \times 10^6) = 8.49 \, \text{N/mm}^2$$

However, to determine stresses in the steel, it is important to remember to reintroduce the modular ratio. This gives for the steel $\sigma_{max} = 8.49 \times 12.5 = 106$ N/mm^2 (tensile). The maximum shear force on the beam occurs adjacent to the supports with a magnitude: $S = \frac{1}{2}\bar{w}\ell = \frac{1}{2} \times 2.24 \times 3.6 = 4.03$ kN. The shear flow across the steel timber interface can be obtained from Eq. (4.7) using $\tilde{A}$ specified as the whole of the original 100 m $\times$ 50 m cross-section whose centroid is situated at 30.56 mm above the neutral axis. Thus

$$\int_{\tilde{A}} y \, dA = 5000 \times 30.54 = 152.8 \times 10^3 \, \text{mm}^3$$

and

$$\tau_{max} = \frac{4.03 \times 10^3 \times 152.8 \times 10^3}{12.6 \times 10^6 \times 50} = 0.98 \, \text{N/mm}^2$$

which is the required answer.

Efficiency of composite action

Ex. 6.7 indicates a type of retrofit strengthening of a timber beam which could be considered if access is only possible to the bottom surface. However, if it is also possible to access the upper surface, a much better structural performance is obtained, using the same amount of steel, by reinforcing the top and bottom equally. By this means, the equivalent timber cross-section becomes a two-flanged symmetric beam giving a large I value (highlighted in discussion of cross-sectional shape in Section 4.4). Table 6.2 compares this sandwich construction (shown in Fig. 6.16.) with the beam analysed in Ex. 6.7. The use of sandwich construction increases the benefit of adding the steel from a factor of 1.87 to a factor of 5.13.

Table 6.2 Comparison of beam performances with the same maximum timber stress

Construction	Permissible loading (kN/m)	Maximum deflection (mm)	Maximum steel stress (N/mm²)	Maximum interface shear stress (N/mm²)
100 mm × 50 mm timber only	1.2	39.4	—	—
With steel below (Fig. 6.15)	2.24	24.2	106	1.0
With sandwich construction (Fig. 6.16)	6.16	39.4	306	1.7

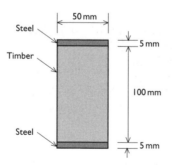

Figure 6.16 Cross-section of a beam with sandwich construction.

An historical note

An early use of iron and steel in beams for building construction occurred in the 1840s when vertical 'flitch plates' were bolted together with timber

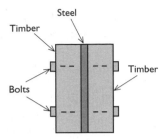

Figure 6.17 Cross-section of a flitch beam.

beams to make a composite flitch beam as shown in Fig. 6.17. Although not as efficient as the sandwich construction above, it is easy to construct, poses less difficulty over shear transfer across interfaces and the iron or steel plate is restrained from buckling by the timber. Because of the fire hazard in multi-storey buildings, such as mills, wrought iron and later steel beams were used in preference to timber for these types of building (Friedman, 1995). However, the benefit of this composite form is evident from their continued use in some forms of timber construction to this day.

Q.6.7
For a simply-supported composite beam of length 3.6 m made from steel ($E_s = 200 \, \text{kN/mm}^2$) and timber ($E_t = 16 \, \text{kN/mm}^2$) having the cross-section shown in Fig. 6.16, determine

a The maximum distributed load which may be supported if the timber stress is to be no greater than $23.3 \, \text{N/mm}^2$.
b The central deflection at maximum load.
c The corresponding maximum steel stress.
d The maximum shear stress across the steel/timber interface.

6.4 Steel and concrete: a productive partnership

Reinforced concrete beams

Reinforced concrete is a widely used form of composite construction in the building industry. Concrete itself is a material which is strong in compression and relatively cheap. Furthermore, because it is cast in moulds, it is easy to shape. However, on its own it would be of little use for beams because of its weakness in tension. In a reinforced concrete beam, internal longitudinally placed steel bars provide the necessary tension carrying capacity that the concrete lacks. The bars need to be separated by more than the maximum aggregate size, so that the concrete can pass round them easily in the mould. Also the complete beam needs to have sufficient cover for the reinforcing bars in order to inhibit corrosion due to ingress of water and chemicals. Concrete cover will also delay high temperature softening of the bars in the event of a fire.

Figure 6.18(a) shows a possible arrangement for a beam simply-supported at its ends, which is designed to carry only sagging bending moment and light shearing forces. The bars are held in place by a non-structural cage of wire. They have ribbed surfaces to enhance shear transfer between the steel and the concrete and their ends are provided with hooks to anchor them in to the concrete. Figure 6.18(b) shows how the design might change if the beam has higher bending moments and shearing forces as well as overhanging ends. Because hogging bending moments will be present over the supports as well as sagging bending moments at centre span, reinforcing bars will be required at the top as well as the bottom of the cross-section. However, the

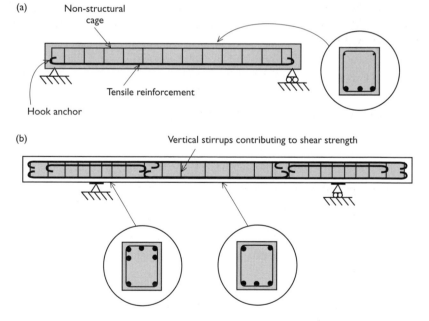

Figure 6.18 Possible design configurations for two simply-supported reinforced concrete beams: (a) for sagging bending moments only and light shear forces; (b) for both hogging and sagging bending moments and large shear forces.

reinforcement can be curtailed where it is not required. If the concrete alone cannot carry the shear force, vertical stirrups in tension work in conjunction with the concrete in diagonal compression to provide the shear strength. In the figure, more stirrups have been used near to the supports where the shearing forces will be largest. Where bars may act in compression, it is important that they cannot buckle outwards bursting the concrete cover. Hence they should not have a long length without any restraint to hold them in position.

Elastic bending of simple reinforced concrete beams

Generally reinforced concrete beams are designed to have adequate collapse strength by using ultimate load theory (considered in Section 9.2). However, the elastic properties are also needed for calculation of deflections, stiffnesses and stresses under working load conditions. Their elastic bending analysis differs from that of other composite beams in so far as it is assumed that the concrete may be cracked and hence incapable to carrying tension.

Figure 6.19(a) shows the cross-section of a rectangular concrete beam of breadth b reinforced to carry sagging bending moments only. It is convention when specifying the beam depth, d, to measure it from the outer fibres of the concrete in compression to the centreline of the outermost reinforcing bars in tension. Furthermore, the stress in each reinforcing bar is assumed to be uniform. With only one line of reinforcing bars, the resultant tensile force they carry will be

$$P_{st} = A_{st}\sigma_{st} \tag{6.8}$$

where A_{st} is the steel cross-sectional area and σ_{st} its stress.

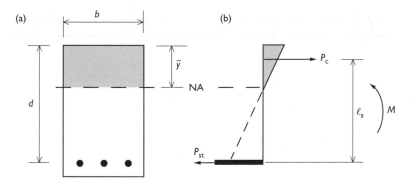

Figure 6.19 Sagging bending moment applied to a reinforced concrete beam: (a) cross-section; (b) assumed stress resultants.

The compressive stress in the concrete will be proportional to the distance above the neutral axis. If σ_c is the compressive stress at the top fibres and the neutral axis is at depth, $\bar{y}$, the resultant compressive force in the concrete will be

$$P_c = \tfrac{1}{2}\sigma_c b\bar{y} \tag{6.9}$$

acting at a depth $\bar{y}/3$ as shown in Fig. 6.19(b). For any given bending moment M, there are three unknowns to determine σ_c, σ_{st} and $\bar{y}$. Two equations are required to satisfy equilibrium:

$$P_c = P_{st} \tag{6.10}$$

and

$$M = P_c \ell_a \tag{6.11}$$

where $\ell_a = d - \bar{y}/3$, called the lever arm, is the distance between the forces of the couple. The other equation comes from the requirement that the strain is a linear function of distance from the neutral axis. If the modular ratio between steel and concrete is $m = E_{st}/E_c$, the strain condition is satisfied when

$$\frac{\sigma_{st}}{m(d - \bar{y})} = \frac{\sigma_c}{\bar{y}} \tag{6.12}$$

Using Eqs (6.8)–(6.10) and (6.12), the following quadratic equation is obtained for $\bar{y}$:

$$\tfrac{1}{2}b\bar{y}^2 - mA_{st}(d - \bar{y}) = 0 \tag{6.13}$$

Once $\bar{y}$ has been determined from this equation, the stresses, strains and curvature of the beam due to M may be estimated.

Ex. 6.8
A rectangular reinforced concrete beam with a span of 6 m carries a uniformly distributed load of 7 kN/m. If $b = 250$ mm, $d = 360$ mm, $E_s = 205.5$ kN/mm^2, $E_c = 13.7$ kN/mm^2 and the reinforcing steel consists of three bars of diameter 16 mm, all placed at the same depth, estimate the maximum stresses in the steel and concrete assuming that the concrete is ineffective in tension.

Answer

From the above information, $A_{st} = 603.2 \, mm^2$ and $m = 15$. Hence Eq. (6.13) gives in mm units

$$\bar{y}^2 + 72.4\bar{y} - 26{,}060 = 0$$

the one positive solution of which is $\bar{y} = 129 \, mm$ giving $\ell_a = 317 \, mm$.

The applied load produces a maximum bending moment at midspan of $7 \times 6^2/8 = 31.5 \, kNm$. Hence from Eqs (6.10) and (6.11): $P_c = P_{st} = 99.4 \, kN$ and from Eqs (6.8) and (6.9), $\sigma_{st} = 164.8 \, N/mm^2$ and $\sigma_c = 6.2 \, N/mm^2$.

Flower pot technology?

Unreinforced concrete (otherwise known as mass concrete) has been in use since Roman times. The Pantheon in Rome built about 120 AD is a celebrated example in which circular mass concrete walls of 43 m internal diameter and 6 m thick support a concrete dome of approximate thickness 1.2 m. The dome is a complete hemisphere except for a central opening of 9 m diameter. The origin of reinforced concrete can be traced back to Joseph Monier, a Parisian gardener, who in 1867 patented the use of wire mesh to reinforce flower pots and baskets made of concrete (Hopkins, 1970). Later he made other items such as pipes and railway sleepers. He appeared to recognise the need to use the reinforcement to take tension when bending forces may be present because, when the concrete thickness was large enough, he put the reinforcement near both faces. Although floor beams he made tended not to be successful, independent experiments by an American lawyer, Thaddeus Hyatt, established that wrought iron bars placed longitudinally near to the tension face of a beam were most effective at increasing its strength. Hyatt ignored possible strength of the concrete on the tensile side of the neutral axis and, as a result, advocated the use of T beams to minimise the volume of concrete required (Fig. 6.20). François Hennebique (1843–1921) and Robert Maillart (1872–1940) are two engineers celebrated for their use of reinforced concrete in many types of structure.

The successful 'marriage' of steel and concrete to form reinforced concrete works well while the ranges of usable strains for the two materials are fairly similar. It is also beneficial that the coefficients of thermal expansion are close. There is an advantage in not making the cross-sectional area of the reinforcing steel of a beam too large, in which case the resulting under-reinforced cross-section fails by the reinforcement stretching (the technical term is 'yielding') so giving warning of its overloaded state. In contrast, if the beam is over-reinforced, the neutral axis will be nearer to the tension

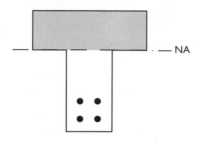

Figure 6.20 Cross-section of a T beam.

steel ($\bar{y} > \frac{1}{2}d$) in which case, the concrete is likely to suddenly crush giving no warning of failure.

Eugene Freyssinet

When high tensile steels became available, it was not possible to use them to their full potential in reinforced concrete beams because they needed to develop large strains to be effective. From 1903 onwards, Eugene Freyssinet developed a solution to this problem which earned him the nickname the 'father of prestressing'.

Viewed from the standpoint of the steel, Freyssinet's solution was to provide it with an initial tensile strain so that its high stress capability could be utilised without the associated large strains needing to be developed by the applied load. From the standpoint of the concrete, prestressing places the concrete initially in compression (by reacting it against the tension in the steel). As a result longitudinal tensile stresses due to bending subtract from the initial stresses in such a way that concrete beams may be designed never to go into tension.

Thus the concrete is fully effective in bending and cracking is avoided. Furthermore, because the concrete is being fully utilised, it is of more benefit to employ quality control measures in its manufacture in order to guarantee high strength. Prestressed concrete beams can thus be more slender than their reinforced concrete counterparts and can be used for larger spans. Hopkins (1970) says 'Consequently sense was restored to concrete construction – the better the materials used for prestressed concrete, the better the resulting structure was likely to be in all aspects'.

The development of prestressed concrete was, however, complicated by shrinkage of the concrete under permanent compressive load and also 'creep' in both the concrete and the steel. The initial prestressing forces needed to be increased to allow for losses because of these factors. Design methods need to ensure safety under all feasible loading cases both initially and also when shrinkage and creep have taken place. In Chapter 2, it was discovered that masonry walls and arches are more stable once they carry large axial compressive forces. This principal also applies to concrete, being itself a form of masonry. Freyssinet's solution, therefore, was to develop a compressive force in the concrete artificially. It has been applied not only to beams, but also to other concrete structures such as bridges, dams and tanks for liquid storage. Although to date prestressing has been mainly applied to concrete, it can also be applied to other forms of masonry.

Pre-tensioning

Prestressed concrete can be either pre-tensioned or post-tensioned. Pre-tensioning involves casting the concrete around prestressing wires (or 'tendons') which have already been tensioned. External clamps tensioning the wire are released when the concrete has set, so transferring the reactive compressive force to the concrete. This is a process which lends itself to manufacture in a factory where beams can be left undisturbed for a while during casting and then, after release, the jacking apparatus can be reutilised within a production cycle (Fig. 6.21).

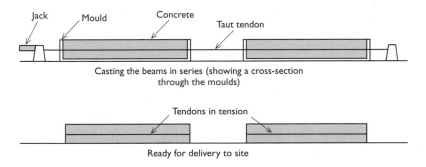

Casting the beams in series (showing a cross-section through the moulds)

Ready for delivery to site

Figure 6.21 The fabrication of two pre-tensioned beams.

Ex. 6.9
A rectangular concrete beam of width b and depth d is to be pre-stressed by pretensioning using a tendon situated along the longitudinal centre-line. If the concrete stress σ is to be limited to $0 \leq \sigma \leq f_c$, determine expressions for the maximum bending moment (either sagging or hogging) which the beam may be designed to carry and the prestressing force required to achieve this. (In this rudimentary investigation neglect the effects of shrinkage and creep.)

Answer

The prestressing force, T, gives rise to a uniform compressive stress of T/bd. A sagging or hogging bending moment M gives rise to outer fibre stresses of $\pm 6M/bd^2$. Hence to keep σ within the required limits

$$\frac{T}{bd} + \frac{6M}{bd^2} \leq f_c \quad \text{and} \quad \frac{T}{bd} - \frac{6M}{bd^2} \geq 0$$

Subtracting these inequalities (remembering to reverse the second inequality when its signs are changed) gives

$$M \leq \frac{bd^2 f_c}{12}$$

To achieve the maximum moment carrying capacity, the inequalities need to be equalities in which case

$$T = \frac{1}{2} bd f_c$$

Hence the concrete compressive stress due to prestressing needs to be half of the limiting stress f_c in order for the beam to develop its maximum bending moment capacity (Fig. 6.22).

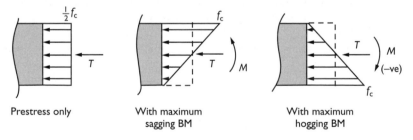

Figure 6.22 Stresses in a centrally prestressed symmetric beam.

Eccentric prestressing

Where the maximum sagging bending moment is higher than the maximum hogging bending moment, it is advantageous to lower the position of the prestressing tendon. For a rectangular beam, if no hogging bending moments are anticipated, placing the tendon with the same prestressing force at $d/3$ from the bottom of the cross-section doubles the maximum possible sagging bending moment (Fig. 6.23). Thus the most important design parameter is the range of bending moments requiring to be carried ($M_{max} - M_{min}$), because this is dependent on the allowable concrete stress, f_c, and the dimensions of the cross-section, whereas the actual values of M_{max} and M_{min} can be adjusted by altering the alignment of the prestressing tendon.

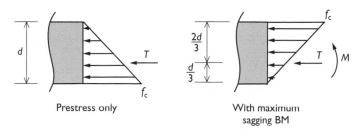

Figure 6.23 Stresses in an eccentrically prestressed rectangular beam.

Post-tensioning

The other procedure for prestressing, called post-tensioning, involves casting tubes (called 'ducts') into the concrete. Once the concrete has set, prestressing tendons are passed through the ducts and stretched by jacking against the ends of the beam. The ducts are then sealed by grouting to protect the prestressing tendons against corrosion. Because strong jacking points are not required, post-tensioning is normally carried out *in situ*, thus avoiding the need to handle the pre-tensioned beams during transport to site. It also affords more flexibility in design by allowing ducts to be curved (Fig. 6.24).

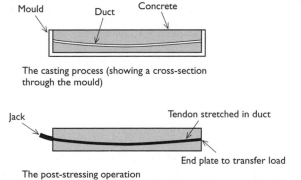

Figure 6.24 The fabrication of a post-tensioned beam.

On the other hand, the concrete does not necessarily need to be cast *in situ* as this may involve expensive shuttering and support structure during the casting process. This is avoided in many modern bridges by using glued segmental construction which is well suited to multispan viaducts. Here bridge segments including ducts are cast off-site in a factory environment. The bridge itself is formed by cantilevering out from the piers by adding segments one at a time. To hold segments in place, the interfaces between them are serrated and glued as well as being clamped by prestressing forces. During construction temporary prestressing is employed until each span is complete, at which time the permanent tendons can be put in place and prestressed (Fig. 6.25). Care needs to be taken to ensure stability and safety throughout the construction process as well as in the completed bridge.

Figure 6.25 A glued segmental prestressed concrete bridge under construction at Belfast in 1994 using a balanced cantilever technique with six segments fitted as part of the cross-harbour bridges.

Q.6.8
Using the computed position, $\bar{y} = 129$ mm, for the neutral axis of the beam of Ex. 6.8, determine I for the equivalent concrete beam and use the modular ratio method to check if the maximum stresses in the

concrete and the steel are correct. Also, estimate the central deflection
of the beam for the specified loading case.

6.5   Statically indeterminate elastic beams

Solution by integration

If a beam has distributed loading and no intermediate supports, the differential relationships of Sections 4.1 and 6.2 relate the curve of loading w to
the curve of deflection v by:

$$w = -\frac{\mathrm{d}^2}{\mathrm{d}x^2}\left(EI\frac{\mathrm{d}^2v}{\mathrm{d}x^2}\right)$$

When the beam is uniform, the constant bending stiffness EI can be taken
outside the differentiation to give

$$w = -EI\frac{\mathrm{d}^4v}{\mathrm{d}x^4}$$

There are always two conditions to be satisfied at each end, thus enabling
the four constants of integration to be determined and the beam behaviour
to be analysed by integration. For instance, the cantilever beam analysed in
Ex. 6.4 has the end conditions:

$$y = 0 \quad \text{and} \quad \frac{\mathrm{d}y}{\mathrm{d}x} = 0 \text{ at } x = 0$$

$$M = 0 \quad \text{and} \quad S = 0 \text{ at } x = \ell$$

Statically indeterminate beam problems may also be solved by integration,
the only difference being that it is no longer possible to split the integration
into two distinct phases, one for equilibrium and the other to determine
deflections.

Ex. 6.10
A uniform beam AB with bending stiffness EI and length ℓ is built-in
at both ends. Analyse the elastic behaviour of the beam when subject
to a uniformly distributed loading of w/unit span, determining the
shapes of the BMD and the deflection curve. Compare critical values
of bending moment and deflection with those of the corresponding
simply-supported beam.

Answer

Integrating the (constant) loading curve four times and including four constants of integration c_1 to c_4, gives

$$S = wx + c_1$$

$$-EI\frac{d^2v}{dx^2} = -M = \frac{1}{2}wx^2 + c_1x + c_2$$

$$-EI\frac{dv}{dx} = \frac{wx^3}{6} + \frac{1}{2}c_1x^2 + c_2x + c_3$$

$$-EIv = \frac{wx^4}{24} + \frac{c_1x^3}{6} + \frac{1}{2}c_2x^2 + c_3x + c_4$$

(6.14)

However, the end conditions are

$$v = 0 \quad \text{and} \quad \frac{dv}{dx} = 0 \text{ at } x = 0 \quad \text{and} \quad x = \ell$$

Substituting for these conditions gives

$$c_1 = -\frac{1}{2}w\ell, \quad c_2 = \frac{w\ell^2}{12}, \quad c_3 = 0, \quad c_4 = 0$$

Hence $M = \frac{1}{2}wx(\ell - x) - w\ell^2/12$ and

$$v = \frac{-wx^2(\ell - x)^2}{24EI}$$

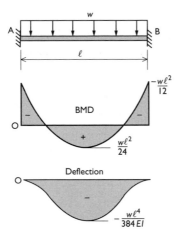

Figure 6.26 Elastic bending of a built-in beam.

The BMD and deflection shape are shown in Fig. 6.26. Critical bending moments are $w\ell^2/24$ at the centre and $-w\ell^2/12$ at the ends. Thus the maximum bending moment is 2/3 of the maximum value for the corresponding simply-supported beam shown in Fig. 4.7. The maximum deflection is $w\ell^4/384EI$ at the centre, one-fifth of the corresponding value for the simply-supported beam shown in Fig. 6.10.

The moment–area method with redundancies

An alternative more versatile technique for solving statically indeterminate beam problems is to define sufficient redundant moments as variables to enable the BMD to be specified in terms of them. Then moment–area equations can be derived and solved to ensure that outstanding constraints are satisfied.

Ex. 6.11
Use the moment–area method to analyse the built-in beam of Ex. 6.10.

Answer
The beam may be made statically determinate by specifying hogging moments M_A and M_B at the ends. However, from symmetry, it may be deduced that $M_A = M_B = \overline{M}$ (say). Hence the bending moment at distance x from A is

$$M = -\frac{1}{2}wx(\ell - x) - \overline{M}$$

with the BMD as shown in Fig. 6.27(a). The outstanding compatibility equation which is not automatically satisfied by the statically

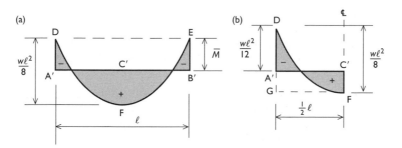

Figure 6.27 BMD for the built-in beam: (a) with unknown end moments; (b) with $\overline{M} = w\ell^2/12$.

determinate system is that the end slopes (equal because of symmetry) are zero. Thus with $i_A = i_B = 0$ and EI constant, Eq. (6.6) yields

$$\int_A^B M \mathrm{d}x = 0$$

Splitting the BMD, Fig. 6.27(a) into two parts, the parabola DFE with base DE has an area $\frac{2}{3} \times$ base $\times$ height $= w\ell^3/12$ and the rectangle A′B′DE has an area $-\overline{M}\ell$. Therefore, for the total area to be zero $\overline{M} = w\ell^2/12$.

Because the end slopes are zero, it is possible to determine the maximum deflection (which will be at the centre C) from the deflection of C above the tangent at A. From Eq. (6.7)

$$EI\delta_A^C = A_m \bar{x}^C$$

The BMD for AC is shown in Fig. 6.27(b). Splitting this into the rectangle A′C′FG with positive area $w\ell^3/48$ and the negative parabola DFG which has an area $\frac{1}{3} \times$ base $\times$ height $= w\ell^3/48$ and a centroid at distance $\ell/8$ from A

$$EI\delta_A^C = \frac{w\ell^3}{48} \times \frac{\ell}{4} - \frac{w\ell^3}{48} \times \frac{3\ell}{8}$$

Hence the central deflection is $-\delta_A^C = w\ell^4/384EI$.
(NB   The same result could have been obtained from δ_C^A.)

Clapeyron and continuous beams

After graduating at the same time as Lamé from the Ecole Polytechnique in 1818 and working in St Petersburg for 10 years, Clapeyron returned to France where he lectured in the Ecole des Ponts et Chaussées (School of Bridges and Roads). Here he met the problem of designing multispan bridges and developed his celebrated 'three-moment equation' for continuous beams.

Consider the case of a beam ABCDEF of uniform bending stiffness EI spanning across six simple supports at spacing ℓ as shown in Fig. 6.28 and compare this with the set of statically determinate simply-supported beams spanning the same supports shown in Fig. 6.29(a). In the latter case, there is a discontinuity in slope across the interior supports and an absence of hogging bending moments there. These discontinuities can be eliminated by introducing hogging moments $\overline{M}_B, \overline{M}_C, \overline{M}_D$ and $\overline{M}_E$ as redundancies at these supports and using moment–area equations to determine their magnitude

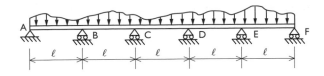

Figure 6.28 A continuous beam over five equal spans.

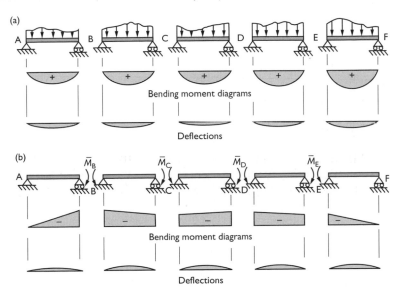

Figure 6.29 Moment–area analysis for the continuous beam: (a) the basic statically determinate system; (b) modifying effects of the redundant forces.

(Fig. 6.29(b)). If i_{CB} and i_{CD} define the slopes at C for the beams BC and CD respectively, continuity at C is satisfied when $i_{CB} = i_{CD}$. However, from the moment–area method using the BMDs of Fig. 6.29, where A_{BC}, $\vec{x}^{B}$, A_{CD} and $\vec{x}^{D}$ relate to the simply-supported beam BMDs, hence

$$EI\ell i_{CB} = EI\delta_{C}^{B} = A_{BC}\vec{x}^{B} - \frac{1}{3}\overline{M}_{C}\ell^{2} - \frac{1}{6}\overline{M}_{B}\ell^{2}$$

and

$$-EI\ell i_{CD} = EI\delta_{C}^{D} = A_{CD}\vec{x}^{D} - \frac{1}{3}\overline{M}_{C}\ell^{2} - \frac{1}{6}\overline{M}_{D}\ell^{2}$$

Hence

$$\frac{\ell^{2}}{6}(\overline{M}_{B} + 4\overline{M}_{C} + \overline{M}_{D}) = A_{BC}\vec{x}^{B} + A_{CD}\vec{x}^{D} \tag{6.15}$$

which is a form of Clapeyron's three-moment equation. Because, for any continuous beam, as many three-moment equations can be derived as there are redundancies, a solution is always possible.

Ex. 6.12
Determine the elastic BMD for a continuous beam having the dimensions shown in Fig. 6.28 if it carries a uniformly distributed load of w/unit span.

Answer
The BMD for the beams when simply-supported each have area $A = \frac{2}{3}\ell \times w\ell^2/8 = w\ell^3/12$ with a moment arm $\bar{x} = \frac{1}{2}\ell$. Noting that the bending moments at the ends of the beam, M_A and M_F, are zero, the set of four three-moment equations yields

$$\begin{bmatrix} 4 & 1 & & \\ 1 & 4 & 1 & \\ & 1 & 4 & 1 \\ & & 1 & 4 \end{bmatrix} \begin{bmatrix} \overline{M}_B \\ \overline{M}_C \\ \overline{M}_D \\ \overline{M}_E \end{bmatrix} = \begin{bmatrix} \frac{1}{2}w\ell^2 \\ \frac{1}{2}w\ell^2 \\ \frac{1}{2}w\ell^2 \\ \frac{1}{2}w\ell^2 \end{bmatrix} \qquad (6.16)$$

which has the solution

$$\{\overline{M}_B\ \overline{M}_C\ \overline{M}_D\ \overline{M}_E\} = \frac{w\ell^2}{38}\{4\ 3\ 3\ 4\}$$

giving the solution for bending moment shown in Fig. 6.30. It should be noted that the largest hogging and sagging bending moments occur near to the ends of the beam because of the lack of rotational constraint there.

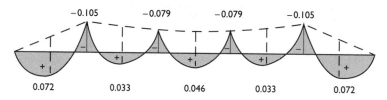

-0.105 -0.079 -0.079 -0.105

0.072   0.033   0.046   0.033   0.072

Figure 6.30 BMD for the continuous beam of Fig. 6.28 with uniformly distributed loading. When multiplied by $w\ell^2$, figures show the bending moments at $\frac{1}{2}\ell$ intervals.

Q.6.9
A uniform beam AB of length ℓ with bending stiffness EI is built-in at A and simply-supported at end B (this is known as a 'propped cantilever'). Specify the four end conditions and determine, by integration, the elastic theory deflection profile and BMD when it is subject to a uniformly distributed load w.

Q.6.10
For the propped cantilever of Q.6.9 determine, assuming elastic theory, expressions for the maximum hogging and sagging bending moments and the deflection under the load, if it is subject to a concentrated load P placed at mid-span instead of the distributed load.

Q.6.11
Apart from multispan girder bridges considered by Clapeyron, suggest some other types of continuous beams.

Q.6.12
Determine the elastic BMD for the continuous beam of Fig. 6.28 if span AB carries a uniformly distributed load of w/unit length with all other spans unloaded. Do not compute any deflections, but sketch what you expect the deflected form to be.

Q.6.13
If the continuous beam shown in Fig. 6.28 can carry a distributed load with a maximum value of w/unit length anywhere on the span, identify where you expect the loading to be applied to obtain:

a   The largest bending moment over support B.
b   The largest bending moment at the centre of span BC.
c   The largest support reaction at C.

NB   This question should be attempted after Q.6.12.

Q.6.14
Derive a more general form of Clapeyron's equation to Eq. (6.15) by allowing BC and CD to have the different spans ℓ_{BC} and ℓ_{BD}.

6.6 Truss deflections

The Williot diagram

Although graphical methods have been superceded by other methods, even a sketch of a Williot or Williot–Mohr diagram can provide an insight into how a truss deflects, or can form a basis for checking results computed by other means. In a Williot diagram, deflections are assumed to be small enough for second order effects to be neglected, and are plotted all on the same diagram. The following example shows how joint displacements can often be much larger than the extensions of the individual members.

Ex. 6.13

The statically determinate crane truss shown in Fig. 6.31 carries a vertical load of 100 kN. An equilibrium analysis has given the member forces shown in the figure. If all the members are made of steel having $E = 200\,\text{kN/mm}^2$ and have a cross-sectional area of 3000 mm, determine the vertical deflection of E.

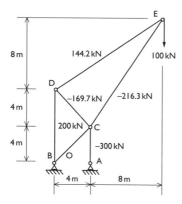

Figure 6.31 A crane truss showing member forces.

Answer

All the members have $EA = 200 \times 3000 = 600{,}000\,\text{kN}$ and their lengths in metres are

$$\ell_{\text{DE}} = \ell_{\text{CE}} = 14.42, \quad \ell_{\text{CD}} = \ell_{\text{CB}} = 5.66, \quad \ell_{\text{BD}} = 8, \quad \ell_{\text{AC}} = 4$$

Hence using Eq. (6.1), the member extensions in mm are

$$e_{\text{DE}} = 3.47, \quad e_{\text{CE}} = -5.20, \quad e_{\text{CD}} = -1.60, \quad e_{\text{BD}} = 2.67,$$

$$e_{\text{BC}} = 0 \quad \text{and} \quad e_{\text{AC}} = -2.00 \tag{6.17}$$

Because joints A and B do not move, they appear on the Williot diagram at the origin O. The movement of C can be determined next because it is connected to both of these joints. On account of member AC, it will move downwards 2 mm and hence lie somewhere on the line *rs* in Fig. 6.32(a). Because of member BC, which does not change in length, C must also lie on the line *pq* at 45° and passing through the origin. The correct position for C is therefore the intersection of *pq* and *rs*. Deriving the position of D from the known positions of B and C by using the extensions of members BD and CD, and then continuing to determine the position of E gives the complete Williot diagram shown in Fig. 6.32(b) (the downwards deflection of E is almost six times the largest extension or contraction of any member despite there only being six members present in the truss).

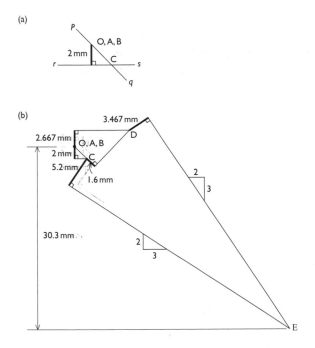

Figure 6.32 The Williot diagram for the crane truss: (a) determining the deflection of C; (b) the complete diagram.

The Williot–Mohr diagram

Where it is not possible to start the Williot diagram because no joint links directly to two fixed joints, Mohr's extension needs to be used. Here an unknown displacement is given an assumed value to start off the Williot diagram. Because of this, there will be a mismatch with a known displacement when the Williot diagram is complete. Mohr corrected this mismatch by subtracting an appropriate rotation of the truss.

Ex. 6.14
Figure 6.33 shows a truss together with computed member extensions. Determine the vertical deflection of C.

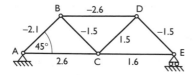

Figure 6.33 Member extensions in mm for a Warren truss.

Answer
Figure 6.34 shows a Williot diagram which has been started by placing C at 2.6 mm to the right of A (thus ignoring any possible unknown vertical movement). From that, the positions of B, D and E have been derived in turn. However, this Williot diagram is incorrect because E

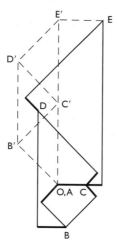

Figure 6.34 The Williot–Mohr diagram for the Warren truss.

is only free to move horizontally. Its vertical component of displacement needs to be eliminated by rotation of the whole truss about A. This rotation involves E in a vertical movement which is catered for on the Williot–Mohr diagram by measuring the deflection of E relative to E′ instead of O. The effect of the rotation on the other joints is to subtract the displacements shown by the ghost frame drawn on OE′ as base. Thus the deflection of C needs to be measured relative to C′ as origin, its vertical component being approximately 7.5 mm downwards.

Displacement compatibility equations

Consider a plane truss in which joint displacements are designated u and v in the x and y directions respectively. If member PQ with direction P $\rightarrow$ Q has an inclination α to the x axis, using small deflection theory, its extension e_{PQ} can be obtained by projecting the joint displacements onto the member axis (see Fig. 6.35). This gives the member constraint:

$$e_{PQ} = (u_Q - u_P)\cos\alpha_{PQ} + (v_Q - v_P)\sin\alpha_{PQ} \qquad (6.18)$$

In Section 5.2, it was established that, for a statically determinate truss restrained against body movements, the number of possible joint movements $(2j - r)$ must equal the number of member constraints (m). Hence, if the extensions to all members are known, the full set of Eq. (6.18) provide sufficient constraints to be able to solve for joint displacements.

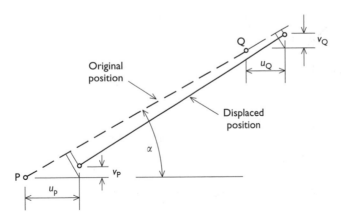

Figure 6.35 Displacement of a truss member PQ.

Ex. 6.15
Use displacement compatibility equations for each of the members of the crane truss Fig. (6.31) to determine the vertical deflection of E.

Answer
The member inclinations are as follows

$$A \to C : 90°, \quad B \to C : 45°, \quad B \to D : 90°, \quad C \to D : 135°,$$

$$C \to E : 56.3°, \quad D \to E : 33.7°$$

Hence, with the displacements of A and B equal to zero and the member extensions given by Eq. (6.17), the member compatibility equations are

$$\begin{bmatrix} 0 & 1 \\ 0.7071 & 0.7071 \\ 0 & 0 & 0 & 1 \\ 0.7071 & -0.7071 & -0.7071 & 0.7071 \\ -0.5547 & -0.8321 & 0 & 0 & 0.5547 & 0.8321 \\ 0 & 0 & -0.8321 & -0.5547 & 0.8321 & 0.5547 \end{bmatrix}$$

$$\times \begin{bmatrix} u_C \\ v_C \\ u_D \\ v_D \\ u_E \\ v_E \end{bmatrix} = \begin{bmatrix} e_{AC} \\ e_{BC} \\ e_{BD} \\ e_{CD} \\ e_{CE} \\ e_{DE} \end{bmatrix} = \begin{bmatrix} -2.00 \\ 0 \\ 2.67 \\ -1.60 \\ -5.20 \\ 3.47 \end{bmatrix} \qquad (6.19)$$

with the solution

$$\{u_C \; v_C \; u_D \; v_D \; u_E \; v_E\} = \{2.0 \; -2.0 \; 8.93 \; 2.67 \; 35.1 \; -30.3\}$$

The vertical deflection of E is therefore 30.3 mm downwards.

Q.6.15
a   If the member extensions for the truss of Fig. 5.18 are in mm,

$$\{e_{AB} \; e_{AD} \; e_{BC} \; e_{CE} \; e_{BD} \; e_{BE}\} = \{-2.4 \; 0 \; 0 \; 0 \; 3.0 \; -3.0\}$$

sketch the Williot diagram and hence determine the displacements of A and C.

b Derive compatibility equations for this truss and use them to check the answer to part (a).

6.7   Stiffness equations for a truss

Significance

This section provides some insight into the most commonly used computer techniques for the analysis of structures by considering their application just to plane trusses.

Equilibrium and compatibility equations

In the previous section, compatibility Eq. (6.19) was derived for a crane truss in order to determine joint displacements from member extensions. However, in order to analyse this truss in its entirety, equilibrium equations are also required. Applying the method described in Section 5.3 to the crane truss gives the following equilibrium equation

$$
\begin{bmatrix}
0 & 0.7071 & 0 & 0.7071 & -0.5547 & 0 \\
1 & 0.7071 & 0 & -0.7071 & -0.8321 & 0 \\
 & & 0 & -0.7071 & 0 & -0.8321 \\
 & & 1 & 0.7071 & 0 & -0.5547 \\
 & & & & 0.5547 & 0.8321 \\
 & & & & 0.8321 & 0.5547
\end{bmatrix}
\begin{bmatrix}
P_{AC} \\
P_{BC} \\
P_{BD} \\
P_{CD} \\
P_{CE} \\
P_{DE}
\end{bmatrix}
$$

$$
= \begin{bmatrix}
H_{C} \\
V_{C} \\
H_{D} \\
V_{D} \\
H_{E} \\
V_{E}
\end{bmatrix}
= \begin{bmatrix}
0 \\
0 \\
0 \\
0 \\
0 \\
-100
\end{bmatrix}
\tag{6.20}
$$

These equations have a remarkable relationship to the compatibility Eq. (6.19). Without reading further, can you identify it? It is that the coefficient matrices are the transpose of each other. A transpose, M^{T}, of a matrix M is one in which the roles of rows and columns have been reversed. Thus if

$$
M = \begin{bmatrix} 3 & 0 \\ 1 & -1 \\ 4 & 2 \end{bmatrix}, \quad M^{T} = \begin{bmatrix} 3 & 1 & 4 \\ 0 & -1 & 2 \end{bmatrix}
$$

In order to obtain this result for the crane truss, it is necessary for the order of the members to be the same in the lists of member forces and member extensions. It is also necessary for the list of joint displacements in the compatibility equations to correspond with the list of joint forces in the equilibrium equations. There is nothing special about the crane truss, however. This transpose relationship is a general property of structures to be discussed further in Chapter 8.

Synthesis of the stiffness equations

The only other information required to formulate a complete elastic analysis of the crane truss is the relationship between the member forces and member extensions which take the form

$$p_{AC} = (EA_{AC}/\ell_{AC})e_{AC} = 150e_{AC} \text{ kN/mm}$$

$$p_{BC} = (EA_{BC}/\ell_{BC})e_{BC} = 1061.1e_{BC} \text{ kN/mm}$$

etc.

which can be written in matrix form as:

$$
\begin{bmatrix} p_{AC} \\ p_{BC} \\ p_{BD} \\ p_{CD} \\ p_{CE} \\ p_{DE} \end{bmatrix} = \begin{bmatrix} 150 & & & & & \\ & 106.1 & & & & \\ & & 75 & & & \\ & & & 106.1 & & \\ & & & & 41.6 & \\ & & & & & 41.6 \end{bmatrix} \begin{bmatrix} e_{AC} \\ e_{BC} \\ e_{BD} \\ e_{CD} \\ e_{CE} \\ e_{DE} \end{bmatrix} \tag{6.21}
$$

Using this equation together with the equilibrium and compatibility equations, it is possible to formulate equations for the joint forces directly in terms of the joint displacements.

To generalise the process so that it is applicable to the elastic analysis of any plane truss, the following designations will be used:

$\mathbf{u}$ = vector of joint displacements
$\mathbf{e}$ = vector of member extensions
$\mathbf{p}$ = vector of member forces
$\mathbf{f}$ = vector of externally applied joint forces
$\mathbf{A}$ = compatibility matrix
$\mathbf{\overline{K}}$ = diagonal matrix of member stiffnesses

The compatibility equations take the form

$$\mathbf{e} = \mathbf{Au}$$

The member stiffness equations are

$$\mathbf{p} = \overline{\mathbf{K}}\mathbf{e}$$

and the equilibrium equations are

$$\mathbf{f} = \mathbf{A}^{\mathrm{T}}\mathbf{p}$$

Substituting for $\mathbf{p}$ and $\mathbf{e}$ gives the external forces directly in terms of the joint displacement as

$$\mathbf{f} = \mathbf{K}\mathbf{u}$$

where $\mathbf{K} = \mathbf{A}^{\mathrm{T}}\overline{\mathbf{K}}\mathbf{A}$ is the stiffness matrix for the truss.

The stiffness matrix

For the crane truss, the stiffness matrix is

$$\mathbf{K} = \begin{bmatrix} 118.87 & 19.20 & -53.03 & 53.03 & -12.80 & -19.20 \\ 19.20 & 284.87 & 53.03 & -53.03 & -19.20 & -28.81 \\ -53.03 & 53.03 & 81.84 & -33.83 & -28.81 & -19.20 \\ 53.03 & -53.03 & -33.83 & 140.83 & -19.20 & -12.80 \\ -12.80 & -19.20 & -28.81 & -19.20 & 41.61 & 38.41 \\ -19.20 & -28.81 & -19.20 & -12.80 & 38.41 & 41.61 \end{bmatrix} \qquad (6.22)$$

The leading diagonal of a matrix is the set of elements having row number i equal to column number j and a symmetric matrix $\mathbf{M}$ is one which is symmetric about the leading diagonal, that is $m_{ij} = m_{ji}$ (hence $\mathbf{M}^{\mathrm{T}} = \mathbf{M}$). It is a property of stiffness matrices that they are symmetric (provided that the vectors $\mathbf{u}$ and $\mathbf{f}$ are similarly ordered and the structure is loaded within its linear elastic range).

To identify what individual elements of the stiffness matrix represent, consider displacing one joint a unit amount in one direction (e.g. $u_j = 1$) and keeping the rest fixed at zero. Multiplying the stiffness matrix by the displacement vector representing this movement (which is null except for $u_j = 1$) indicates that the joint forces required to maintain this displacement are given by column j of the stiffness matrix.

For the crane truss example, a unit displacement v_D (with the other joint displacements restrained) requires the joint forces to be the elements of the fourth column of the stiffness matrix (6.22) as shown in Fig. 6.36. It is easy to carry out spot checks on individual elements. For instance, the forces acting at C in Fig. 6.36 are due to the extension $1/\sqrt{2}$ mm of member CD. The force in this member of $106.1/\sqrt{2} = 75$ kN has components in the x and y directions of $75/\sqrt{2} = 53.03$ kN as shown. Taking account of their directions of action, this justifies the first two elements of the fourth column of the stiffness matrix.

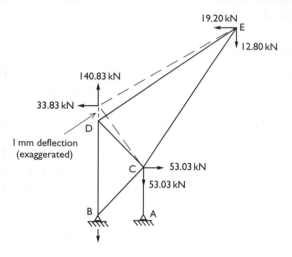

Figure 6.36 Joint forces required to produce a deflection v_D of 1 mm for the crane truss.

The stiffness matrix for statically indeterminate trusses

A statically indeterminate truss having m members and n joint displacements, will have $m > n$. In this case the matrix dimensions will be

$$\mathbf{A}^{\mathrm{T}}(n \times m), \quad \overline{\mathbf{K}}(m \times m), \quad \mathbf{A}(m \times n), \quad \mathbf{K}(n \times n)$$

Thus the determination of its stiffness matrix $\mathbf{K}$ from the multiplication $\mathbf{A}^{\mathrm{T}}\overline{\mathbf{K}}\mathbf{A}$ is as straightforward as it is for a statically determinate truss.

The direct stiffness method

In the direct stiffness method, the stiffness equations for a structure

$$\mathbf{Ku} = \mathbf{f}$$

are formulated and (given the applied forces $\mathbf{f}$) are directly solved for the displacements $\mathbf{u}$. It is then possible to evaluate the member forces from the multiplication:

$$\mathbf{p} = \overline{\mathbf{K}}\mathbf{Au}$$

Such a technique is cumbersome using hand calculation for the crane truss and other statically determinate trusses. However, on a computer, it has the advantages of being straightforward and versatile, one aspect of the versatility being that it can be used to solve statically indeterminate structures just as easily as statically determinate ones.

Stiffness contributions of individual members

Consider member i of a truss linking joints P and Q. If the truss suffers displacements $\mathbf{u}$, from Eq. (6.18), the force in the member will be

$$p_i = k_i \left[(u_Q - u_P) \cos \alpha_i + (v_Q - v_P) \sin \alpha_i \right] \tag{6.23}$$

where α_i is the inclination in the direction P $\to$ Q and k_i is the member stiffness. Designating the end forces which need to act on the member to keep it in equilibrium as those shown in Fig. 6.37, they relate to the member force according to

$$-H_{PQ} = H_{QP} = p_i \cos \alpha_i \quad \text{and} \quad -V_{PQ} = V_{QP} = p_i \sin \alpha_i$$

Substituting for p_i this gives

$$
\begin{bmatrix} H_{PQ} \\ V_{PQ} \\ \\ H_{QP} \\ V_{QP} \end{bmatrix}
=
\begin{bmatrix}
k_i c_i^2 & k_i c_i s_i & -k_i c_i^2 & -k_i c_i s_i \\
k_i c_i s_i & k_i s_i^2 & -k_i c_i s_i & -k_i s_i^2 \\
-k_i c_i^2 & -k_i c_i s_i & k_i c_i^2 & k_i c_i s_i \\
-k_i c_i s_i & -k_i s_i^2 & k_i c_i s_i & k_i s_i^2
\end{bmatrix}
\begin{bmatrix} u_P \\ v_P \\ \\ u_Q \\ v_Q \end{bmatrix}
\tag{6.24}
$$

where c_i and s_i have been used as a shorthand notation for $\cos \alpha_i$ and $\sin \alpha_i$, respectively.

Because the external forces acting on each joint must be the sum of the forces acting on each impinging member, the stiffness matrix for the truss can be formed by adding the stiffness contributions of all the members. However, the contribution of member PQ needs to be separated into the four submatrices so that each can be added into the appropriate position in the full matrix as defined by joints P and Q. As an example, the contribution

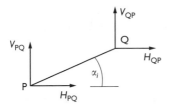

Figure 6.37 End forces acting on member PQ.

of member CE to the stiffness matrix of the crane truss is

$$
\begin{array}{c}
\begin{array}{cccccc} u_C & v_C & u_D & v_D & u_E & v_E \end{array} \\
\begin{array}{c} H_C \\ V_C \\ H_D \\ V_D \\ H_E \\ V_E \end{array}
\left[
\begin{array}{cccccc}
12.80 & 19.20 & 0 & 0 & -12.80 & -19.20 \\
19.20 & 28.81 & 0 & 0 & -19.20 & -28.81 \\
0 & 0 & 0 & 0 & 0 & 0 \\
0 & 0 & 0 & 0 & 0 & 0 \\
-12.80 & -19.20 & 0 & 0 & 12.80 & 19.20 \\
-19.20 & -28.81 & 0 & 0 & 19.20 & 28.81
\end{array}
\right]
\end{array}
\tag{6.25}
$$

The off-diagonal blocks appear unaltered in the fully compiled matrix, but the diagonal blocks have contributions added in from other members meeting at the particular joints. For members AC, BC and BD, only one of the four blocks is active because no displacement variables have been included for joints A and B.

Trusses with a large number of joints will only have non-zero off-diagonal submatrices when the relevant joints have a linking member. In Fig. 6.38, for instance, there is a non-zero submatrix in position H–C because there is a member HC in the lattice tower. However, no member HB exists and hence the submatrix in position H–B is null. Efficient programs take advantage of the presence of zero coefficients in the stiffness matrix to reduce computer time and also storage requirements (Jennings and McKeown, 1993).

Figure 6.38 The pattern of the non-zero submatrices in the stiffness matrix for the lattice tower of Fig. 5.30(a).

Q.6.16
Figure 6.39 shows a triangular truss with three possible joint movements.

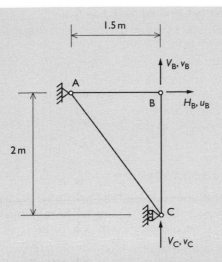

Figure 6.39

a   Determine the compatibility equations relating the member forces $\{e_{AB}\ e_{BC}\ e_{AC}\}$ to the joint displacements $\{u_B\ v_B\ v_C\}$.
b   Determine the equilibrium equations relating the external joint forces $\{H_B\ V_B\ V_C\}$ to the member forces $\{P_{AB}\ P_{BC}\ P_{AC}\}$.
c   Identify the transpose relationship between the matrices involved in (a) and (b) above.

Q.6.17
The truss shown in Fig. 6.40 is being used as a trial example to test a new structural analysis program. All the members have $E = 70\,\text{kN/mm}^2$ and $A = 1700\,\text{mm}^2$.

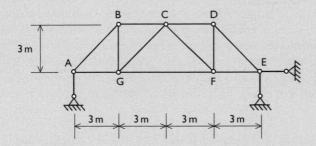

Figure 6.40

a What is the size of the stiffness matrix you expect the computer to formulate?

b Determine which 2×2 submatrices of the stiffness matrix contain non-zero elements.

c Derive all the submatrices on the two rows relating to joint D.

6.8 Computer analysis and modelling

As soon as digital computers became available, the benefit of direct stiffness methods became apparent, as recognised by Livesley (1953) for framed structures and Argyris and Kelsey (1960) for aircraft structures.

Key requirements for effective computational methods are simplicity, versatility and efficiency. Furthermore, with the speed of modern computers, efficiency is only important when equations to be solved have very many variables. The direct stiffness method has gained its prominence because of its extreme versatility. It is straightforward to program the construction of the stiffness matrix by adding together the various member contributions (as illustrated by matrix (6.25)). A computer program based on this technique is able to analyse any plane truss, whether it is statically determinate or indeterminate and whatever the shape. A discussion of how the versatility of the direct stiffness method can be extended, whilst retaining simplicity of technique, is given next.

Inclusion of restraint variables

If stiffness equations are formulated as for the crane truss of Ex. 6.16, deflections and member forces can be determined from their solution, but not the support reactions. A way of remedying this situation is to include extra members to simulate the restraints. Figure 6.41 shows how this may be done for the crane truss. The four artificial restraining members at A and B need to be given axial stiffnesses which are an order of magnitude larger than the stiffnesses of the other members so that any deflections induced at A and B will be negligible. Whereas this increases the number of stiffness equations requiring to be solved, it has the following advantages:

• The restraints are now member forces and will be computed automatically alongside the other member forces.

• There will always be two displacement variables per joint irrespective of the positions of the restraints, thus making it easier to program the construction of the stiffness matrix (e.g. if the simply-supported truss in

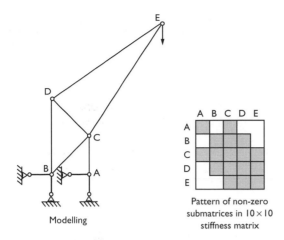

Figure 6.41 Inclusion of restraining members in the crane truss (Fig. 6.36).

Fig. 6.40 had not been presented in this way, joint A would have had only one variable).

NB   A user friendly computer program will have facilities for determining restraints without the need for the user to input dummy members.

Ex. 6.16
The three bar truss shown in Fig. 6.42 has members with $E = 200\,\text{kN/mm}^2$ and $A = 560\,\text{mm}^2$. It is to be used to check the workings

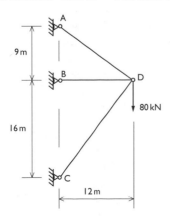

Figure 6.42 A statically indeterminate three bar truss.

of a computer program in which the restrained joints are provided with restraining members having axial stiffnesses of 10^6 kN/mm. Construct the stiffness matrix and determine the member forces and reactions when a vertical load of 80 kN is applied at D (as shown).

Answer

For all members $EA = 112,000$ kN

Hence for AD: $\ell_1 = 15$ m, $k_1 = 7.47$ kN/mm, $\cos\alpha_1 = 0.8$, $\sin\alpha_1 = -0.6$

for BD: $\ell_2 = 12$ m, $k_2 = 9.33$ kN/mm, $\cos\alpha_2 = 1$, $\sin\alpha_2 = 0$

for CD: $\ell_3 = 20$ m, $k_3 = 5.60$ kN/mm, $\cos\alpha_3 = 0.6$, $\sin\alpha_3 = 0.8$

With restraining members included, the stiffness equations are

$$
\begin{bmatrix}
10^6 & -3.584 & & & & & -4.779 & 3.584 \\
-3.584 & 10^6 & & & & & 3.584 & -2.688 \\
& & 10^6 & 0 & & & -9.333 & 0 \\
& & 0 & 10^6 & & & 0 & 0 \\
& & & & 10^6 & 2.688 & -2.016 & -2.688 \\
& & & & 2.688 & 10^6 & -2.688 & -3.584 \\
-4.779 & 3.584 & -9.333 & 0 & -2.016 & -2.688 & 16.128 & -0.896 \\
3.584 & -2.688 & 0 & 0 & -2.688 & -3.584 & -0.896 & 6.272
\end{bmatrix}
$$

$$
\times
\begin{bmatrix}
u_A \\ v_A \\ u_B \\ v_B \\ u_C \\ v_C \\ u_D \\ v_D
\end{bmatrix}
=
\begin{bmatrix}
0 \\ 0 \\ 0 \\ 0 \\ 0 \\ 0 \\ 0 \\ -80
\end{bmatrix}
\tag{6.26}
$$

Here the large elements in the diagonal positions of the first six equations ensure that, to three figure accuracy at least, the joints A, B and C do not move. The solution in mm units is

$$
\{u_A \ldots v_D\} = \{42.7 \times 10^{-6}, -32.0 \times 10^{-6}, -6.7 \times 10^{-6}, 0,
$$

$$
- 36.0 \times 10^{-6}, -48.0 \times 10^{-6}, -0.714, -12.857\}
$$

Using Eq. (6.23), the member forces and reactions are found to be as shown in Fig. 6.43.

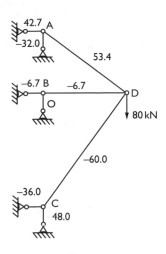

Figure 6.43 Computed forces in kN for the three bar truss including restraining members.

2D rigidly jointed frames

Where bending of the members of a frame needs to be taken into account, a rigidly jointed frame analysis is employed. A rigid joint is one in which all the members linking to it have the same rotation where they meet. Hence for a plane frame, there are three displacement variables per joint, u, v and θ, with θ being the in-plane rotation. Because the members carry bending moment, there is an extra equilibrium equation per joint to ensure that the end moments of the impinging members must be in equilibrium with the externally applied moment. Thus the member stiffness equations corresponding to Eq. (6.24) involve a 6×6 stiffness matrix which separates into four 3×3 submatrices. Elements of these matrices depend on the bending stiffness EI of the member as well as its axial stiffness EA.

Apart from increasing the number of variables to three, the main computational difference between truss and rigidly jointed frame analyses is concerning the lateral loading applied to frame members. This is catered for by using the theory of superposition to split a problem into the following two cases:

Case (a) has the actual loading applied, but has all the possible joint movements restrained. Each member is therefore considered as built-in.

Case (b) is to correct the joint forces introduced in Case (a) by allowing the joints to move. Thus the stiffness equations need to be solved, but with the loading being only at the joints.

Figure 6.44 shows how this is achieved for a particular example. Here the end forces and moments for the built-in portal member have been evaluated from the answer to Ex. 6.10. Once the stiffness equations have been solved, the results for cases (a) and (b) need to be added to obtain the correct member forces, bending moment distributions and reactions. However, these operations will all be part of any user friendly computer program.

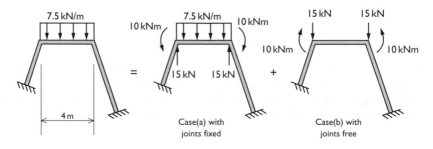

Figure 6.44 Use of the theory of superposition in the analysis of rigidly jointed frames.

3D trusses and frames

If a truss is to be analysed in three dimensions, three variables, u, v and w need to be allocated per joint to represent displacements in the x, y and z directions. If a member i with length ℓ_i connects joints P and Q, its length is given by

$$\ell_i = [(x_Q - x_P)^2 + (y_Q - y_P)^2 + (z_Q - z_P)^2]^{1/2}$$

Projecting the movements of joints P and Q on to the axis of the member gives the following formula for its extension:

$$e_i = (u_Q - u_P)\ell_i + (v_Q - v_P)m_i + (w_Q - w_P)n_i$$

in place of Eq. (6.18) for a plane frame. Here ℓ_i, m_i and n_i are the direction cosines of the member such that:

$$\ell_i = \frac{x_Q - x_P}{\ell_i}, \quad m_i = \frac{y_Q - y_P}{\ell_i}, \quad n_i = \frac{z_Q - z_P}{\ell_i}$$

It is not difficult to program the direct stiffness method to analyse 3D trusses in this way.

On the other hand, the analysis of 3D rigidly jointed frames is much more complicated to program than even 2D frames. Three displacement and three rotation variables are required per joint, and hence the member stiffness equations corresponding to Eq. (6.25) have a 12×12 coefficient matrix with elements dependent on the axial stiffness, bending stiffnesses in two directions and the torsional stiffness of the member. Furthermore, unless the program is restricted only to deal with members having doubly symmetric cross-sections, it will be necessary to cater for a non-zero product moment of area (see Section 4.10) and a shear centre which does not coincide with the centroid (see Section 4.9). Not only are 3D frame analysis programs more complicated, but also greater care needs to be taken in inputting data and interpreting results.

Computer modelling of rigidly jointed frames

There are many reasons why computer results for frame analysis might be inaccurate. Apart from more obvious possible sources of error due to inaccuracies in magnitude and positions of load, overall dimensions, cross-sectional dimensions and material properties, significant errors may arise from

a   ignoring the thickness of members;
b   assuming that joints are either pinned or rigid.

The following example explores these types of inaccuracies for a frame which, being statically determinate, has a relatively simple structural action.

Ex. 6.17
A crane rail is to be supported on a bracket BC which itself cantilevers from a column AB. Both members are to comprise I sections with the following properties:

Bracket: length 700 mm, depth 400 mm, $A = 7600\,\text{mm}^2$, $I = 240 \times 10^6\,\text{mm}^4$
Column: length 5200 mm, depth 340 mm, $A = 25,000\,\text{mm}^2$, $I = 500 \times 10^6\,\text{mm}^4$

The joint B is to be welded, with stiffeners welded into the column to prevent distortion at the joint. A plate is to be welded to the column base at A so that it can be bolted to a pad footing. Hence the elevation of the structure will be as shown in Fig. 6.45. Lateral bracing is present to ensure that neither the columns or bracket deflect out of plane.

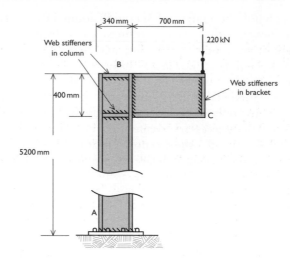

Figure 6.45 A crane rail support.

If the structure is to be modelled using two members located on the centre-lines of the actual members, and having the cross-section properties as given here, shown in Fig. 6.46(a), discuss possible modelling errors.

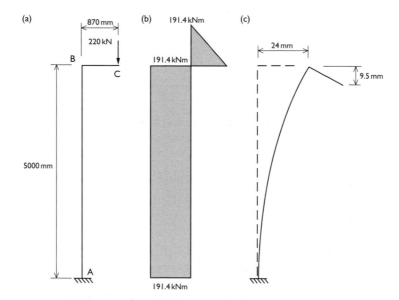

Figure 6.46 Frame analysis of the crane rail support: (a) computer model; (b) BMD drawn on tension face; (c) deflection (exaggerated).

Answer
For a crane load of 220 kN with $E = 200$ kN/mm², the computer
model will predict the bending moments and deflections shown in
Fig. 6.46(b) and (c). However, if the bracket and stiffened top part
of the column are considered as a 1040 mm long beam supported by
the column flanges, its loading and BMD will be approximately be as
shown in Fig. 6.47 involving a reduction of over 18% in the maximum
bending moment for the bracket. The bending moment in the column
for most of its height, however, will be as computed. The computed
vertical deflection of 9.5 mm at the crane rail comprises 8.3 mm due
to bending of the column, 1.0 mm due to bending of the bracket and
0.2 mm due to axial compression of the column. Modelling of joint B
will not have caused much error in this deflection because its largest
component (due to bending of the column) is for the most part correctly modelled. However, two other factors increase the deflection.
These are

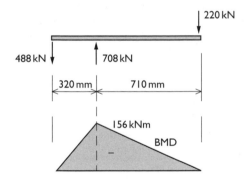

Figure 6.47 Alternative modelling of the top of the crane rail support.

- shear deflections in the bracket and the joint B;
- rotation of the column base.

Since the base pad cannot be infinitely stiff, there is bound to be some
small rotation of the column base which may or may not be significant.
There is also the possibility that the left-hand side of the column is trying to lift up (if the column is only supporting the eccentric 220 kN
load, this will be the case, but not necessarily if other loads are also
being supported). Although the bolts are there to prevent lift off, bending of the base plate is likely to cause some rotation which would be
difficult to estimate accurately with or without a computer.

Alternative modelling strategies

Different modelling techniques could be used for this problem. Some account could be taken of the different behaviour in the region of joint B by stopping the members at the edge of the joint and introducing short stiffer connecting members to simulate the behaviour of the joint (as shown in Fig. 6.48(a)). Alternatively, a truss model such as that shown in Fig. 6.48(b) could be used. This would enable the flanges, diaphragm stiffeners and baseplate to be located in their correct positions, but would not model the web behaviour properly. Although this form of model could be improved by reducing the spacing of the web bracing, 2D or 3D finite elements could be used which model all types of structural components more effectively. However, finite elements are a major study in themselves and are beyond the scope of this book.

General guidelines for computer modelling are as follows:

- There will be errors in all computer modelling.
- Do not dismiss simple computer models, but try to assess their applicability.
- When more accuracy is required, more complex forms of modelling may be used. However, much more computing power will be required. Also, more time will be needed to formulate the model and interpret results.

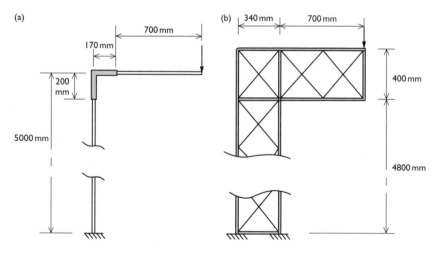

Figure 6.48 Alternative computer models for the crane rail support: (a) using a frame analysis; (b) using a truss analysis.

Q.6.18
If you have access to equation solving facilities (e.g. MATLAB):

a   Check the solution to Eq. (6.26).
b   Change the right-hand side to one in which the third element is 10^6
   with the rest zero and solve the equations again. What does this
   solution represent?

6.9   Rigidly jointed frames

Importance

Programs for the elastic analysis of rigidly jointed frames are extensively used
for building frames and many other types of structures. When using such
programs, it is essential for engineers to appreciate how the structures they
are analysing are likely to behave. Otherwise there is considerable scope for
incorrect modelling or misinterpretation of computer results. For this reason,
this section is concerned with identifying expected behaviour, but more in
a qualitative than a quantitative way. This can be done from sketches of
the deflection shapes (exaggerated) for the members and their BMDs, which
should be adjusted to satisfy the following criteria.

Structural characteristics

Regarding the position of joints:
 (i)   Axial extensions of members are so small relative to bending deflections
      that they can be ignored.
(ii)   Hence joint movements are negligible except where sway can take place
      (see Section 5.2).

Along each member:
(iii) The BMD must conform with the lateral loading, being linear if there
      is no lateral loading.
(iv)   The curvature must be proportional to M (or M/EI where EI is not
      constant).

At the joints:
 (v)   All forces acting at a joint must be in equilibrium.
(vi)   End moments for all members meeting at a joint must sum to zero (or
      balance the externally applied moment if one exists).
(vii) End rotations of all members meeting at a joint must be the same.

End moment applied to a member

Because moments will be transferred across joints, it is necessary to establish what effect an end moment will have on an otherwise unloaded member. If a member AB is uniform, two important cases for the response to an end load M_{AB} are as shown in Fig. 6.49. Where end B is free to rotation, $\theta_B = -\frac{1}{2}\theta_A$ and the rotational stiffness $M_{AB}/\theta_A = 2EI/\ell$. However, where B is fully restrained against rotation, the stiffness increases to $M_{AB}/\theta_A = 4EI/\ell$ and $M_{BA} = \frac{1}{2}M_{AB}$.

Moment applied to a joint

If a moment is applied to a joint of a frame in which there are no sway modes, the joint will rotate with end moments developing in the impinging members in proportion to their rotational stiffnesses. The members which connect to this active joint will develop deflections and bending moments somewhere between the two cases shown in Fig. 6.49, because the remote ends of these members, being connected to passive joints, will be themselves only partially restrained against rotation.

Figure 6.50 shows how a frame may respond to a moment applied to joint A. Small amounts of rotation will be induced in the other joints B–E by the lateral deflections of members AB, AC, AD and AE, but the effects on members not directly connected to the active joint (namely BC, CD and DE) will be relatively small.

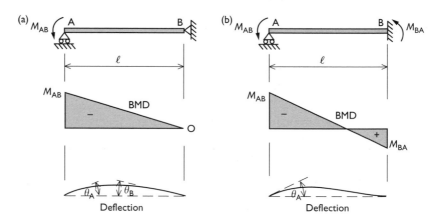

Figure 6.49 Response of a uniform member to an end moment M_{AB}: (a) with end B pinned; (b) with end B built-in.

Frames without sway modes

Visualisation of the response of rigidly jointed frames to lateral loads acting on members is helped by calling on the theory of superposition. Consider firstly a case of a frame which does not have any sway modes. Figure 6.51 shows how lateral loading applied to one member of such a frame may be broken down, via the theory of superposition, as the sum of three simpler loading cases. The bending moment and deflections may be sketched by

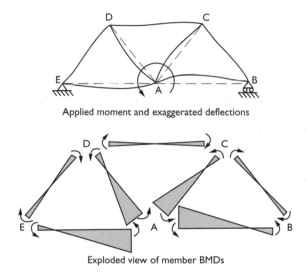

Applied moment and exaggerated deflections

Exploded view of member BMDs

Figure 6.50 A sketch of the response to an applied moment at a joint.

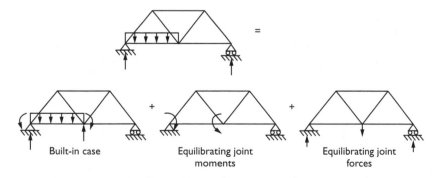

Built-in case   Equilibrating joint   Equilibrating joint
                       moments   forces

Figure 6.51 Use of the theorem of superposition in the analysis of rigidly jointed frames.

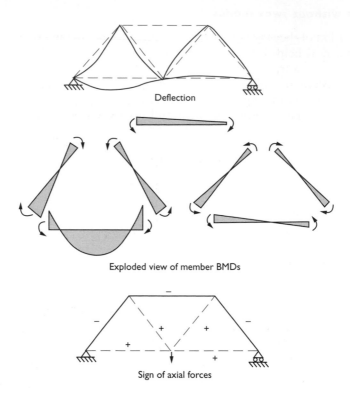

Figure 6.52 A sketch of the response for the frame loading of Fig. 6.51.

considering the first two of these loading cases with the axial forces mainly arising through the third case as shown in Fig. 6.52.

Sway frames

When load is applied to sway frames, it is usually necessary for bending moments to develop in the members in order to resist sway movements. These bending moments can often dominate the structural action. Consider a portal frame to which a horizontal load is applied at the eaves level. Figure 6.53 shows how a deflection diagram can be predicted which allows movement of the sway mechanism yet maintains rigidity in all the joints. Also shown is a prediction for the bending moments in which, to satisfy equilibrium, the bending moments at the ends of the beam must equal the bending moments at the tops of the columns.

The amount of rotation of the eaves joints depends on the relative rotational stiffness of the column and beam members. Thus if the beam is very

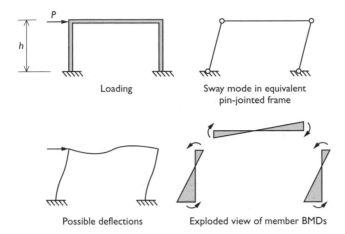

Figure 6.53 Sway of a portal frame.

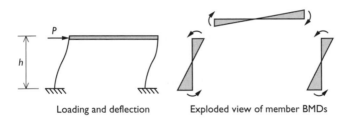

Figure 6.54 Sway of a portal frame with a stiff beam member.

much stiffer than the columns, it will remain almost horizontal with most of the bending deflection occurring in the columns (Fig. 6.54).

Figure 6.55 shows different possible sway displacements for two sway frames both of which have two sway mechanisms. If lateral loading is applied along members of a frame, it may be easier to predict the structural response if the theory of superposition is first applied as in Fig. 6.51.

Significance of points of inflection

In these examples it was important to ensure that, if the bending moment in a member changes sign, the curvature does also at the same spanwise location. Such a position is called a 'point of inflection'. Furthermore, if imaginary hinges are introduced at all places where points of inflection are expected to occur, it is usually possible to use equilibrium principles to estimate bending moments and other internal forces or reactions.

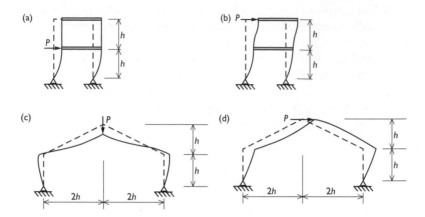

Figure 6.55 Possible deflections of frames having two sway modes.

Ex. 6.18

a For the portal frame shown in Fig. 6.54 having an infinitely stiff beam, obtain an expression for the magnitudes of the critical bending moments.

b Discuss how the critical bending moments are affected by the relative rotational stiffnesses of the beams and columns.

Answer

a Because the tops of the columns do not rotate, their points of inflection will be at height $\frac{1}{2}h$ (as indicated on their BMDs). Furthermore their shear forces, being the slopes of their BMDs, will be the same, say S. Hence, from equilibrium of the top part of the frame as shown in Fig. 6.56, their shear force $S = \frac{1}{2}P$ and the critical bending moment, occurring at both eaves joints and both column bases, will be $M = Ph/4$.

Figure 6.56 Equilibrium of the portion of the portal frame, Fig. 6.54, above the points of inflection in the columns.

b   When the beam is not as stiff, the eaves joints rotate and the points
    of inflection migrate up the columns as illustrated in Fig. 6.53.
    Because the shear forces in the columns will still be $S = \frac{1}{2}P$, the
    critical bending moment in each column will be at its base with
    a magnitude $\frac{1}{2}Py$ where y is the height of the point of inflection
    ($\frac{1}{2}h < y < h$). On the other hand, the critical bending moment
    for the beam will occur at the eaves joints with a value $\frac{1}{2}P(h - y)$,
    which must be less than $Ph/4$.

Ex. 6.19

a   For the pitched roof portal frame and loading shown in Fig. 6.55(c),
    determine the critical bending moment if points of inflection in the
    rafter members occur half way along their length.
b   What affects the position of the points of inflection in the rafter
    members and what effect do they have on the critical bending
    moments?

Answer

a   Figure 6.57(a) shows the forces acting on the position of the
    frame outboard of one of the points of inflection (which is drawn
    as a hinge). Since, from symmetry, the vertical component of
    reaction at the base must be $\frac{1}{2}P$, taking moments about the
    point of inflection gives the horizontal component of base reac-
    tion as $P/3$. Hence it follows that the bending moment at the
    eaves joint is $Ph/3$ (with the rafter member hogging) and at
    the apex is $Ph/3$ (with the rafter member sagging). Hence the
    critical bending moment is $Ph/3$ in both the columns and the
    rafters.
    NB   The same result could have been obtained by drawing the
    thrust line (which is triangular for this loading case) through
    the hinges at the bases and the points of inflection (as shown in
    Fig. 6.57(b)).
b   Increasing the ratio of k_R/k_C (where k_R is the rotational stiffness
    for a rafter member and k_C is that for a column) will move the
    points of inflection closer to the eaves. This will decrease the bend-
    ing moment at the eaves joints (and hence the maximum bending
    moment in the column), but will increase the bending moment

at the apex, which will be the maximum bending moment in the rafter members.

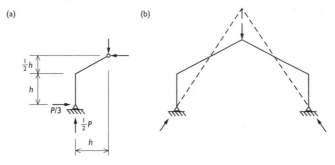

Figure 6.57 The pitched roof portal frame, Fig. 6.55(c), with points of inflection at the quarter chord positions: (a) equilibrium of outboard portion; (b) position of the thrust line.

Q.6.19
Why do the end moments in Fig. 6.49, which are both anticlockwise, have opposite signs in the BMD?

Q.6.20
Show that, for a uniform elastic member AB with length ℓ, bending stiffness EI and subject to an end moment M_{AB}, the end moment required to prevent rotation of joint B is $M_{BA} = \frac{1}{2}M_{AB}$. Also show that the rotational stiffness of the member M_{AB}/θ_A is $4EI/\ell$ (this is the case shown in Fig. 6.49(b)).

Q.6.21
Use a structural analysis program to analyse a frame similar to that of Fig. 6.50 and check whether deflections and bending moments conform reasonably closely to those of diagrams of Fig. 6.52.

Q.6.22
Treating the continuous beam shown in Fig. 6.28 as a frame with no sway modes, sketch the expected deflection form and bending moment distribution when span AB carries a uniformly distributed load and no other span is loaded.

Q.6.23
For the frame and loading shown in Fig. 6.55(d), sketch the member BMDs and determine the magnitude of the maximum bending moment.

Q.6.24
Using a structural analysis program, analyse loaded frames similar to Fig. 6.51 and Fig. 6.55(b) to determine axial forces, bending moments and deflections. Investigate whether these solutions satisfy the structural characteristics for joints and members specified in this section of the book.

6.10 Symmetric structures

Reflective symmetry

It is difficult to think of man-made structures which do not have any form of symmetry. Very regularly structural components have symmetric properties (as noted in Section 4.4 for beam cross-sections) and entire structures are often symmetric. Symmetric structural forms are useful because of savings in design time, construction costs and also because their structural behaviours are more easily understood. This section is concerned with structures which have reflective symmetry about one plane. Many examples have already been considered in this book (e.g. the frames of Figs 6.53–6.55). Simply-supported structures such as the truss of Fig. 6.50 and also the continuous beam of Fig. 6.28 can be treated as having reflective symmetry because the position of the horizontal restraint is of no significance to any analysis when the loading does not induce any horizontal reaction.

When analysing linear elastic structures, engineers can use symmetry in one of two ways. Either they can analyse the complete structure, reserving knowledge of the symmetric properties to provide a partial check on the validity of the results. Alternatively knowledge of the behaviour of symmetric structures may be used to condense the analysis. When only hand solution techniques were available, the time saving gained from condensing the analysis was particularly useful. When using computer programs, it may still be advantageous to consider condensing analyses for symmetric structures in order to save data preparation time and, when structures are large, to keep down the time and capacity requirements of the computer.

Symmetric loading applied to 2D symmetric structures

When symmetric loading is applied to a symmetric structure, the internal forces, stresses, reactions, deflections and strains will all be symmetric. Also parts of the structure, originally on the plane of symmetry, must deflect only within the plane of symmetry.

Figure 6.58 shows four symmetric structures with vertical planes of symmetry and subject to symmetric loading (it is necessary for complete symmetry that the member cross-sections and material properties are also symmetric). Figure 6.59 shows how they may be modelled so that the correct behaviour is obtained for the half frames.

Notes

- Joints on the planes of symmetry all have horizontal movement restrained.
- Joints in rigidly jointed frames which are on the planes of symmetry have rotation restrained as well.
- Where an extra joint has been included, at the mid-span position of the lower chord of the roof truss, this has been prevented from moving vertically as well as horizontally to avoid a mechanism being created.
- Loads acting on the plane of symmetry have been halved.
- The cross-sectional area of members lying along the plane of symmetry have been halved.

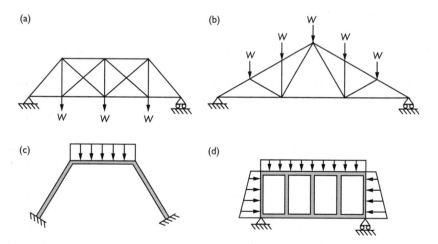

Figure 6.58 Symmetric structures with symmetric loading: (a) a truss girder; (b) a roof truss; (c) a fixed base portal frame; (d) a Vierendeel truss.

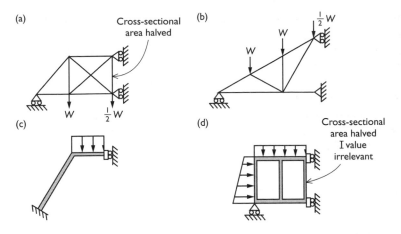

Figure 6.59 Alternative modelling for analyses represented in Fig. 6.58.

- Where a member of a rigidly jointed frame lies in the plane of symmetry, it will not be subject to bending moment and so its I value is irrelevant.
- Because horizontal restraints are provided at the plane of symmetry, no other horizontal restraint should be present where the structure was originally simply-supported.

Antisymmetric loading applied to 2D symmetric structures

Antisymmetric loading is where it is a mirror image on opposite sides of the plane of symmetry except for a sign change. In this case, the internal forces, stresses, reactions, deflections and strains will all be antisymmetric. The displacements and forces which can occur at a plane of symmetry are the complete reverse of the situation for symmetric loading as shown in Table 6.3.

Figure 6.60 shows two symmetric structures with vertical planes of symmetry subject to antisymmetric loading. The roof truss can be classified as symmetric because the loading does not generate any horizontal reaction. (This would not be the case if the loads acted normal to the rafter members instead of vertically.) Figure 6.61 shows how the structures may be modelled so that the correct behaviour is obtained for the half frames. The portal frame of Fig. 6.55(d) is symmetric with antisymmetric loading. It is therefore possible to use Table 6.3 to show that there can be no moment transfer across the plane of symmetry, thus indicating the apex joint as a point of inflection.

Table 6.3 Deflections which can occur at a vertical plane of reflective symmetry and internal forces which can act across it

Variable	Symmetric loading	Antisymmetric loading
Horizontal displacement	0	✓
Vertical displacement	✓	0
Rotation	0	✓
Horizontal force	✓	0
Vertical force	0	✓
Moment	✓	0

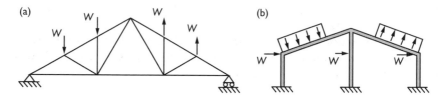

Figure 6.60 Symmetric structures with antisymmetric loading: (a) a roof truss; (b) a rigidly-jointed frame.

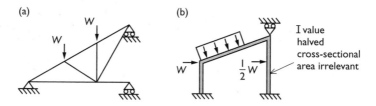

Figure 6.61 Alternative modelling for loading cases shown in Fig. 6.60.

Sway frames with lateral load

Figure 6.53 is an example of a symmetric frame which would have anti-symmetric loading if half of the load were to be transferred to the opposite eaves joint as shown in Fig. 6.62. However, such a transfer would only affect the axial force in the beam. The bending moments and lateral deflections of the members of the original frame are antisymmetric as seen in Fig. 6.53. Similarly the two storey portal frames in Fig. 6.55 exhibit antisymmetric responses.

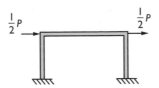

Figure 6.62 A modification of the loading on the portal frame of Fig. 6.53.

Plane structures subject to lateral loading

In 3D, let u, v and w be displacements in the x, y and z directions and let θ, ϕ and ψ be rotations about the x, y and z directions respectively. If a structure lies entirely in the x–y plane, that plane must be a plane of symmetry. Hence, any loading applied in the z direction amounts to an antisymmetric loading. Therefore the response, being antisymmetric, will involve only displacements w and rotations θ and ϕ with $u \equiv v \equiv \psi \equiv 0$. Examples of laterally loaded plane structures are shown in Fig. 6.63.

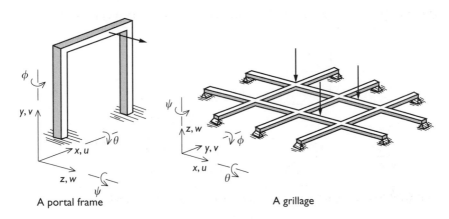

A portal frame   A grillage

Figure 6.63 Two laterally loaded plane structures.

Warning

It is important to remember that the results in this section can only be relied upon when the structures being investigated behave in a linear elastic manner. In Chapter 7, the possibility of unsymmetric buckling modes is discussed for symmetric structures loaded symmetrically, but beyond their linear range. Also it is well known that a flat plate, although being symmetric about its own plane, will develop in-plane (and therefore symmetric) forces when loaded laterally once the deflections exceed (approximately) its own thickness. This is also due to non-linear behaviour.

Q.6.25
'A symmetric beam subject to symmetric loading has symmetric diagrams of loading, shear force, bending moment, slope and deflection'. Discuss the validity of this statement.

Q.6.26
Figure 6.64 shows a simply-supported beam which is strengthened underneath by truss bracing (assumed to carry axial load only).

a   For a uniform load w acting across the whole span, show how an analysis for the half structure could be formulated to provide the correct modelling.
b   Would it be possible to obtain a half-frame solution if the loading on the left-hand side was $2w$/unit span downwards with no loading on the right-hand side? If so, derive any computer model(s) to be used.

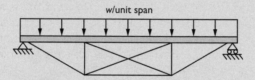

w/unit span

Figure 6.64

Q.6.27
Choose values for dimensions, member properties and loading for any one of the frames shown in Fig. 6.58. Carry out a computer solution of the whole frame and also one of the corresponding half frame using the modelling technique given in Fig. 6.59. Check whether the answers correspond.

Q.6.28
For the case of the simply-supported structures shown in Fig. 6.58, explain why the displacements will not completely agree between the full analysis and the half-frame analysis.

6.11   Redundancy and its importance

Alternative methods of analysis

When a structure is statically indeterminate, the strain and deflection characteristics need to be considered alongside equilibrium in order to determine the internal forces. Methods of analysis divide into two types, those which use redundant forces as the basic variables and those which use displacements. Methods employing redundant forces, generally classified as 'force' or 'flexibility' methods, were well suited for hand calculations when the number of redundancies were small (e.g. most trusses). Arches, portal frames and statically indeterminate beams were also frequently analysed using redundant forces. Clapeyron's method for continuous beams (section 6.5) is the only flexibility method given in this book.

Methods which use displacements as the basic variables, generally classified as 'displacement' or 'stiffness' methods, have been used for hand solution of some of the more highly redundant structures. Rigidly jointed frames, in particular, could be analysed with just joint rotations and sway modes as basic variables, making the number of equations to be solved often less than the number of redundancies. However, with computer solution there is no longer a need to make the number of basic variables as small as possible and hence direct stiffness formulations, as described in Section 6.8, have been found to be most effective because of their versatility. Direct stiffness methods are distinguishable in so far as the displacement form of every member is entirely defined by the displacements of the joints to which it is attached. When sufficient joint displacements are introduced to make this possible, the structure is described as being 'kinematically determinate'.

Why bother about redundancy?

Because flexibility methods are rarely used in computer solutions, the question could be asked: is it any longer necessary to know anything about redundancy? The answer is an emphatic yes. It is important because of its influence on structural behaviour. A statically determinate structure has only one load path, whereas a statically indeterminate structure has just as many additional load paths as there are redundancies. Several factors can influence the magnitudes of the redundant forces and hence which load paths are most active. These are

- stresses arising through manufacture;
- movement of supports;
- slippage within joints;
- non-linear material behaviour;
- non-linear geometric response;
- temperature strains.

Stresses due to manufacture

Stresses can be caused when prefabricated parts of a structure do not fit together properly and therefore have to be forced into place. If a fixed-base pitched roof portal frame is manufactured in four parts to be bolted together on site, conventional analysis methods would suggest that there will be three redundancies and therefore three independent stress systems which could be induced if force is required to close the joints. If the frame is to be erected by cantilevering with the ridge joint the last to be bolted, misalignments could occur horizontally, vertically and in rotation as shown in Fig. 6.65. However, in practice, the situation could be worse than that because a steel fabricator is necessarily working in 3D rather than 2D. With lateral deflection, lateral rotation and twist also possible, there are actually six possible independent force systems which may need to be induced in order to correct misalignments (corresponding to six redundancies for the completed structure when analysed in 3D).

Even when a structure is prefabricated in such a way that no forces are required to close the joints, the internal forces due to self-weight will depend on the construction sequence. For instance, if a bridge, which is to be in the form of a continuous beam, is erected by cantilevering half spans from the supports, the self-weight will be carried entirely by hogging bending moments which will be maximum over the supports. If, on the other hand, full span beams are lifted onto bearings at the supports and afterwards neighbouring beams are connected over the supports, the self-weight will be carried entirely by sagging bending moments which will be maximum at the centre-span positions.

Although unwanted internal forces can arise through correcting for lack of fit, particularly when the manufacturing process has not been well planned, it is also possible to use lack of fit deliberately to induce beneficial stress regimes. One example is the pretensioning of tie bracing in trusses to

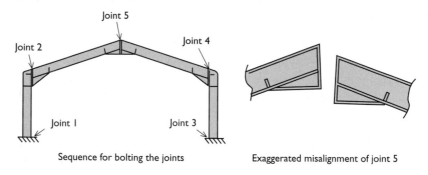

Sequence for bolting the joints   Exaggerated misalignment of joint 5

Figure 6.65 A poor erection procedure for a pitched roof portal frame.

avoid ties going slack under working load (see Section 5.6) and another is fabrication techniques used on the Britannia Tubular Bridge and also on more recent box girder bridges to reduce the sagging bending moments at the centres of their main spans (see Section 4.11).

Whereas simply-supported beams and cantilevers are considered to be statically determinate, they are only statically determinate as far as internal forces are concerned. The distribution of stress within the cross-section cannot be found from statics alone (in Chapter 4, elastic behaviour was assumed in order to predict stresses). Hence it is possible to generate self-equilibrating stress systems within such beams. Examples are longitudinal stresses arising from manufacture of steel sections by rolling, stresses arising through cooling of weldments and the stresses caused in span 14–15 of the West Gate bridge due to the construction sequence (discussed in Section 4.12).

Temperature strains and stresses

Temperature effects occur mainly because of

- sunshine and the daily temperature cycle;
- the yearly temperature cycle;
- industrial activity;
- fire.

During careful monitoring of the Tower of Pisa, it has been observed that heating by the sun causing expansion on one side, produces a small variation in the angle of lean. Similar diurnal (daily) temperature variations will be occurring in all kinds of structures on a regular basis. The other items on this above list have the potential of creating much larger temperature strains. When a structure is statically determinate, expansion or contraction due to temperature effects will produce deflection, but will not affect stresses. However, with statically indeterminate structures, such movements may not be able to take place freely, in which case quite large temperature stresses may develop.

Ex. 6.20

a   In a chemical plant, a simply-supported steel beam of span 9 m, depth 550 mm and second moment of area 700×10^6 mm^4, could be subject to a temperature differential in which the top flange is as much as 45°C warmer than the bottom flange. Assuming that the temperature varies linearly between the two flanges and that the coefficient of thermal expansion is $3.6 \times 10^{-6}/°$C, estimate the maximum central deflection this could cause in the beam.

b   If the beam is given an additional support at its centre, estimate the change in the support reactions and the maximum bending stress in the beam due to the temperature differential reaching its maximum (it may be assumed that none of the supports move).

Answer

a   The maximum strain difference between the outer fibres is $45 \times 3.6 \times 10^{-6} = 0.162 \times 10^{-3}$. This will create curvature in the beam with the strain at the outer fibres, distance 275 mm from the central axis, differing from the strain at the centre by $\varepsilon = 0.081 \times 10^{-3}$. From Eq. (6.2), the curvature

$$\frac{1}{R} = \frac{0.081 \times 10^{-3}}{0.275} = 2.945 \times 10^{-3}/\text{m}$$

It is a well-known geometric theorem that any two interesting chords, AEB and CED, of a circle have the property that $AE \times EB = CE \times ED$ (see Fig. 6.66). With a circle of radius R, if the chord AEB is the undeflected position of the beam and CED bisects AB at right angles, CE is the central deflection δ and $ED \simeq 2R$. Hence, if ℓ is the span, $(\frac{1}{2}\ell)^2 \simeq 2R\delta$ giving

$$\delta = \ell^2/8R$$

which, for the example, gives the central deflection as 3.0 mm.

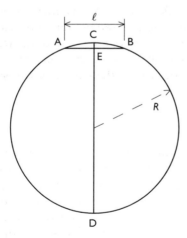

Figure 6.66 Property of intersecting chords of a circle.

b   The central support will need to pull the beam downwards by a deflection of $\delta = 3.0$ mm. From the third case of Fig. 6.10, it follows that the downwards force P requiring to be applied at the centre to prevent this deflection is governed by the equation

$$P = 48EI\delta/\ell^3 = 48 \times 200 \times 700 \times 10^6 \times 3.0/9000^3$$

$$= 27.5 \, \text{kN}$$

The end reactions on the beam will be 13.75 kN and the maximum bending moment at the centre: $M = P\ell/4 = 61.9$ kNm giving a maximum bending stress of $\sigma = My/I = 61.9 \times 10^6 \times 275/700 \times 10^6 = 24.3 \, \text{N/mm}^2$.

The yearly temperature cycle causes longitudinal expansion and contraction of the decks of bridges which is not generally matched by any changes to the distance between the supports. To avoid large tensile or compressive forces arising in the decks, it is normal practice to include an expansion joint.

Structures subject to specified displacements

Sometimes it may be necessary to consider what happens when structures are required to withstand specified displacements (e.g. due to ground subsidence). Even when a structure was originally statically determinate, when one of its joints is constrained to a specified displacement, that effectively adds a redundancy. The internal forces that need to be developed to accommodate the displacement can be classified as a self-equilibrating system. The stiffer is the structure, the higher will be the forces induced into it. Hence, if stresses are too high, increasing the strength of the structure in a way which increases also the stiffness is likely to be exactly the wrong way of solving the problem.

Load tests on structures

If increasing load is placed on a structure till it fails (as was done with the tests shown in Fig. 5.59), sudden catastrophic collapse often takes place with no chance of carefully examining what really initiated failure. A better procedure for conducting experiments is to apply loading through a hydraulic jack which is reacting against a loading frame. This is easier to achieve when model structures are tested in a laboratory than with larger structures. In this case, when a part of the structure starts to give way, the additional displacement increases the jack extension, so reducing the force being applied to the structure. Thus the structure is held in the partially collapsed state till more fluid is pumped into the jack. Effectively the structure is being given an applied displacement rather than an applied load. For this procedure to work, it is important that the loading frame is both strong enough and stiff enough otherwise it will be the loading frame rather than the test structure which will actually be tested. The test on the masonry arch shown in Fig. 2.40 was carried out by jacking and that is why the partially collapsed state was able to be photographed easily.

Q.6.29
It would be difficult to assemble the portal frame of Fig. 6.65 in the sequence described. Can you suggest a better procedure?

Q.6.30
The reinforced concrete deck of a road bridge of length 80 m has a coefficient of thermal expansion of $3.4 \times 10^{-6}/°C$. The equivalent concrete cross-section of the deck has an area of $2.8\,m^2$ and a modulus of elasticity of $23\,kN/mm^2$.

a   Determine the width of the expansion joint necessary to cater for annual temperature variations from $-30°C$ to $40°C$.
b   If the expansion joint seizes up due to lack of maintenance, determine the longitudinal compressive force that would be generated in the deck if the temperature were then to rise a further $40°C$.

Q.6.31
A bridge over a motorway is to have an intermediate column support as shown in Fig. 6.67. Discuss options available for catering for thermal expansion of the deck.

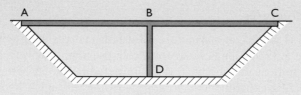

Figure 6.67

Q.6.32
It is required to formulate a method of obtaining deflections and stresses due to temperature in a truss having m members and n unknown joint displacements where $m > n$ (hence it is statically indeterminate). The vector of member extensions $\mathbf{e}$ will be such that

$$\mathbf{e} = \mathbf{e}_s + \mathbf{e}_t$$

where $\mathbf{e}_s$ contains the member extensions due to their axial forces and $\mathbf{e}_t$ contains the member extensions due to the specified temperature regime.

Formulate a matrix method of determining the deflections of the joints expressed as the vector $\mathbf{u}$, of order $n \times 1$ where $n \le m$ using the matrices $\mathbf{A}$ and $\overline{\mathbf{K}}$ given in Section 6.7. Also indicate how the internal forces $\mathbf{p}$ of order $m \times 1$ may be determined.

Q.6.33
A bridge has been designed, using elastic theory, as a uniform steel beam continuous over three supports. The design adequately caters for all the loading cases. However, before construction takes place, concern arises over the possibility of ground subsidence due to mining activity. An investigation has revealed that, if a support sinks a specified amount (considered to be possible), the bridge will be too highly stressed. Of the following possibilities which are likely to alleviate the problem?

(i)   Increasing the depth of the cross-section to cater for higher bending moments.
(ii) Keeping the cross-section the same depth, but increasing the flange areas to cater for more bending moment.
(iii) Reducing the depth of the cross-section.
(iv) Making the beam discontinuous over the central support and ignoring the subsidence case in the redesign.

6.12 Why did the roof fall in?

The Kemper Arena

Increasingly during the second half of the twentieth century, sporting events attracted large crowds and became big business. This provided structural engineers with an opportunity to display their ingenuity and initiative in providing roofs covering the arenas themselves as well as the areas for spectators. There have been a multitude of different schemes which have included various types of domes, trusses, cable nets and cable-stayed structures. Some of the most ingenious have involved a retraction facility to provide organisers with the option of open air or indoor events.

The R Crosby Kemper Jr Memorial Arena at Kansas City was recognised as one of the USA's finest buildings after its construction in 1974. Built originally for the local basketball team, it had a seating capacity of 17,600 and was used for very many kinds of events including a Republican Party national convention in 1976. Unusually for such a large span, the 98 m × 94 m roof (covering 4 acres) was flat. The roof membrane was of concrete placed on a corrugated steel deck to form a composite reinforced concrete slab. This was supported in turn by three forms of truss. The tertiary structure comprised 'open web joists' running north–south at 2.7 m spacing. These were formed from steel angles as chords with bent rods to act as bracing in such a way that the resulting joist had the profile of a Warren truss. These joists carried the roof loading to the secondary structure, which was seven deeper trusses running east–west at 16.4 m spacing. These trusses acted as continuous beams, being each supported at six places where they intersected with the two lower chords of each of the three large space trusses acting as the primary structure and running north–south. Unlike the primary and secondary structures which were internal, these space trusses were external to the shell of the building, giving it an immediately recognisable appearance.

Collapse

On the evening of 4 June 1979, approximately 11 mm of rain fell in a short space of time on Kansas City. With no-one present inside, one-third of the roof fell in, although the primary structure was still intact (see Fig. 6.68). Only 17 months earlier, in January 1978, two other large span roofs had fallen in the USA (a space truss at the Hartford Coliseum in Connecticut and a truss dome at the C.W. Post College of Long Island University). Then the trigger was snow loading. Hence the public trust in large span roof structures was at a low ebb. No one had been killed, but if any one of these buildings had been hosting an event, there could have been a large death toll.

Bolt failure

It was discovered that collapse of the Kemper Arena roof started with failure of two of four bolts connecting a secondary truss to short hangers separated from the central primary space truss. This failure overloaded the two other bolts and then neighbouring hangers thus causing a chain reaction. The hangers were there to separate the primary space trusses from the roof itself

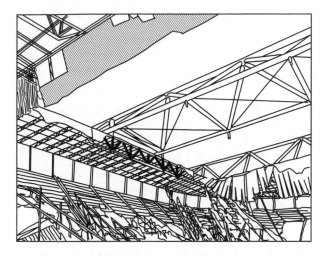

Figure 6.68 The Kemper Arena after the roof collapse.

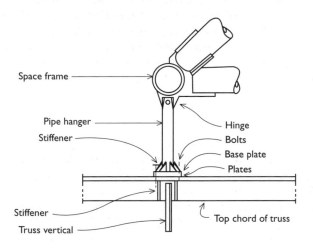

Figure 6.69 A hanger assembly.

(Fig. 6.69). They had not only to carry vertical gravity loading, but were subject to sideways movements when wind forces acted on the walls. Each bolt was designed to carry 700 kN, but they were found to fail when only about one quarter of this load was applied. Being made of high strength steel, they had a good load carrying capacity under steady load, but were not suitable for dynamic loads. Variable loading arising through wind, rain and snow made the choice of this type of bolt unsuitable.

The drainage system

When large building developments occur on greenfield sites, there is often a problem with undercapacity of the sewers. In order not to overload them, the Kemper Arena roof had been provided with only eight 130 mm diameter drainpipes when 55 would be required to satisfy the normal requirements of the Kansas City code for a roof of that area. In the event of a heavy downpour, it was anticipated that water might accumulate on the roof up to a depth of 50 mm, at which stage it would spill out of holes called 'scuppers' and down the walls. The situation with the Kemper Arena roof, therefore, was somewhat bizarre, although not completely without precedent, in that it was designed to retain water temporarily rather than shed it as quickly as possible.

The failure initiated below the southern side of the central space frame where it was discovered that water had accumulated to a depth of 225 mm ($4\frac{1}{2}$ times the design level). Why should this build up of water have occurred?

Wind effect

On 4 June, gusts of up to 110 kph were recorded which would have pushed water to the southern side by wind friction. However, that affect alone could not have produced this degree of ponding, particularly in view of the presence of the scuppers.

Structural deflections

Another wind effect would have been to create a suction on the northern side of the roof as the air swept over the northern parapets (called the 'Bernoulli effect'). Both this upward loading on the northern side and the additional downward loading due to wind blown water would have caused the roof to tilt south side down so encouraging further ponding there. Furthermore, since the scuppers would not deflect downwards, downwards deflection of the roof structure would explain why the local water level could have built up so high.

The likelihood of ponding can be described by a parameter g which is the depth of water whose weight will cause a 1 mm downwards deflection of the roof. If this is less than 1 mm, any ponding action will be self-perpetuating thus making collapse virtually inevitable. A g value of 0.627 has been quoted for the Kemper Arena (Levy and Salvadori, 1992). If this is correct, collapse could have easily occurred without wind effects being present.

Since also lateral deflections of the roof due to wind loading on the walls had a major influence on the failure of the bolts, lack of appreciation of the effects of deflections on the performance of the structure could be considered as the most important basic factor contributing to the collapse.

Aftermath

The roof of the arena was rebuilt using the primary space frame which was still intact. The structural support system for the roof was improved and it was no longer required to have an auxiliary role as a storage reservoir.

> Q.6.34
> What advantages and disadvantages can you identify for having the primary structure external to a building?

6.13   Footbridges for the twenty-first century

Pedestrian access – a modern requirement?

Changes in employment, transportation and social requirements have led to some areas of inner cities around major rivers falling into disuse. Regeneration plans usually involve different forms of leisure facilities in completely or partly traffic free zones, included in which are often specifications for footbridges across the rivers. Because footbridges are narrower and not as heavily loaded as road or rail bridges, there is much scope for innovation in design. The London Millennium Bridge is one solution which uses shallow suspension cables for the main support (discussed in Section 7.4). Some other designs entail cable stay supports from masts or hanger supports from arches.

An advantage of torsionally stiff decks

Traditionally most girder, truss or suspension bridges have the main structure situated in two planes, one supporting each side of the deck. Simplification can be achieved if only one support plane is required. For a straight footbridge supported from a suspension cable or an arch, the first two cross-sections in Fig. 6.70 have the advantage of retaining symmetry, but have hangers which interfere with the space requirements for pedestrians. The third cross-section keeps the hangers clear of the pedestrian space by using a torsionally stiff deck to transfer the resulting torsion loads to the abutments. The first footbridge with this type of configuration built in 1991 was La Devesa Bridge at Ripoll in Spain, designed by the celebrated Spanish Architect/Engineer Santiago Calatrava (Frampton *et al.*, 1966).

Merchants Bridge

The first footbridge to emmulate Calatrava's idea and employ one-sided deck support in the UK was Merchants Bridge at Manchester (Fig. 6.71) built in

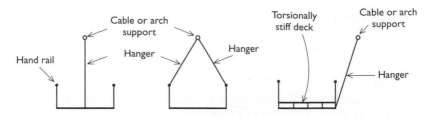

Figure 6.70 Possible cross-sections for straight footbridges.

Figure 6.71 Merchants Bridge, described as a 'sickle arch'.

1995. Adequate torsional stiffness in the deck was achieved by welding plates to edge tubes in order to form a closed box. Internal stiffening was included in the box in order to avoid distortion of the cross-section (Fig. 6.72). Whereas the Ripoll Bridge had a straight deck, the deck of Merchants Bridge was curved in both elevation and plan. Furthermore by inclining the arch away from the deck, tension in the hanger supports introduced a certain amount of compression force into the deck. The hangers were provided with bending stiffness in order to resist lateral forces acting on the arch and also stabilise it against lateral buckling. Whilst the curves of both the deck and the arch complicated design and construction, they helped to create a structural form which appears both simple and elegant.

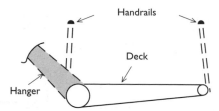

Figure 6.72 Cross-section of the deck of Merchants Bridge.

The Gateshead Millennium Bridge *(Clark and Eyre, 2001; Johnson and Curran, 2003; Wells, 2002)*

Coupled with regeneration of the south bank waterfront of the Tyne at Gateshead was a need for a low level pedestrian crossing of the river downstream of the swing bridge built in 1876. Like the swing bridge, however, it needed to be movable to allow the passage of ships. The accepted design involved a torsionally stiff deck of span 105 m supported on the upstream side by stays from a single arch. In order to achieve a 4.5 m clearance over the 30 m wide shipping lane when the bridge is closed, whilst avoiding steep gradients for pedestrians, the deck plan was provided with a substantial curve.

However, the main innovative feature is that both the deck and arch join to hinge about common springing points so that the deck can be raised to provide a 25 m clearance for ships as shown in Figs 6.73 and 6.74. Both the arch and deck are parabolic in shape to match the neighbouring 1928 high level Tyne Bridge, a two-pinned trussed arch which was forerunner to the Sydney Harbour Bridge (as seen in the cover photograph). The way the Gateshead Bridge opens to allow the passage of ships has resulted in it being nicknamed the 'blinking eye'. In receiving a Structural Steel Design Award in 2002, it was described as having 'visual daring and elegance in its closed position, giving way to theatre and power in operation'.

Structure of the Gateshead Bridge

In order to make the bridge easy to open and close, it needed to be as light as possible particularly near the centre of the span where the greatest movements need to take place. Hence both the arch and the deck structure have cross-sections which are lighter nearer the centre span than they are near to the hinges. Although the main purpose of the stays is to relieve the deck of bending moment by transferring forces to the arch, with the given 3D geometry it is inevitable that both the deck and the arch need to carry some bending moments. When the bridge is closed, each end of the deck

Figure 6.73 The Gateshead Millennium Bridge open for river traffic.

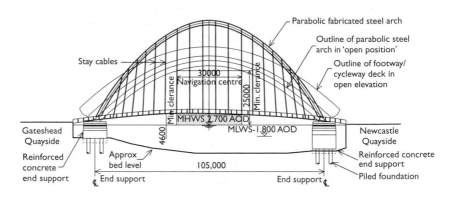

Figure 6.74 Elevation of the Gateshead Bridge.

rests on a pedestal and is also connected via an arm to the nearby springing point of the arch where the hinge is situated. The bridge is rotated by means of rams located below the level of the hinges (Fig. 6.75). As the bridge is opened, because it is not completely counterbalanced, its centre of gravity moves from downstream to upstream of the hinges. Allowing for the possibility of high wind loads, ram forces could change from 10,000 kN push

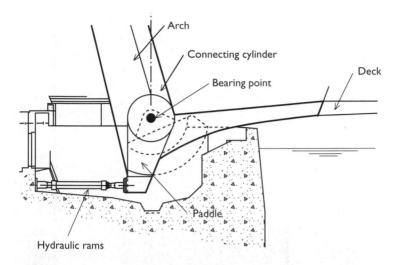

Figure 6.75 Hinge and ram control of the Gateshead Bridge.

to 45,000 kN pull. Figure 6.76 shows how moments due to out of plane forces acting on both the arch and deck react at the springings against the combined moment from the counterbalance weight and the ram forces.

The main deck comprises an unsymmetric stiffened steel box supporting the footway, to which is attached a cycleway at a slightly lower level by means of an aluminium cantilever structure (Fig. 6.77). The arch is also of stiffened steel box construction whose cross-section reduces in size near to the centre span (Fig. 6.78). The structural analysis has therefore been highly complex on account of

- the curvature and non-uniformity of the members;
- the combined compression, bending and torsion loading on the deck;
- the degree of redundancy including the possibility of prestress in the hangers;
- the changes of attitudes of the structure during the opening and closing movements.

Use of computers

The successful completion of such highly 3D innovative structures would be a much more risky task without the aid of sophisticated 3D computer analysis facilities. Computer analysis is not only required for linear static analysis of the structures due to the various loading cases, but to predict

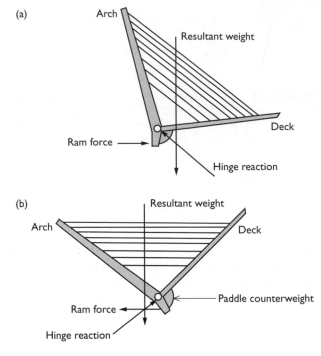

Figure 6.76 Equilibrium of the Gateshead Bridge subject to gravity loading, showing end views: (a) with the bridge just starting to open; (b) with the bridge fully open.

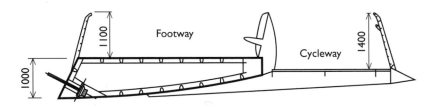

Figure 6.77 Cross-section of the Gateshead Bridge deck (mm units).

non-linear effects relating to buckling and also dynamic behaviour. Computers are increasingly being utilised for other tasks within the design and manufacturing process as well.

However in discussing the design of some innovative modern footbridges, Mairs and Lomax (2002), say that 'Irrespective of the final complexity of the

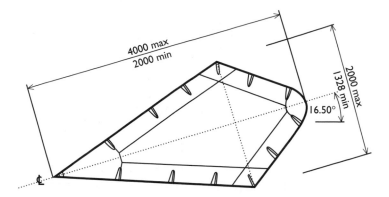

Figure 6.78 Cross-section of the Gateshead Bridge arch (mm units).

structures, hand calculations were carried out to determine the initial sizing of the bridge elements, and hand calculations were performed throughout the design to assist in the validation of computer output'. There is always a need for engineers to appreciate load paths and understand structural actions themselves.

Chapter 7

Stability

Stability can be defined as the power to recover equilibrium and is an essential requirement for all structures. The stability of structures and structural members may increase or decrease with the addition of loading. For instance, in Chapter 2, it was found that masonry structures generally become more stable with increasing dead weight. However, when iron and steel became available in quantity, elastic buckling due to loss of stability of slender members appeared as a particular hazard. Problems arose determining how close it was safe to go to buckling loads for compression members whilst, for tension members, too much reliance was sometimes placed on the stabilising effect of their tensile forces. These considerations are as important for today's structural engineers as they were for the pioneering developers of the great structures of the past.

On completion of this chapter, you should be able to do the following:

- Predict elastic buckling loads for struts and columns with standard end conditions.
- Appreciate the danger of having insufficient stiffness in structural components, which are subject to compression forces.
- Understand that mathematical models to examine buckling characteristics lead to eigenvalue analyses.
- Recognise vibration as an important stability-related phenomenon.
- Understand the stabilising effect of large tensile forces and its influence on the development of suspension bridges.

7.1 Elastic buckling

The Euler buckling load

Consider a pin-ended, uniform and initially straight strut of length ℓ carrying an axial compressive load P (with sign compression positive rather than negative). If the strut deflects laterally such that its curve is $v(x)$ as shown in Fig. 7.1, the horizontal end reactions must remain zero and the bending moment at lengthwise coordinate x can be determined from equilibrium as

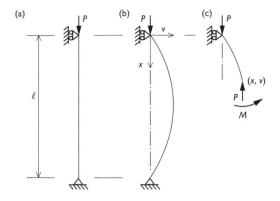

Figure 7.1 Bending of an Euler strut: (a) unbent shape; (b) with lateral deflection (exaggerated); (c) equilibrium of a segment.

$M = -Pv$. Substituting this into the elastic bending formula, Eq. (6.4), gives

$$EI\frac{d^2v}{dx^2} + Pv = 0 \tag{7.1}$$

Whereas this differential equation always has a trivial solution $v = 0$, there are specific values of P for which non-zero solutions exist. With the known end conditions $v = 0$ at $x = 0$ and $x = \ell$, the smallest such load is

$$P = \pi^2\frac{EI}{\ell^2} \tag{7.2}$$

in which case $v = a\sin(\pi x/\ell)$ where a is an indeterminate constant. This load is known as the Euler buckling load and designated P_E.

A historic note

Euler discovered this critical load whilst investigating elastic curves for beam bending. In his 1770 publication, Euler used a constant to describe the linear relationship between curvature and bending moment because the significance of the second moment of area in relation to bending (as specified in Eq. (6.4)) was not known at the time. He stated: 'unless the load P to be borne be greater than (the buckling load), there will be absolutely no fear of bending; on the other hand, if the weight be greater, the column will be unable to resist bending' (Timoshenko, 1983). Figure 7.2(a) shows Euler's envisaged (theoretical) load path for a strut in which lateral deflections remain at zero until the axial compressive force reaches the Euler load. At the buckling load, any magnitude of lateral displacement is in equilibrium (subject to small deflection theory being still applicable) and so the strut proceeds to buckle. The unstable equilibrium situation with $P > P_E$ is thus never reached.

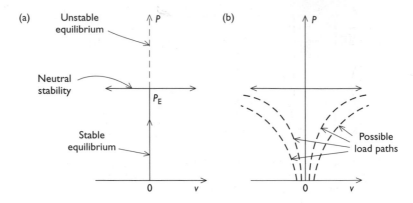

Figure 7.2 Lateral deflection of an Euler strut: (a) ideal; (b) in practice.

In practice, ideal struts cannot be found. They tend not to be perfectly straight, there may be side loads (e.g. due to wind) and the end conditions will not be exactly as assumed. When axial compressive force is increased, lateral deflections due to any extraneous effects will increase to be asymptotic at the buckling load. In a design situation, P_E may be easy to determine, but it is much more difficult to identify which of the load paths shown in Fig. 7.2(b) are likely to be relevant and hence how close it might be safe to go to the buckling load. Thus even this dramatic statement by Euler, a renowned Swiss mathematician, understates its importance. The elastic buckling load is something which engineers only approach close to at their peril. Despite its importance, this and related buckling formula received little attention until large iron structures were being built (for instance the Britannia Tubular Bridge, 1850, mentioned in Section 4.11).

Different end conditions

Pinned-end conditions are used frequently. However, there are several possible end conditions with theoretical buckling loads ranging from $P_E/4$ to $4P_E$ as shown in Fig. 7.3. Mathematical analysis can be used to show that in each case the displacement mode is sinusoidal.

Effective length

By identifying points of inflection for cases (a) and (b) of Fig. 7.3, it is possible to determine the length of strut which acts like an Euler strut. Such a length is called the effective length (see Fig. 7.4). To make a similar comparison for cases (c), (e) and (f), it is necessary to create mirror images of the struts as shown in Fig. 7.5. Using this effective length concept allows the same

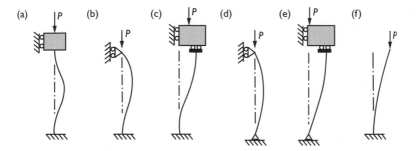

Figure 7.3 Buckling of uniform struts with different end conditions: (a) fixed–fixed, $P_{cr} = 4P_E$; (b) fixed–pinned, $P_{cr} = 2.046P_E$; (c) fixed–fixed allowing sway, $P_{cr} = P_E$; (d) pinned–pinned, $P_{cr} = P_E$; (e) pinned–fixed allowing sway, $P_{cr} = P_E/4$; (f) fixed–free, $P_{cr} = P_E/4$.

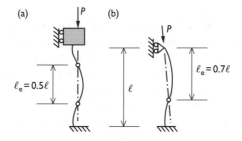

Figure 7.4 Struts with effective length $< \ell$: (a) fixed–fixed; (b) fixed–pinned.

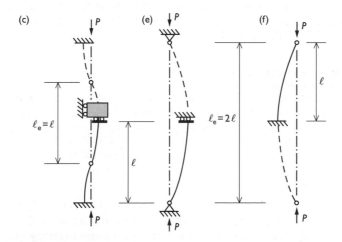

Figure 7.5 Struts with effective length $\geq \ell$: (c) fixed–fixed allowing sway; (e) pinned–fixed allowing sway; (f) fixed–free.

buckling formula to be used for all struts:

$$P_{cr} = \pi^2 \frac{EI}{\ell_e^2} \tag{7.3}$$

where P_{cr} is the critical load and ℓ_e is the effective length.

Ex. 7.1

An overhead crane gantry in the form of a portal frame with the dimensions shown in Fig. 7.6 is to have columns with $E = 200\,\text{kN/mm}^2$ and $I = 650 \times 10^6\,\text{mm}^4$ and a very deep beam which can be considered to be fully stiff in bending. Estimate the magnitude of the load W which, if centrally placed on the beam, would cause in-plane elastic buckling of the whole frame:

a   if the column bases are built-in as shown;
b   if they are both pinned.

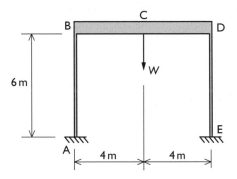

Figure 7.6 An overhead crane gantry.

Answer

Because of the large bending stiffness of the beam, joints B and D will have negligible rotation due to normal portal frame action. The columns will therefore act as struts, effectively built-in at the top but with sway allowed, each supporting a load of $\frac{1}{2}W$.

a   With both ends of each column prevented from rotating, buckling will be by case (c) of Fig. 7.3 with both columns swaying sideways in unison (Fig. 7.7(a)). Thus $\ell_e = \ell$ with the buckling load for each column equal to

$$P_{cr} = \frac{\pi^2 \times 200 \times 650 \times 10^6}{6000^2} = 35,600\,\text{kN}$$

Hence $W_{cr} = 2P_{cr} = 71,200\,\text{kN}$

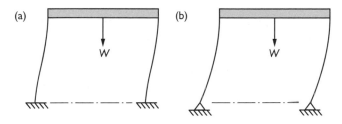

Figure 7.7 Buckling modes for the overhead crane gantry: (a) with fixed bases; (b) with pinned bases.

b If the bases are pinned, the columns will buckle according to case (e) of Fig. 7.3 as shown in Fig. 7.7(b). With $\ell_e = 2\ell$:

$$P_{cr} = \frac{\pi^2 \times 200 \times 650 \times 10^6}{12,000^2} = 8900\,\text{kN}$$

Hence W_{cr} becomes 17,800 kN.

Slenderness ratio

The term radius of gyration r is used to relate second moment of area, I, to cross-sectional area, A, according to $I = Ar^2$ (for the elementary beam, whose concept was introduced in Section 4.5, r is half the distance between the flanges). Writing $P_{cr} = A\sigma_{cr}$ and substituting for I in Eq. (7.3) gives the critical compressive stress at which buckling occurs as being defined by the equation:

$$\sigma_{cr} = \pi^2 E / (\ell_e/r)^2 \qquad (7.4)$$

The ratio ℓ_e/r is a measure of the slenderness of the column and is called the 'slenderness ratio'. A large slenderness ratio results, from Eq. (7.4), in the theoretical buckling load occurring at a low compressive stress. Substituting the approximate E values given in Section 6.1 for different materials into Eq. (7.4) gives the curves shown in Fig. 7.8. When the theoretical buckling stress is of the same order of magnitude or less than the stress at which the material itself fails, buckling will be an overriding design consideration for members loaded in compression.

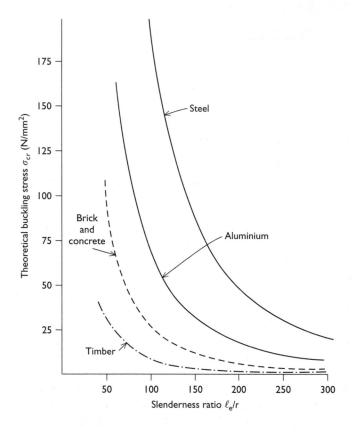

Figure 7.8 Theoretical elastic buckling stress as a function of slenderness ratio for different structural materials.

Ex. 7.2

A member of a pin-jointed truss has length 8 m and is to be fabricated from steel tube of average diameter 200 mm. Determine what thickness the tube needs to be to ensure that its elastic critical buckling load is no less than 900 kN.

Answer
Assuming that the thickness of the tube, t, is small compared with the radius, the second moment of area, I, in mm units (from Fig. 4.22) is $100^3 \pi t$.

However, $A = 200 \pi t$. Hence the radius of gyration is

$$r = (100^3 / 200)^{\frac{1}{2}} = 70.7 \, \text{mm}$$

The slenderness ratio $= 8000/70.7 = 113.2$.
Hence $\sigma_{\text{cr}} = \pi^2 E / 113.2^2 = 154 \, \text{N/mm}^2$ and the critical load is $200 \pi t \times 154 \, \text{N} = 96.8 t \, \text{kN}$.
For this load to be greater than $900 \, \text{kN}$, $t \geq 900/96.8 = 9.3 \, \text{mm}$.

In-plane buckling of trusses and frames

When a truss has all of its joints pinned, there will be no interaction between the lateral bending deflections of individual members. Hence the Euler buckling load of the truss will be the load at which the weakest compression member buckles as an Euler strut. However, when a truss is rigidly jointed, there will be interaction between the lateral deflections of neighbouring members through rotation of the common joint in the manner discussed in Section 6.9. This will stiffen the truss and thereby increase the buckling load. However, it is important to note that two compression members meeting at a joint (e.g. as part of a continuous chord) will not provide mutual support if they are both reaching their own buckling load simultaneously. It is therefore unsafe to assume much enhancement of the buckling load of a truss due to rigidity of the joints unless a computer analysis is used to establish the correct theoretical buckling load.

The lateral stiffness of a frame tends to be reduced significantly by the presence of sway modes. Hence, where the equivalent pin-jointed frame contains one or more mechanisms, buckling will normally involve sway movements. An example of this is the buckling of the overhead crane gantry shown in Fig. 7.7.

Buckling of symmetric structures

When examining the elastic stability of symmetric structures loaded symmetrically, it is essential to recognise that buckling modes will not necessarily be symmetric. However, they must be either symmetric or antisymmetric (e.g. the buckling modes shown in Fig. 7.9). On account of this, it would not normally be a good idea to condense the buckling analysis of a symmetric frame as shown (for equilibrium analysis) in Section 6.10. Instead,

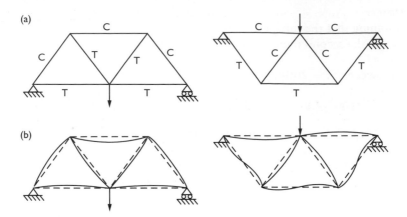

Figure 7.9 Buckling modes for rigidly jointed trusses: (a) loading and internal forces; (b) anticipated buckling modes. (Key: C= compression; T=tension.)

knowledge that the buckling modes should be either symmetric or antisymmetric can be used to check the validity of any analysis involving the whole frame.

Remember the third dimension

A column lying along the x axis can buckle in any direction in the y–z plane, with the actual direction being dictated by whichever offers least bending resistance to possible buckling modes. If the end conditions do not favour buckling in any particular direction (e.g. both ends are built-in), the column will buckle in the direction having the smallest I value, identified by means of the principal second moments of area discussed in Section 4.10. However, analysis for the least principal second moment of area would only be necessary in the unusual event of the column cross-section not having a plane of symmetry. Where a cross-section is symmetric about the y or z axis, the least of I_y and I_z needs to be used. Tubes are ideal for compression members of space trusses because they have good two-way bending characteristics with $I_y = I_z$ (e.g. the space truss supporting the Kemper Arena roof, Figs 6.68 and 6.69).

Where joint restraints are different in two lateral directions and particularly where joints are unrestrained in one of the directions, it is essential to ensure that there is sufficient elastic stability in both directions. Long compression chords of plane trusses rarely have sufficient lateral bending stiffness to stand without some form of lateral bracing. For example, the horizontal members linking the two trusses in Fig. 7.10(a) would do little to restrain lateral buckling of the top (compression) chords. Effective restraint of lateral buckling is achieved by including diagonal bracing into the horizontal panels, thus restraining horizontal joint movements by means of truss action (Fig. 7.10(b)).

(a)   (b)

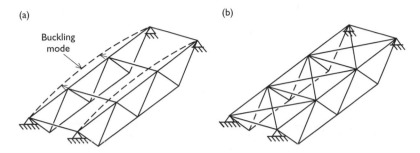

Figure 7.10 Lateral buckling of the compression chords of a girder bridge: (a) inadequate bracing; (b) adequate bracing.

Q.7.1

Figure 7.11 shows the tower, AB, of a cable-stayed bridge supporting cables which link its top to various points along the centre-line of the horizontal deck. If the base of the tower at deck level is built-in and the deck is very stiff when loaded laterally, discuss what effective lengths might be used to predict the buckling characteristics of the tower:

a   in the plane of the cables;
b   laterally.

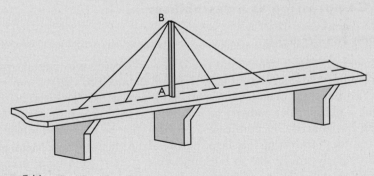

Figure 7.11

Q.7.2

The rigidly jointed Warren truss shown in Fig. 7.12 has members AB, BF and FG fabricated as tubes with outer diameter 140 mm, thickness 16 mm and $E = 200\,kN/mm^2$.

a   What can you say about its elastic critical load when the lower chord joints carry equal loads as shown in the figure.
b   Discuss how the elastic critical load might change if the load applied to joint E is removed.

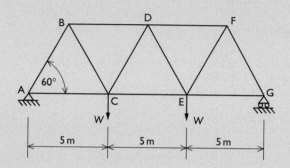

Figure 7.12

Q.7.3   Observation question

Examine the compression chords of any trusses you can find and identify how lateral buckling is prevented.

7.2   Compression as a destabiliser

Stiffness requirements

The Euler buckling of struts is just one of several reasons why lateral stiffness is required to avoid structures side-stepping the need to carry compressive load. The buckling phenomenon can be observed by, for instance, compressing a plastic ruler endwise or extending a steel tape measure as far as possible horizontally. In both cases, the item may be recovered intact after buckling, thus illustrating the elastic nature of the phenomenon. Metal plates are particularly susceptible to buckling when subject to in-plane compressive loading, which is one of the reasons for including stiffeners in panels of thin-walled beams (Section 4.11) and was one of the contributory causes of problems at the Yarra (Section 4.12).

Unlike the differential equation solution for the Euler load, most other buckling analyses are carried out using a discrete set of variables. Some simple problems requiring only one or two variables are discussed here.

Ex. 7.3
Two aligned pin-ended struts AB and BC each of length ℓ are simply-supported at A and C and have a lateral support at B as shown in Fig. 7.13(a). If the struts have a high bending stiffness, determine an expression for the buckling load, P, of the system in terms of ℓ and the stiffness, k, of the support at B.

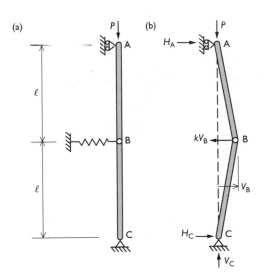

Figure 7.13 Buckling of a system of two struts: (a) undisplaced position; (b) forces acting when displaced.

Answer
If B displaces laterally a small distance v_B, the restoring force supplied by the support is kv_B (Fig. 7.13(b)). From overall equilibrium $V_C = P$ and $H_A = H_C = \frac{1}{2} kv_B$. Taking moments about B for the strut AB gives

$$M_B = H_A\ell - Pv_B$$

However, because of the pin at B, $M_B = 0$.
    Thus $\frac{1}{2} kv_B\ell = Pv_B$ with a displaced position being in equilibrium when $P = \frac{1}{2}k\ell$ (which is the required buckling load).

A multi-degree of freedom problem

Because Ex. 7.3 only had one degree of freedom, it was too simple to be illustrative of typical buckling problems.

Ex. 7.4

Figure 7.14(a) shows a similar problem to the one considered before, but with three pin-ended struts placed end to end and with lateral supports having stiffness k for both internal joints. Determine the buckling load of the system.

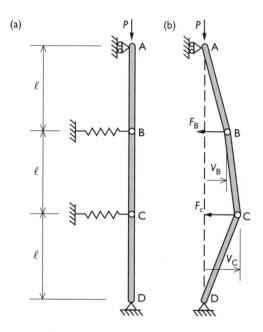

Figure 7.14 A system of three struts: (a) undeflected position; (b) deflected position.

Answer

Figure 7.14(b) shows the system subject to a small lateral movement in which joints B and C displace by v_B and v_C, respectively. The restraining forces at B and C required to maintain equilibrium in this position are

$$F_B = P(2v_B - v_C)/\ell \quad \text{and} \quad F_C = P(-v_B + 2v_C)/\ell$$

Equating these forces to kv_B and kv_C respectively gives

$$\frac{P}{\ell}\begin{bmatrix} 2 & -1 \\ -1 & 2 \end{bmatrix}\begin{bmatrix} v_B \\ v_C \end{bmatrix} = k\begin{bmatrix} v_B \\ v_C \end{bmatrix}$$

Being a 2×2 eigenvalue problem, this has two non-zero solutions for P as follows:

(i)   With $\{v_B\ v_C\}$ in the ratio $\{1 -1\}$, the equation is satisfied when $P = k\ell/3$.

(ii) With $\{v_B\ v_C\}$ in the ratio $\{1\ 1\}$, the equation is satisfied when $P = k\ell$.

Of these two buckling modes, the important one is (i) in so far as it corresponds to the lowest buckling load (see Fig. 7.15).

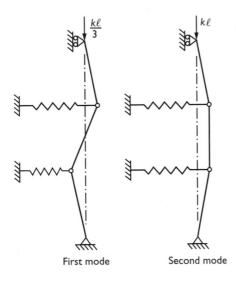

First mode   Second mode

Figure 7.15 Buckling of the system of three struts.

Comment

Many engineers would probably expect the symmetric second mode to be the most critical because of its similarity to the Euler buckling mode. Thus particular care needs to be taken when investigating buckling.

Higher buckling modes

In virtually all structural problems involving buckling, there will be possible buckling modes which correspond to higher loads than the critical one. For the system of three struts, there was just one such mode. The Euler strut also exhibits additional modes. The differential Eq (7.1) is satisfied with the correct end conditions whenever $P = n^2\pi^2 EI/\ell^2$ where n is an integer. Thus the Euler load is one solution (with $n = 1$) of an infinite series of possible solutions as shown in Fig. 7.16. Because a structure becomes unstable when the magnitude of the loading reaches the lowest of these, it may be argued that the higher modes are of no real interest. However, they may be useful in a design context. If the truss shown in Fig. 7.17(a) is

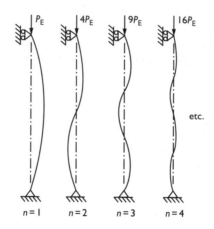

Figure 7.16 Possible buckling modes for a simply-supported column.

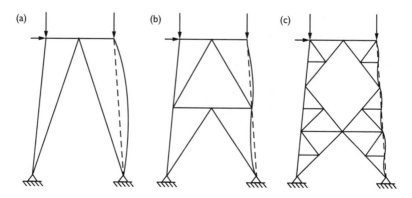

Figure 7.17 Effect of lateral restraints on column buckling: (a) with unbraced columns; (b) with one lateral restraint; (c) with multiple restraints.

considered to be pin-jointed and the buckling load of the column is too low, restraints may be included to prevent the lower buckling modes forming. With one restraint, this is most effectively placed at mid-height, in which case the second mode becomes the most critical (Fig. 7.17(b)). But if this is still not satisfactory, additional restraints could be added (as shown in Fig. 7.17(c)).

Foundation stability

If a tall building rests on soft ground, there is a possible instability mode in which the building is in equilibrium in a tilted position. In the critical state, the moment about the base due to the weight of the building will equal the restoring moment due to the change in ground reaction (see Fig. 7.18). Because soil comprises particles, there is less tendency for elastic recovery than with man-made materials. Hence any small movements caused by wind loading or changes in the ground water level, for instance, are likely to produce a small amount of permanent tilt where the stability of the foundation is poor.

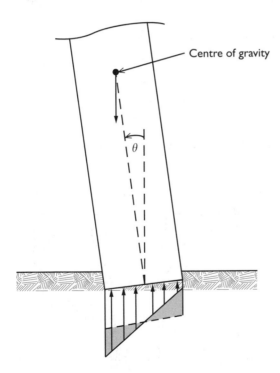

Figure 7.18 Cross-section of a tall building at an angle of tilt θ showing the change in ground pressure due to tilt as shaded.

Q.7.4
Although a straight pin-ended strut AC of length 24 m is very stiff in bending, it contains a weak joint at its centre B with a rotational stiffness (moment/rotation) of 580 kNm. By examining the equilibrium of the strut when B is displaced laterally a distance v_B, as shown in Fig. 7.19, estimate the buckling load.

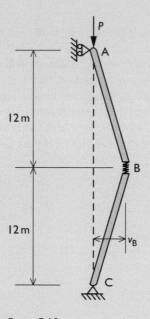

Figure 7.19

Q.7.5
Figure 7.20 shows a similar strut to that of Q.7.4, but with a further length added. The joints at B and C both have a rotational stiffness of 580 kNm. By examining the equilibrium when the lateral displacements of B and C are v_B and v_C respectively, determine the buckling mode and corresponding buckling load.

Q.7.6
A building is to be designed for a maximum total weight of 24,000 kN with the centre of gravity at a height of 32 m above the base, which has dimensions of 12 m × 28 m. By investigating a tilted position as shown in Fig. 7.18, determine the critical value of soil stiffness in kN/m³ (assuming it to be elastic) below which the building will topple.

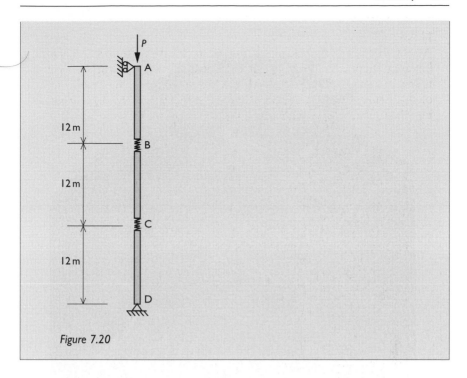

Figure 7.20

7.3 Bridging the St Lawrence

Collapse

The collapse of the partially erected Quebec Bridge was a disaster which had repercussions throughout the engineering profession. At 5.30 p.m. on 29 August 1907, the south arm cantilevered out over the St Lawrence a distance of 223 m with the anchor arm spanning back over the shore a distance of 153 m. The north arm superstructure was only just starting to be erected. When completed, it was to be the largest span bridge in the world. However, in the space of 15 seconds, the south arm lost its fight against gravity. Of 86 workers on or under the arm, only 11 survived from amongst the 17,000 tons of twisted steel (Fig. 7.21).

Countdown to disaster *(quotations from Schneider, 1908)*

The Quebec Bridge Company was set up by act of parliament in 1887 with a capital of one million dollars and powers to raise more by issuing bonds. However, despite the economic benefit that the bridge would provide for Quebec City, progress was severely hampered for 16 years by lack of funds.

Figure 7.21 Onshore wreckage of the south arm of the first Quebec Bridge.

During this period, in 1900, Theodore Cooper was engaged as consulting engineer and a tender by the Phoenix Bridge Company was accepted. However, very little happened till 1903, when more public funding was allocated. At this stage, design and production went into full swing to materialise the Phoenix Bridge Company's plan, which had been deemed to be the best and cheapest of those submitted.

During 1906, there was concern over alignment. Norman McLure, a young site engineer appointed by Cooper, wrote of a number of compression members: 'in sighting from end to end, the webs in places are decidedly crooked, and show up in wavy lines apparently held that way by the lacing angles. This makes a very bad appearance, for a person seeing the member like that, and knowing it to be in compression, would at once infer that it had been overstrained sufficiently to bulge the webs'. In 1907, he had more concerns about bent compression members and there was correspondence about how some of these bends may be eliminated. The Phoenix Bridge Company continuously maintained that these bends arose during manufacture and that they were of no concern. However, on 27 August, McLure reported: 'Only a little over a week ago, I measured one rib of the 9-L chord of anchor arm here shown, and it was only $\frac{3}{4}$ inch (19 mm) out of line. Now it is $2\frac{1}{2}$ inches (57 mm)'.

Erection was at first halted, but then resumed because, as Edward Hoare of the Quebec Bridge Company said: 'the moral effect of holding up the work would be very bad on all concerned, and might also stop the work for this season on account of losing men'. McLure went to New York to report the situation to Cooper, but just before the end of the shift, with the request from Cooper barely opened to 'Add no more load to bridge till after due consideration of facts', the collapse occurred.

Theodore Cooper

The Quebec Bridge Company's own chief engineer, Edward Hoare had not worked on a bridge with a span over 100 m and so it was essential that they engage a bridge engineer of considerable experience. Therefore Cooper, one of the most experienced bridge engineers in North America was appointed as the company's consulting engineer in May 1890. At the age of 60, he expected the bridge to be his last work, the crowning achievement of an elegant career (Tarkov, 1986). However, because of the company's weak financial position, it was not prepared to pay him the full costs of design and supervision. Reluctantly he agreed to be paid a lump sum for preparing a general specification, reporting on plans and tenders submitted for a design and build contract and acting in an advisory capacity during construction.

Although the tenders were submitted for a span of 488 m (1600 ft), Cooper decided to increase the span to 549 m (1800 ft). His reasons were that the piers, being closer to shore would be quicker to build and less vulnerable to damage by heavy ice flows (this may have had nothing to do with the fact that the change would make it the world's largest span bridge). Despite the huge size of this venture, preliminary tests and research studies were not carried out. This was partly due to the financial difficulties of the Quebec Bridge Company, but also because of Cooper's pride and confidence in his ability.

When more finance was available and work on the superstructure was pressing ahead at a great pace, it was discovered in 1906 that, in the rush to get going, the magnitude of the dead load due to self-weight of the bridge had not been re-estimated after the design phase either by the Phoenix Bridge Company or the Quebec Bridge Company. It is normal procedure to re-evaluate the dead load once more information is known about the design details of the individual members and this was particularly important in this case because of the increase in span. It was found that the dead load was approximately 10% more than allowed for. When faced with this problem, and realising the difficulty and cost of any major modifications once much superstructure had been manufactured and some had been erected, Cooper accepted an even higher working stress than had previously been adopted, using as justification the fact that the dead load/live load ratio was higher than for shorter span bridges (modern practice is to put a lower load factor on dead load than on live load, the latter normally being less well defined, giving rise to dynamic effects and creating possibilities of fatigue).

A criticism levelled at Cooper was that he objected to anyone checking his work. When Collingwood Schreiber, Chief Engineer of the Department of Railways and Canals suggested that a competent engineer be employed to approve or correct detailed drawings as necessary, Cooper said, 'This puts me in a position of subordinate, which I cannot accept'. A further criticism was his reluctance to pay site visits. He only came to the site three times, the last being in May 1903. After that he refused to visit on health grounds. This omission was particularly unfortunate because no engineer on site was of sufficient calibre to make authoritative decision.

Overall design

Figure 7.22(a) shows the intended truss configuration in which the central 206 m comprised a suspended span. Design requirements at the time were to keep the stresses within the elastic limits allowing a significant margin for unknown factors and taking particular precautions with compression members. The practice prevailing then was to use compressive stresses of no more than half of the elastic limiting values. If it had not been for the mistake in estimating the dead load, stresses would have been close to the elastic limit leaving little margin for error, these high stresses being most in conflict with the practice then, as regards compression members.

Although the suspended span was to be much larger than those of the Forth Rail Bridge, it was being erected in the same way, that is, by cantilevering out until the two halves met in the centre. This was imposing additional forces on the members of the main trusses until such time as the closure took place and the central 206 m was converted to being simply-supported. At the collapse configuration shown in Fig. 7.22(b), the dead load stresses may have been at least as high as they would have been in the completed bridge. However, the traveller and crane are likely to have produced lower stresses than the design live loading of trains and vehicular traffic. The wind was not strong at the time of the collapse.

(a)

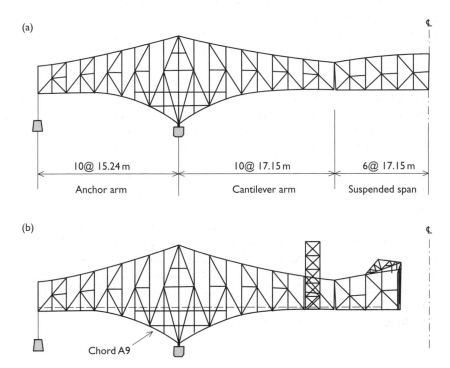

| 10@ 15.24 m | 10@ 17.15 m | 6@ 17.15 m |
| Anchor arm | Cantilever arm | Suspended span |

(b)

Chord A9

Figure 7.22 The first design of the Quebec Bridge: (a) design configuration; (b) situation at collapse of the south arm.

Compression chords

Buckling of both lower chord members A9 (see Fig. 7.22(b)) in the anchor arm was thought to be the prime cause of the collapse. Not only were they found in a buckled condition, but also one cross-bracing member in this panel was broken and the other was buckled. Movement of the shoes supporting the main vertical posts shorewards from the pier suggest that the anchor span rather than the cantilever span was first to give way. Figure 7.23 shows the general arrangement of one of the two chord members. Each chord comprised four plates 1.37 m deep with a made up thickness of over 76 mm (called ribs) in parallel vertical planes with angle members attached at the top and bottom. These longitudinal angles were braced together by a lattice system of angles to create an open box structure.

The area of the cross-section was $504 \times 10^3 \, \text{mm}^2$ with the largest slenderness ratio (corresponding to bending in a vertical plane) equal to approximately 42. From Eq. (7.4), this gives a critical stress estimate of $1100 \, \text{N/mm}^2$ (about 10 times the axial stresses likely to be occurring at the

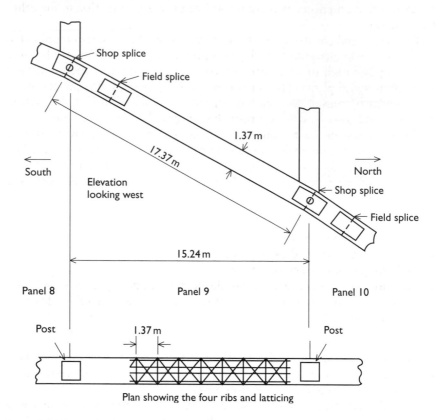

Figure 7.23 First Quebec Bridge design: chord A9.

time of failure). The lattice bracing which comprised light members with a minimum of rivets has been described as 'woefully inadequate'. If, for instance, local buckling occurred in one of the outer plates because of failure of some of the rivets connecting it to the bracing, the theoretical buckling load of the rest of the cross-section would be reduced to 0.4 of its previous value and the neighbouring plate would receive 7/3 times its expected share of any additional load, thus rapidly bringing it to its elastic limit. Once the elastic limit is reached, load–deflection properties deteriorate so reducing the effective buckling load and increasing lateral deflections due to lateral load or eccentricity. Thus the integrity of compression members require all three of the following:

1   They should act as integrated units.
2   Stresses should not be close to the elastic buckling stress.
3   Stresses should not be close to the elastic limit.

It appears that criterion 2 was compromised by lack of attention to the other two criteria.

The design and construction of the compression chords were complicated by the change in direction at each panel point. Here the individual ribs were butted against each other to provide continuity in the load path and held in place by splice plates. The vertical posts were connected to the compression chords by pins. Cover plates, top and bottom, were included across the joint and also across an offset diaphragm. If there was only a very slight difference in the manufactured lengths of the ribs, there would be a tendency for the four ribs in each chord to take up different loads when compressed and for the most heavily loaded to want to deflect laterally. Reasons for bending moments developing in the compression members were as follows:

- lack of straightness in fabrication;
- moments transmitted through the joints;
- self-weight of the member itself;
- manufacturing tolerances in lengths of ribs and positions of rivet holes;
- magnification of lateral displacements as the buckling load is approached.

Hence using significantly higher working stresses for compression members of such size and complexity, without adequate experimental tests, was a form of pioneering not to be recommended.

After the collapse, a one-third scale model of a compression member was tested and it was found to fail by buckling in the horizontal direction. This was despite the radius of gyration being slightly larger than for bending in the vertical direction, but specifically because the lattice bracing and associated riveting was not sufficiently strong and stiff to ensure that the four ribs acted as a single unit.

Re-design

A new superstructure was designed and built utilising the existing pier supports. It was significantly heavier than the original. A K-truss configuration was used with a slightly shorter suspended span of 195 m. Design and fabrication of the compression and tension chords was simplified by making them straight (Fig. 7.24). The suspended span was no longer fabricated by cantilevering out from the ends of the cantilever spans. Instead it was raised as one unit. However, at the first attempt, a casting broke plunging it into the estuary with the loss of 11 lives. There was a delay of one more year clearing the site, manufacturing a new suspended span and lifting it into place.

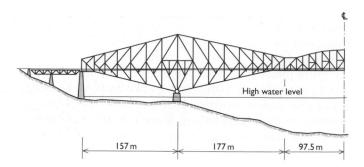

Figure 7.24 The Quebec Bridge completed in 1918.

Never to forget

When civil engineering students graduate at Canadian universities, they are presented with a steel ring to remind them of the lessons of the Quebec Bridge disaster. For those who can, another way of appreciating the problems faced by the bridge engineers and the design solution which eventually proved to be successful, is to walk across the bridge noting its features first hand.

> Q.7.7
> Discuss the statical determinacy or otherwise of the original and revised designs for the Quebec Bridge.

7.4 Sources of instability and vibration

Significance

Compression buckling of straight members is just one of many flexibility-related structural problems which may arise. Ponding instability, mentioned in Section 6.12, is another. Not only are there other forms of instability, but also excessive vibration can render structures unfit for purpose. Theoretical analysis of these phenomenon is often complicated and will not be undertaken here. Instead some qualitative descriptions are given so that readers will be aware of their existence.

Snap-through buckling

Snap-through buckling can occur with shallow arches, thin domes and plates which are almost flat, when inward loading causes a reversal of curvature over all or part of the structure (Allen and Bulson, 1980; Bazant and Cedolin, 1991).

Inward loading on the shallow arches shown in Fig. 7.25 produces large axial compressive forces and hence axial compressive strains. These strains make the arches even shallower, so increasing the ratio of inward movement/increment of load. When the load can no longer be supported in this way, there is snap-through. Snap-through of a dome can be illustrated by applying localised pressure to a table tennis ball. Snap-through of an almost flat plate is illustrated by safety lids on some jars. These are held in the snap-through position by the internal vacuum until such time as the jar is opened.

Load–deflection diagrams for such behaviour are shown in Fig. 7.26. A sudden jump in deflection from A to B occurs when the snap-through load is reached. If after buckling, the load is reduced, the return path is towards the hollow of the load–deflection diagram. In some cases, the buckle will remain even when the load is completely removed and in other cases it will snap-back.

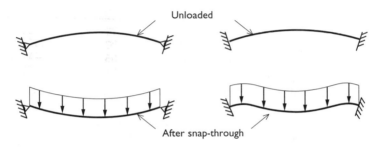

Figure 7.25 Snap-through buckling of shallow arches.

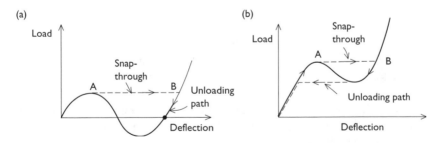

Figure 7.26 Load–deflection curves for different types of snap-through buckling: (a) buckle remains when loading is removed; (b) snap-back occurs when loading is removed.

Lateral–torsional buckling

When a large bending moment is applied to a beam, a way in which it can avoid carrying the load is to twist. The resulting form of instability,

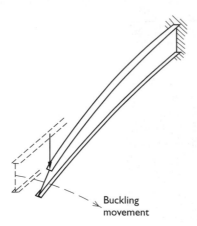

Buckling
movement

Figure 7.27 Lateral–torsional buckling of a cantilever beam.

called lateral–torsional buckling, requires the product of both the torsional and lateral bending stiffnesses of the beam to be low. The twisting action throws a component of the bending moment on to the weak bending axis, thus increasing the lateral displacement and, with it, the twisting moment. When a beam is simply-supported with a vertical loading applied, the top (compression) flange will suffer the greatest lateral deflection. Hence this form of instability might appear to classify as lateral buckling of the compression flange. However, when lateral–torsional buckling of a cantilever beam occurs, it is the tension flange which has the largest lateral deflection (Fig. 7.27).

The vulnerability of deep slender beams to toppling over sideways was witnessed at Barton, near Manchester, in 1959 when a set of long parallel slender beams (depth 5.5 m and flange width 0.7 m at the support), being cantilevered across the Manchester Ship Canal, had inadequate cross-bracing and buckled sideways under their own self-weight. The resulting collapse killed two construction workers and injured others, as well as delaying the work schedule (Fig. 7.28) (Short, 1962). This is an example of how structures are often much more at risk from instability when they are being erected than when they are complete.

Aeroelastic divergence

Up to the stalling angle, the lift provided by an aircraft wing is proportional to its angle of incidence relative to the airstream. If the wing is given a nose up twisting disturbance, the centre of pressure of the resulting increase in lift acts through the centre of lift, a point just behind the quarter chord position, as shown in Fig. 7.29. Because this is normally in front of the shear centre of the wing, its effect is to help to maintain the twist. Also since aerodynamic

Figure 7.28 Collapsed beams at Barton, Manchester.

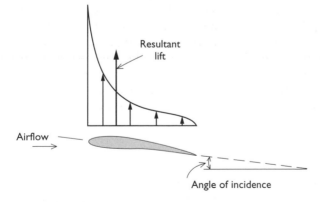

Resultant
lift

Airflow

Angle of incidence

Figure 7.29 Pressure distribution due to increase in incidence of an aircraft wing in subsonic flow.

forces increase with the square of the airspeed, there is normally a speed, called the divergence speed, at which the wing twists off.

Divergence speeds of wings (and also tailplanes and fins) are increased by providing them with a high torsional stiffness and by keeping the shear centre well forward. This type of instability would also be possible for cable-supported bridge decks, although flutter (discussed later in this section) is the form of instability that they seem most vulnerable to in practice.

Resonant vibration

There are normally different frequencies, called 'natural frequencies', at which a structure oscillates about its equilibrium position once it is disturbed. The analysis of undamped vibration, like buckling analysis, can be reduced to an eigenvalue problem. However, in this case, the different eigenvalues correspond to different natural frequencies, with the eigenvectors identifying the corresponding mode shapes of oscillation. A ruler held horizontally, so that it cantilevers out over the edge of a table, will oscillate in a vertical plane at its lowest frequency of vibration as shown in Fig. 7.30(a). There are higher modes of vibration such as those illustrated in Fig. 7.30(b) and there will also be more modes of vibration involving torsion and horizontal vibration (although they will be difficult or impossible to identify with a ruler).

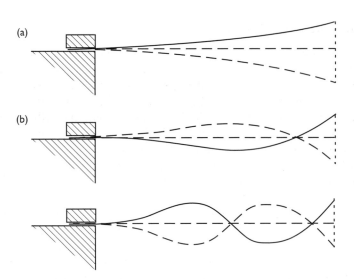

Figure 7.30 Vertical vibration of a cantilever: (a) mode of lowest frequency vibration; (b) higher modes of vibration.

The frequency of any particular vibration mode depends on the ratio of modal stiffness/modal mass. For the cantilever of Fig. 7.30, less force is required to bend the beam into mode (a) than other modes of vibration and hence it has the lowest modal stiffness and the lowest associated vibration frequency. Another important property is the damping. This governs the rate at which an oscillation dies away once the initial disturbance is removed. Even when the material remains within its elastic limit, there will be some damping due to 'hysteresis' (i.e. because the loading and unloading paths of the load–deflection diagram tend to be slightly different). Different materials exhibit different amounts of hysteresis as can be witnessed, for example, by comparing the rate of decay of oscillations for wooden, plastic, aluminium and steel rulers. Damping also comes from friction in joints, so that riveted or bolted structures tend to be more highly damped than welded ones.

Vibration modes are of concern when there is a pulsating force whose frequency is the same as or close to one of the lower natural frequencies of the structure. The amplitude of the resulting resonant response could, in situations where there is little damping, be much larger than the deflection resulting from steady application of the force. Resonant vibration is of concern when machinery is causing pulsating forces close to a natural frequency of a structure. This can be a problem in power stations and in transportation vehicles such as ships and helicopters. Other examples arise when people move in synchronisation with each other, for example at concerts taking place in concert halls, arenas or even out of doors, where they are on temporary tiered stands.

The London Millennium Bridge

Footfall forces may accidentally or deliberately synchronise with a vibration frequency of a footbridge if it is in the region of 2 Hz or less. The effect that vertical footfall forces might have on bridges has been recognised since the nineteenth century when marching soldiers were required to break step on bridges. They were taken into account in the design of London's Millennium Bridge (Fig. 7.31) which is a novel form of shallow suspension bridge with the cables situated mainly below deck level. However, when it was opened on 10 June 2000, it had to be closed again because of horizontal oscillations.

There has been very little prior evidence of horizontal oscillations taking place in resonance with pedestrian footfalls and no research on the topic prior to year 2000. Tests were carried out which revealed that pedestrians are sensitive to horizontal movements and will synchronize the way they walk to the movements of a bridge. The bridge reopened after viscous and tuned dampers were installed to prevent lateral oscillations. These were designed to ensure that the bridge's damping always exceeded any destabilising effects from the people on it. Dallard *et al.* (2001) state that 'The phenomenon of synchronous lateral excitation is not linked to the technical innovations of

Figure 7.31 London's Millennium Bridge.

the Millennium Footbridge' and it 'could occur on any future or existing bridge with a lateral frequency below approximately 1.3 Hz loaded with a sufficient number of pedestrians'.

Vortex shedding

When a circular cylinder is exposed to fluid flow, the air flow separates from the surface on both sides of the cylinder. Circulation of fluid in the wake of the cylinder is in the form of vortices which alternate as shown in Fig. 7.32. Formation of the vortices correspond with movements of the points of separation, so causing pulsating forces to act on the cylinder whose frequency is dependent on the speed of flow. Whereas these forces have components in the direction of flow, the largest components are at right angles to the flow. These forces can produce oscillations of the cylinder or, where the cylinder wall is not very stiff, ovalling oscillations of the cross-section.

Examples in which vortex shedding oscillations occur are telephone wires (which can produce a musical note), submarine periscopes, television antennae, stays and hangers in cable-supported structures, chimneys and the tubes of heat exchangers. The characteristics of this type of vibration are affected by the surface geometry and Scruton found that vibration of circular metal chimney stacks could be virtually eliminated by attaching helical strakes to stabilise the points of separation of the boundary layer (Fig. 7.33) (Simiu and Scanlan, 1996). Vibration caused through vortex

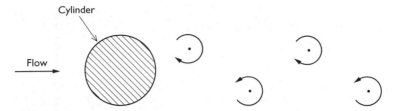

Figure 7.32 A Von Karman 'vortex street.'

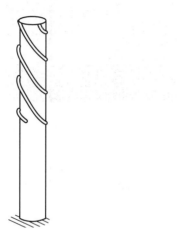

Figure 7.33 Strakes attached to a chimney to inhibit wind-induced oscillations.

shedding is not confined to cylindrical structures. It can also occur with rectangular cross-section and other types of bluff bodies including aerofoils in the stalled condition.

Flutter

Flutter is a form of dynamic instability more dangerous than resonant vibration because it can lead to immediate catastrophic failure. It was encountered in the pioneering days of manned flight and has been a major consideration in aircraft design ever since. It is also a phenomenon which has afflicted power transmission cables and long span cable-supported bridges.

In its simplest form, flutter entails the coupling of a bending and a torsional mode of vibration which oscillate out of phase with each other. The four phases of a complete wing oscillation, shown in Fig. 7.34, constitute

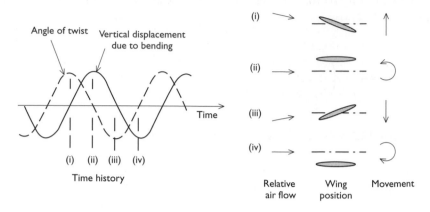

Figure 7.34 Sequence of events in a bending/torsion flutter motion.

a porpoising motion which is self-sustaining as follows:

(i) When the wing moves upwards, its incidence is nose-up to such an extent that the nett force on it is upwards rather than downwards. Hence the aerodynamic lift assists rather than damps the motion.

(ii) When the wing is at the top of its oscillation, it starts to twist nose down on account of the torsion motion being out of phase with the bending oscillation.

(iii) When the wing moves downwards, its incidence is nose-down such that the aerodynamic force again assists the motion.

(iv) When the wing is at the bottom of its oscillation, it starts to twist nose-up, thus being aligned for a repeat (or more violent) oscillation.

Unlike aeroelastic divergence, which is only influenced by the way aerodynamic forces interact with stiffness characteristics, flutter is also affected by aerodynamic and structural damping and the inertia properties. Indeed the position of the centre of gravity of the wing mass relative to the centre of lift is very important, as this affects the way in which the wing twists. Flutter characteristics are improved by

a   sweeping the wings backwards;
b   keeping the mass of the wing as far forwards as possible.

This latter point is a reason why engine pods tend to be mounted in front of wings rather than further back. Darts, having their centre of mass far in front of their tails (their only lifting surfaces), are the most stable of all projectiles. Unfortunately, a bridge deck is exposed to winds coming from any direction and so it is not possible to adopt either of these devices to improve its flutter performance.

The 'galloping' of transmission lines is a form of flutter seen from time to time in regions subject to severe winter weather. It arises when ice builds up on electric cables to form an approximately elliptical cross-section. If this happens in a sleet storm, violent wind-induced oscillations involving vertical amplitudes of 6–10 m may occur.

> *Q.7.8* Observation question
> Clamp one end of a long ruler in a vice so that it becomes a horizontal cantilever with weak lateral bending characteristics. See if lateral–torsional buckling can be induced by hanging a load onto the free end.

7.5 Tension as a stabiliser

Stiffness due to tension

The converse of the destabilising effect of compressive forces is the stabilising effect of tensile forces. This may be illustrated by means of the pendulum shown in Fig. 7.35. If the length of its rod is ℓ and the weight of its bob is W, it may be held with a lateral displacement u by means of a horizontal force P where

$$P = W \tan \theta$$

with $\sin \theta = u/\ell$ and the tension $T = W/\cos \theta$.

However, for small movements $\sin \theta \simeq \tan \theta \simeq \theta$ and $\cos \theta \simeq 1$. Hence $T \simeq W$ and the pendulum exhibits a lateral stiffness of

$$\frac{P}{u} = \frac{T}{\ell}$$

Thus the lateral stiffness of the pendulum is proportional to the tension in its rod.

The hanging chain or cable

Hooke's anagram for the arch, given in Section 2.4, implies that the geometry of chains in equilibrium held no mysteries in the seventeenth century. If a chain supports a set of parallel vertical forces whose lines of action are known, it is possible to use the condition of zero bending moment throughout the chain to identify possible equilibrium positions.

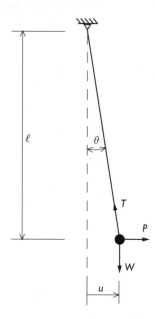

Figure 7.35 Displaced equilibrium of a pendulum.

Ex. 7.5
A cable is to span a horizontal gap of 24 m and support three 10 kN concentrated loads at 6 m spacing as shown in Fig. 7.36. If the maximum sag is to be 4 m, determine the geometry and forces involved.

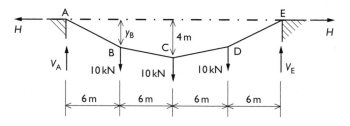

Figure 7.36 A hanging cable.

Answer
Taking moments about E for the whole cable gives $V_A = 15 \, \text{kN}$. Since C will be the lowest point of the cable, it will be 4 m below A. Hence taking moments about C for the segment ABC of the cable gives

the bending moment at C as

$$M_C = 12V_A - 10 \times 6 - 4H = 0$$

Hence $H = 30\,\text{kN}$. Taking moments about B for cable segment AB gives

$$M_C = 6V_A - Hy_B = 0$$

Hence $y_B = 3\,\text{m}$.

For segments AB and DE: length $= (6^2+3^2)^{\frac{1}{2}} = 6.708\,\text{m}$, tensile force$=(30^2 + 15^2)^{\frac{1}{2}} = 30.15\,\text{kN}$ and for segment BC and CD: length $= (6^2 + 1^2)^{\frac{1}{2}} = 6.082\,\text{m}$, tensile force $= (30^2 + 5^2)^{\frac{1}{2}} = 30.41\,\text{kN}$. The support reactions are the same as the forces in segments AB and DE.

Comment

This example was easily solved because the line of action of the applied loads, rather than the lengths of the cable segments were specified. Thus it can be classified as a design problem. If instead, the lengths of the cable segments were to be specified, an analysis to find the alignment of the applied loads would be more difficult, requiring the adoption of an iterative technique.

Cable shape

A useful property to know is that the shape of a cable subject to vertical loading is proportional to the BMD of a simply-supported beam having the same span and loading. If, at the same spanwise position, the cable is distance y below the line joining the supports and M is the bending moment of the simply-supported beam.

$$M = Hy$$

where H is the horizontal component of tension in the cable. This may be verified for the cable of Fig. 7.36 by reference to the BMD shown in Fig. 7.37.

Loss of stiffness

One of the most important design aspects of cable-supported structures is that stiffness is lost when cables go slack. Consider, for instance, a flat roof supported by cables. If wind load could create a larger upwards pressure than the self-weight of the roof, the cables would lose their stiffness and would be unable to prevent the roof distorting grossly or even lifting off (see Fig. 7.38).

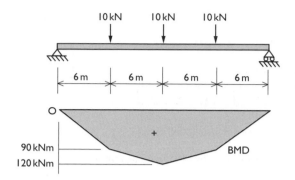

Figure 7.37 A simply-supported beam corresponding to the cable of Fig. 7.36.

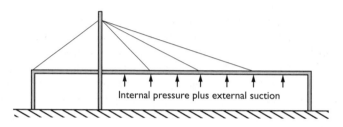

Figure 7.38 A loading case to beware of.

Structural characteristics of suspension bridge cables

The decks of suspension bridges are usually uniform and supported from the cables by vertical hangers. In the 2D analysis given here, it is assumed that the hangers are sufficiently closely spaced and the hanger inclinations sufficiently small that the dead load acting on the cables is both uniform and vertical. By analogy with the equivalent simply-supported beam subject to a uniformly distributed load, it follows that the shape of the cable will be parabolic when only the dead load is acting. If w is the weight/unit span, ℓ is the span and d is the dip of the cable measured from equal height tower tops, taking moments about the centre of the cable for the left-hand cable segment (Fig. 7.39) gives the horizontal component of cable tension as

$$H = \frac{w\ell^2}{8d}$$

When $\ell > 8d$ (which is normally the case) the tension in the cables is greater than the total load they have to support. Because of the very large tension in them, suspension bridge cables have a high inherent stiffness which has been called gravity stiffness (Jennings, 1983).

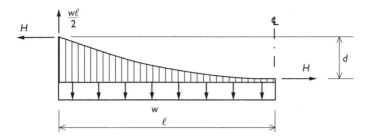

Figure 7.39 Dead load equilibrium of a suspension bridge cable.

To examine the stiffness of suspension bridge cables, consider the addition of a concentrated load P, acting at distance x from the left-hand tower, whose magnitude is small compared with the dead load (Fig. 7.40(a)). The BMD of the corresponding simply-supported beam is now the original parabola plus a triangle with maximum value $Px(\ell - x)/\ell$ at the position of the concentrated load. If H remains unaltered by the addition of the concentrated load, the distance of the cable below the line joining the tower tops will be the original parabola shown in Fig. 7.39 plus a triangle of maximum depth $Px(\ell - x)/H\ell$. However, because the hanger lengths vary across the span to match the parabolic profile of the cables, the only movement transmitted to the deck will be the triangle resulting from the concentrated load as shown in Fig. 7.40(b). This figure also indicates that the downwards movement of the cable draws the tower tops inwards.

If, on the other hand, H increases causing the curvature of the cable to decrease, the deck will develop a uniform hogging curvature in addition to the movement shown in Fig. 7.40(b). The extreme situation is when movements of the tower tops are fully restrained and the cable itself has negligible axial strain, in which case it can be shown that the total area of the deck deflection diagram must be zero as shown in Fig. 7.40(c). Where there is a single main span with the cables continuing over the tower tops to on-shore anchorages, the cable stiffness characteristics are much closer to those of Fig. 7.40(c) than Fig. 7.40(b) and most suspension bridges are one of the two types shown in Fig. 7.41. However, the French have built suspension bridges with more than one main span by adding tie cables to restrict deflection of the tower tops (Fig. 7.42).

When the tower tops are fully restrained, the lowest stiffness occurs when a concentrated load is placed at $x = 0.21\ell$, giving a value of $12H/\ell$. Hence using Eq. (7.5), the stiffness $P/v = 1.5w\ell/d$, from which it follows that the gravity stiffness is increased by increasing the dead weight or by increasing the span/dip ratio. The Severn Suspension Bridge has a main span of 990 m, suspended side spans of 330 m, a span/dip ratio of 12 and a dead load when

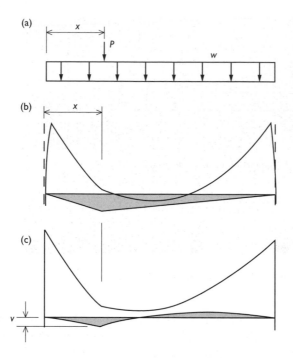

Figure 7.40 Deflections of a suspension cable subject to a concentrated load (exaggerated vertical scale): (a) loading; (b) with *H* unaltered (tower tops being pulled inwards); (c) with tower tops fully restrained.

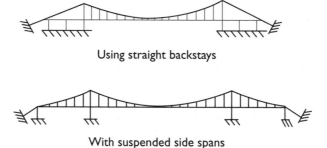

Using straight backstays

With suspended side spans

Figure 7.41 Use of cables to restrain tower top movements of suspension bridges.

it was built of 123 kN/m. Hence for the dead load case, $H = 182,700$ kN and the maximum deflection for a 100 kN concentrated load calculated using gravity stiffness is approximately 50 mm (which is 1/20,000 of the span). Other significant influences are the extensibility of the cables which

Figure 7.42 Use of 'cables de tête' over the Rhone at Condrieu.

increases deflections and the bending stiffness of the deck which decreases them. However, for this particular bridge, gravity stiffness is by far the most important factor in determining stiffness (Jennings, 1983). The most severe deflections usually arise when distributed loading acts over a half of the main span, as shown in Fig. 7.43, unless horizontal tie connections are provided between the cables and the deck at centre span.

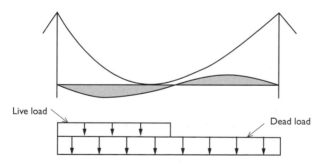

Figure 7.43 Deflections (exaggerated) due to half-span loading.

Q.7.9
A cable ABCDE is to span a gap of 32 m with the support E being at height 8 m above the support A. It is required to carry concentrated

loads of 5 kN, 10 kN and 5 kN at spanwise distances of 8 m, 16 m and 24 m, respectively.

a   If the cable is not to drop below the level of A, determine the smallest horizontal component of cable tension which can be adopted, the resulting geometry and also the maximum tensile force in the cable.

b   Determine the BMD for the equivalent simply-supported beam (having supports at the same level) and show how this is related to the profile of the cable.

Q.7.10
Prove that the shape of a cable AB subject to vertical loads is proportional to the BMD of a simply-supported beam having the same span and loading. If you can manage this with supports at the same level, can you prove it when supports are at different levels?

7.6   The suspension bridge saga

Early vehicular bridges

Although there is a long history of rope foot bridges (some using iron chains) in China and the Himalayas in particular, the birth of the modern suspension bridge is attributed to James Finlay at Jacob's Creek, USA in 1801. His 21 m span bridge had a level deck with stiffening girder supported by vertical hangers from two parallel parabolic chains passing over two towers and held at the ends by ground anchors. Except for the use of chains rather than cables, this matches the principal features of modern long span suspension bridges. The benefit of the stiffening girder to the chains was that it prevented the large movements that would otherwise occur from heavy loads. The benefit of the chains to the stiffening girder was that they removed its need to carry dead load and also reduced the bending moments arising from live load considerably. This 'belt and braces' form of support seemed to be effective with the chains and stiffening girder enhancing each other's performance and with large compressive forces restricted to the tower supports.

Captain Samuel Brown in Britain and the Seguin brothers in France were the earliest pioneers in their respective countries. Whereas a few of the early bridges have survived (Fig. 7.44), there were many which came to grief because of poor engineering or corrosion on the one hand, or vibration arising from traffic, marching soldiers or wind on the other. Furthermore, they were found to be unsuitable for railways because of the heavy live

Figure 7.44 Samuel Brown's Union Bridge across the Tweed, 1820. At 137 m it was once the longest span bridge in the world and it is the only one of his to survive.

loading and the sensitivity of the locomotives to changes in gradient, loading cases such as the one shown in Fig. 7.43 being particularly onerous.

The Roeblings

The most successful of the pioneers in the later part of the nineteenth century was John Roebling. He used cables made from parallel galvanised steel wire strands strung in place and spirally wrapped with wire for corrosion protection. Continuity in the wire strands of the cable proved ideal to carry large tensile forces as it avoided the need for linkages with their possibility of fracture. Figure 7.45 shows a cable built to support a canal aqueduct at Lackawaxan on the Delaware River in 1845. After the canal fell into disuse, the bridge was used for local vehicular traffic. Since this photograph was taken, the bridge has been renovated as a historic monument.

Figure 7.45 A Roebling suspension cable after over 130 years of neglect.

On his larger spans, Roebling supported the stiffening girder by means of diagonal stays radiating from the tower tops in addition to the hangers linking to the cable. Even this belt and two braces type of approach did not work well for a railway bridge across the Niagara Gorge, but was successful for major bridges carrying road and light rail traffic. The most celebrated of these was the 486 m span Brooklyn bridge completed in 1883, the towers of which dominated the New York skyline at the time of its construction (Fig. 7.46). The story of its building is one of dogged determination against adversity with John Roebling dying from a site accident and his son Washington Roebling, who carried on the work, being invalided through contracting what was then the mysterious 'caisson disease' (now familiar to divers as 'the bends') (McCullough, 1982).

American long span bridges of the twentieth century

The success of the Roeblings' bridges was one springboard for the development of much longer span suspension bridges for vehicular traffic in the

Figure 7.46 On the Brooklyn Bridge.

twentieth century. Another springboard was Melan's deflection theory published in 1888 and translated from German to English by Steinman in 1906 (Steinman, 1929). Melan developed a differential equation from which the deflection characteristics of suspension bridges could be predicted if they had a stiffening girder supported from cables by vertical hangers. Ammann's George Washington Bridge over the Hudson River at New York, completed in 1931, was designed using the deflection theory and, at 1067 m, was almost double the span of any previous bridge. Other bridges followed including, six years later, the Golden Gate at San Francisco with a span of 1280 m. Diagonal stays were no longer thought to be necessary and, for these very long spans, the deck girder played little part in resisting the gravity movements of the cables. Thus the weight of the deck girder appeared to be more important than its bending stiffness and relatively shallow girders attracted less bending moment when gravity movements of the cables took place.

Repercussions from Tacoma

The situation changed in 1940 when vibrations caused by a steady wind of only 68 kph (42 mph) brought down the Tacoma Narrows Bridge in Washington State completed earlier in the same year. The bridge had been well designed to carry all the anticipated loads in a static sense, but the designers had not considered the possibility of dynamic instability. If they had appreciated that several of the early suspension bridges had collapsed due to wind-induced vibration, they did not think that there could be a similar problem for their much longer span and heavier bridges. Nor did they appear to be aware of problems of flutter which, by then, aeronautical engineers were mastering.

A suspension bridge deck is effectively a long wing, which is most liable to flutter if vertical and torsional modes of vibration exist which are not only similar in shape to each other, but also have similar frequencies. The lowest frequency of bending vibration for classical forms of suspension bridge normally has a mode in which the main span deforms in a full sine wave. Furthermore, if the deck girder stiffness is provided by two beams or trusses in the same planes as the cables, there is likely to be a similarly shaped torsional mode whose frequency is also similar. In such a situation, a coupled bending/torsion flutter motion is easy to excite. The first Tacoma Narrows Bridge had plate girders on either side of the roadway which complicated the aerodynamics of the deck in wind and caused buffeting. Despite this complication, the mode of vibration which caused failure was a full sine wave with a very strong torsional component (Fig. 7.47).

Since 1940, the importance has been recognised of providing decks with good torsional stiffness. This improves the flutter characteristics by increasing the torsional frequencies, so separating them from the vertical bending frequencies which have similar mode shapes. Several bridges existing at the time of the Tacoma Narrows failure were found to vibrate and had to be modified (Fig. 7.48). Vertical movements of 3.3 m and horizontal swinging through 3.7 m were recorded in the deck of the Golden Gate Bridge during a 111 kph (69 mph) gale in 1951. Since then extra bracing has been added between the lower chords of the deck trusses to convert it into a torsionally stiff closed section.

Taking account of the lessons from Tacoma, the Severn Bridge was designed in the UK to have a shallow box girder deck which was both torsionally stiff and aerodynamically streamlined (Fig. 7.49). The top surface of the girder was used as a platform for the roadway, thus making the bridge both lighter and cheaper than the Forth Road Bridge completed with a very similar span two years previously (Roberts, G., 1968). Since then, box girder decks have been used for many major suspension bridges.

Figure 7.47 Flutter vibration of the first Tacoma Narrows Bridge.

Figure 7.48 The Bronx-Whitestone Bridge showing truss additions to the original stiffening girders and also diagonal wind stays connecting the deck to the tower tops.

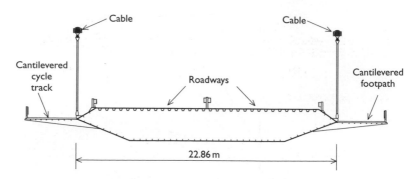

Figure 7.49 Cross-section of the Severn Bridge deck (1966).

Cable-stayed bridges

German engineers faced with rebuilding 4808 bridges following the Second World War, not only pioneered box girder construction, but also developed cable-staying as a method of extending spans across the Rhine and other broad rivers (Walther *et al.*, 1988). To the general public, there may not seem a lot of difference between cable-stayed and suspension bridges, but they are quite different structurally. Cable-stayed bridges are forms of cantilever bridge in which their decks form the compression chords and the tension chords are separated into individual tie cables linking the deck to the towers. Some advantages relative to suspension bridges are as follows:

- Once the towers are constructed, deck erection can take place without waiting for cables to be spun.
- Cable anchorages are not required.
- There is more scope for varying the geometry to suit site requirements and to control stiffness characteristics.

And some disadvantages:

- The towers need to be taller in comparison with the span.
- There is no advantage to be gained from gravity stiffness (the stabilising effect of cable tension is negated by the destabilising effect of deck compression).
- Whereas suspension bridges are generally acclaimed for their graceful lines, cable-stayed bridges appear more utilitarian.

Although some bridges have been constructed with a geometry matching that of classical suspension bridges, that is, having equal height towers at either end of the main span which carry two planes of cables supporting each side of the deck, others support the deck along its centre-line by means of one plane of cables carried on one tower (Fig. 7.50). Another innovation in

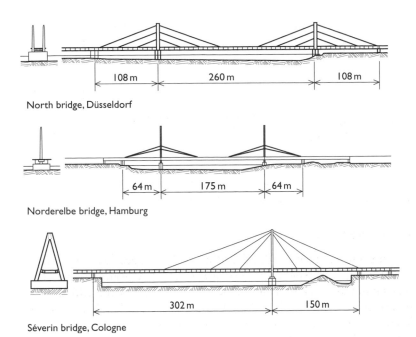

North bridge, Düsseldorf

Norderelbe bridge, Hamburg

Séverin bridge, Cologne

Figure 7.50 Three early cable-stayed bridges showing cross-sections at the towers.

Figure 7.51 A cable-stayed swing bridge at Glasgow.

construction has been to build a bridge on the bank of the river and swing it across the river by rotating the towers once the cantilever is complete. Indeed cable-stayed bridges have been developed as swing bridges to allow the passage of ships (Fig. 7.51).

Shirley-Smith (1964) said that cable-stayed bridges proved to be the most economical on the Rhine for spans of 200–300 m with suspension bridges being the most economical beyond that. Since then cable-stayed bridges of larger span have been built (e.g. the Pont de Normandie, 856 m). Although there has been much recent innovation in the development of cable-stayed bridges, suspension bridges may still have a future, even for shorter spans as indicated by the London Millennium Bridge with a main span of 144 m. It is risky to predict future trends in bridge building, the prime illustration being Drewry (1832) who said 'it may be safely pronounced that bar chains are better adapted than wires for anything beyond the size of a footbridge'.

Chapter 8

Energy concepts

Energy methods are useful tools in the kit of the structural analyst. They were particularly valuable for hand solution of elastic structures in determining deflections and solving for redundancies. Although their use for this purpose is now less important, they are invaluable in the formulation of computer methods, especially for more complex analyses such as those which include non-linear effects and 2D and 3D finite elements. One of their attributes is that they can often provide a direct means of obtaining specific results and another is in facilitating approximate solutions. They also help by providing insight into structural behaviour through energy-related concepts such as Maxwell's reciprocal theorem and Müller–Breslau's principle.

On completion of this chapter, you should be able to do the following:

- Use energy methods to obtain equilibrium equations for both statically determinate and indeterminate systems.
- Determine deflections of linear elastic statically determinate structures using energy methods.
- Use approximate methods for the analysis of buckling.
- Appreciate the connection between virtual work and potential energy methods of analysis.
- Understand energy-related properties of stiffness and flexibility matrices and their relationship to structural behaviour.
- Determine or sketch influence lines and appreciate their significance in design.

8.1 Virtual work

The basic concept

Virtual work, as a concept, derives principally from the work of Jean Bernoulli in the early eighteenth century. When applied to a mechanical system (i.e. one in which the components do not distort), it is found that the forces acting on a system in equilibrium do no nett work when the system suffers a small movement. This result is conditional on the forces not changing their magnitude or direction.

A simple illustration is a block resting on an inclined plane, Fig. 1.11. If the block is given a small displacement du' up the plane as shown in Fig. 8.1, the reaction displaces by $du' \sin 20°$ and the weight by $-du' \sin 30°$. Hence, for the nett work done to be zero,

$$P\,du' + R\,du' \sin 20° - 60\,du' \sin 30° = 0$$

Furthermore, if the block is given a small displacement dv' perpendicular to the plane and the forces remain unaltered:

$$R\,dv' \cos 20° - 60\,dv' \cos 30° = 0$$

Cancelling du' and dv' from these two equations gives two valid equilibrium equations from which the unknown forces may be determined. Here du' represents a feasible movement of the block but dv' does not. The force system is thus considered without reference to its physical origin and the work done is termed 'virtual'.

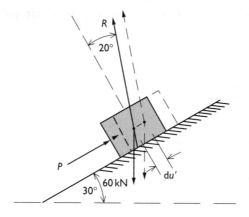

Figure 8.1 Movement of a system of forces in equilibrium.

Application to truss equilibrium

Otto Mohr extended the virtual work method to trusses in the nineteenth century at a time when iron and steel trusses were starting to be used extensively.

If member i of a truss carries an axial force p_i, the amount of work required to stretch it by a distance de_i is $p_i de_i$. For the whole truss, the work required is thus $\sum p_i\,de_i$. It is found that, for the truss to be in equilibrium, the external forces need to do precisely this amount of nett work. If a typical external force f_j is moved a distance du_j in its direction of action, the virtual work equation is

$$\sum f_j \, du_j = \sum p_i \, de_i \tag{8.1}$$

The power of this technique for obtaining equilibrium equations is that virtual displacements can take any form provided that compatibility conditions are always satisfied.

Ex. 8.1
Figure 8.2 shows a Warren truss with a single applied load. Obtain equilibrium equations from the following three virtual displacements:

a   A downwards vertical displacement Δ at C with no other joint moving.
b   A bodily rotation of the two panels ABDC about A such that C drops a distance Δ with joint E not moving.
c   A rotation of the whole truss about A such that C drops a distance Δ.

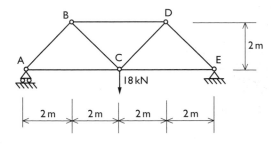

Figure 8.2 A seven bar truss.

Answer
The three virtual displacements (exaggerated) are as shown in Fig. 8.3.

a   Members BC and CD both extend by 0.707Δ with no other member changing length.
    Hence the virtual work equation is

$$18\Delta = (P_{BC} + P_{CD}) \times 0.707\Delta$$

giving the equilibrium equation

$$P_{BC} + P_{CD} = 25.46\,\text{kN}$$

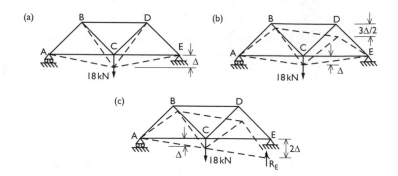

Figure 8.3 Different virtual displacements of the seven bar truss.

b   The two panels ABDC must rotate through an angle $\Delta/4$ (where Δ is in metres). Taking account of the coordinates of D relative to A, it will drop a distance 1.5Δ and move to the right a distance $\frac{1}{2}\Delta$. Hence member DE reduces in length by $(1.5 + 0.5) \times 0.707$ with no other member changing length. The virtual work equation becomes

$$18\Delta = -P_{DE} \times 1.414\Delta$$

giving $P_{DE} = -12.73\,\text{kN}$.

c   The internal work is zero, but joint E contributes to the external virtual work. The virtual work equation is $18\Delta - R_E \times 2\Delta = 0$. Hence, $R_E = 9\,\text{kN}$.

Comment
Thus, by appropriate choice of the virtual displacement, it is possible to obtain, not only joint equilibrium equations and reactions, but also member forces. In case (b), the member force was obtained directly without first needing to determine either of the support reactions. However, it is only possible to obtain member forces directly where the truss (or the relevant part of it) is statically determinate.

Application to beam and frame equilibrium

At a cross-section where the bending moment is M, the amount of work required to give a virtual rotation $d\theta$ to a beam is $M\,d\theta$. Hence, if equilibrium equations are required which involve bending moments at specific cross-sections, these may be obtained by using virtual deflections which involve changes of slope there. A more general equation to Eq. (8.1) covering different types of frame would be

$$\sum f_j\,du_j = \sum p_i\,de_i + \sum m_k\,d\theta_k \qquad (8.2)$$

Ex. 8.2

For a beam ABCD which is simply-supported at A and D, determine an expression for the bending moment at B due to a load P applied at C, where AB, CD and AD are x, b and ℓ respectively.

Answer

If a virtual displacement is given to the beam such that the displacement is Δ at B and the only beam deformation is a sagging rotation θ_B at B as shown in Fig. 8.4,

$$\theta_B = \theta_{AB} + \theta_{BD} = \frac{\Delta}{x} + \frac{\Delta}{\ell - x} = \frac{\Delta\ell}{x(\ell - x)}$$

and the deflection of C is $\Delta' = \Delta b/(\ell - x)$.

Hence substituting in the virtual work equation $P\Delta' = M\theta_B$ gives $M = Pbx/\ell$.

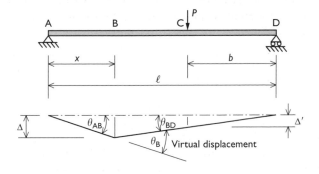

Figure 8.4 Use of virtual work to obtain a bending moment.

Ex. 8.3

For statically indeterminate structures, it is possible to use virtual work to obtain relationships between internal forces. For the pin-based pitched roof portal frame shown in Fig. 8.5(a), determine a relationship between the bending moments at joints C and G for the given loading.

NB   Where sway movements of frames occur, it is normal in hand calculations to ignore axial extensions of members and associated virtual work terms because their effect is likely to be negligible.

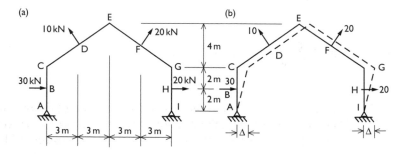

Figure 8.5 A statically indeterminate frame: (a) geometry and loading; (b) a virtual displacement.

Answer

Figure 8.5(b) shows a virtual displacement in which the only distortions of the members are rotations at C and G. With C and G each deflecting horizontally a distance Δ, the rotations of members AC and GI, and hence of joints C and G are $\Delta/4$. The displacements, in their directions of action, of the applied forces are as follows: $F_B, \frac{1}{2}\Delta$; $F_D, -0.555\Delta$; $F_F, 0.555\Delta$; $F_H, \frac{1}{2}\Delta$. Hence with M_C and M_G defined as being positive when the adjacent rafter members have a sagging bending moment, the virtual work equation is

$$M_C\Delta/4 - M_G\Delta/4 = 30 \times 0.5\Delta - 10 \times 0.555\Delta + 20 \times 0.555\Delta$$
$$+ 20 \times 0.5\Delta$$

giving the equilibrium equation:

$$M_C - M_G = 122.2\,\text{kNm}$$

Linear elastic deflections

Instead of applying a virtual displacement to an actual force system, it is also possible to apply a virtual force system to an actual displacement pattern in order to obtain compatibility equations or displacements. This inversion of roles is only valid, however, if the structure being analysed is loaded within its linear elastic range.

In order to determine a specific deflection, u, of a statically determinate structure, a virtual force (conveniently of unit magnitude) needs to be applied in the alignment of the required deflection. If, due to the unit force, member i carries an axial (virtual) force $\bar{p}_i$ and a (virtual) bending moment $\bar{m}_k$ at position x_k (whose magnitude will vary along its length), equating internal and external work gives the following replacement for Eq. (8.2):

$$u_j = \sum \bar{p}_i e_i + \sum \int \bar{m}_k \frac{d^2 v}{dx_k^2} dx_k \qquad (8.3)$$

Here e_i is the actual extension of member i and the integration of the product of the virtual bending moment with the actual curvature is necessary for every member subject to bending action. Because the member extensions of bar members and the curvatures of beam members will normally be obtained from the forces and bending moments induced by the actual loading, a more convenient form of Eq. (8.3) is

$$u_j = \sum \frac{\bar{p}_i p_i \ell}{EA_i} + \sum \int \frac{\bar{m}_k m_k \, dx}{EI} \qquad (8.4)$$

Ex. 8.4

The truss shown in Fig. 8.6(a) has members having $E = 200 \, \text{kN/mm}^2$ and $A = 5000 \, \text{mm}^2$. Assuming linear elastic behaviour, use virtual work to predict the deflection of C.

Answer

From joint equilibrium, member forces are as shown in Fig. 8.6(b) for the given loading and in Fig. 8.6(c) for the virtual loading. With $EA = 10^6 \, \text{kN}$ for all members, the contributions to the internal virtual work are as shown in Table 8.1, from which the vertical deflection of C is found to be 8.2 mm.

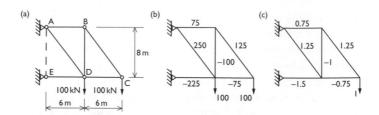

Figure 8.6 Deflection calculations for a statically determinate truss: (a) geometry and loading; (b) member forces (kN); (c) member forces for a unit (virtual) force.

Table 8.1 Internal virtual work for the truss of Fig. 8.6

Member	ℓ/EA (mm/kN)	p_i (kN)	$\bar{p}_i$	$p_i\bar{p}_i\ell_i/EA$ (mm)
AB	0.006	75	0.75	0.34
BC	0.010	125	1.25	1.56
CD	0.006	−75	−0.75	0.34
DE	0.006	−225	−1.5	2.03
AD	0.010	250	1.25	3.13
BD	0.008	−100	−1	0.80
Σ				8.20

Ex. 8.5

Use virtual work to obtain an expression for the central deflection of a uniform beam ABC of length ℓ and bending stiffness EI which is simply-supported at A and C and carries a central load W at B.

NB   The product integrals shown in Table 8.2 are useful for calculating bending moment contributions to Eq. (8.4) when members are uniform.

Answer

The BMDs for the actual loading and the virtual load are shown in Fig. 8.7. Length AB of the beam has BMDs to match row 2 and column 2 of Table 8.2. Hence with $\ell' = \frac{1}{2}\ell$, the virtual work contribution of AB is

$$\frac{\ell}{2} \times \frac{\ell}{4} \times \frac{W\ell}{4} \times \frac{1}{3EI}$$

Adding in the similar contribution from length BC gives the total virtual work (and hence the required deflection) as $W\ell^3/48EI$, which agrees with the result quoted in Fig. 6.10.

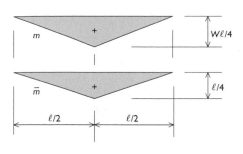

Figure 8.7 BMDs for Ex. 8.5.

Table 8.2 Product integrals for use in the virtual work method

$\bar{m}$	m		
	t ▭ ℓ'	t ◣ ℓ'	t ⌒ ℓ'
s ▭ ℓ'	$\ell'st$	$\ell'st/2$	$\ell'st/3$
s ◣ ℓ'	$\ell'st/2$	$\ell'st/3$	$\ell'st/3$
◢ s ℓ'	$\ell'st/2$	$\ell'st/6$	$\ell st/3$

Ex. 8.6
Use virtual work to estimate the deflection of the free end of the uniform simply-supported timber beam shown in Fig. 8.8 if $E = 16\,\text{kN/mm}^2$ and $I = 12 \times 10^6\,\text{mm}^4$.

Answer
From overall equilibrium $R_C = 5.4\,\text{kN}, R_E = 1.8\,\text{kN}$ and from the method of sections $M_B = -0.6\,\text{kNm}$, $M_C = -2.4\,\text{kNm}$ and $M_D = 1.2\,\text{kNm}$ giving the BMD shown in Fig. 8.8. Due to a unit download applied at A, the BMD is as specified for $\bar{m}$ in Fig. 8.8. By separating the parabolic and triangular parts of the actual BMD, the virtual work integral can be determined as shown in Table 8.3 by using the integrals given in Table 8.2. Dividing the integral shown in the table by EI gives the required deflection of A as 12.5 mm (downwards).

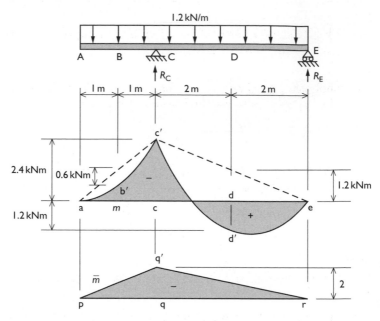

Figure 8.8 A statically determinate beam with BMDs for Ex. 8.6.

Table 8.3 Internal virtual work for the beam of Fig. 8.8

$\overline{m}$	m	$\ell(m)$	$s(m)$	$t(kNm)$	Integral (kNm^3)
$\Delta pqq'$	$\Delta acc'$	2	−2	−2.4	3.2
$\Delta pqq'$	$ac'b'$	2	−2	0.6	−0.8
$\Delta qq'r$	$\Delta cc'e$	4	−2	−2.4	6.4
$\Delta qq'r$	$c'd'e$	4	−2	2.4	−6.4
Σ					2.4

Q.8.1
For the crane truss shown in Fig. 6.31 use virtual work to determine

a   the vertical equilibrium equation for joint D;
b   the force in member BC;
c   the vertical reaction of A.

Q.8.2
Devise a check for the answer to Ex. 8.3 without using virtual work.

Q.8.3
For the simply-supported bent beam with overhangs shown in Fig. 8.9, use virtual work to obtain the bending moment acting at C.

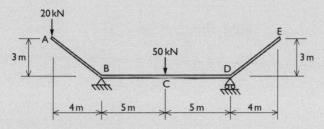

Figure 8.9

Q.8.4
For the truss shown in Fig. 8.6, use virtual work to estimate the vertical deflection of D.

Q.8.5
Use the member forces for the crane truss shown in Fig. 6.31 to determine, using virtual work, an estimate for the horizontal deflection of D based on an *EA* value of 600,000 kN.

Q.8.6
Figure 8.10 shows a uniform simply-supported beam with overhanging ends. Use virtual work to predict its maximum downwards deflection as a function of the bending stiffness *EI*, given the dimensions and loading shown.

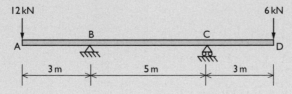

Figure 8.10

Q.8.7
Use virtual work to find an expression for the horizontal spread of joint
D of the frame of Fig. 8.11 due to the specified uniformly distributed
load. The members of the frame may be assumed to be uniform with
bending stiffness *EI*.

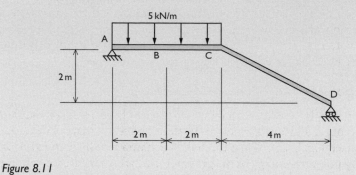

Figure 8.11

8.2   Potential energy

Mechanical forms of energy

Consider a mass *m* subject to a force *F*. If, at time *t*, its displacement is *x*
and velocity is *v*, Newton's second law of motion indicates that

$$F = m\frac{\mathrm{d}v}{\mathrm{d}t}$$

Multiplying both sides of this equation by *v* (which equals d*x*/ d*t*) gives

$$F\frac{\mathrm{d}x}{\mathrm{d}t} = mv\,\frac{\mathrm{d}v}{\mathrm{d}t}$$

However

$$v\frac{\mathrm{d}v}{\mathrm{d}t} = \frac{\mathrm{d}}{\mathrm{d}t}\left(\frac{1}{2}v^2\right)$$

hence

$$F\frac{\mathrm{d}x}{\mathrm{d}t} = \frac{\mathrm{d}}{\mathrm{d}t}\left(\frac{1}{2}mv^2\right)$$

Thus the velocity of a body increases when work is done by a force on it.
Conversely, to slow down, the body must do work by pushing an opposing
force backwards.

The technical definition of energy is the capacity for doing work. Kinetic energy (which for a body in linear motion is $\frac{1}{2}mv^2$) is the energy of movement which must always be non-negative. Potential energy is the energy due to position. Whereas forces have potential energy, it is not often feasible to specify the total amount of energy available. A gravity force, for instance, has an absolute potential energy dependent on how far it could fall (which is to the centre of the earth if a suitable hole could be made) and also how it would vary on the way there. For the calculation of potential energy, the datum position of forces is set arbitrarily (usually their original position is chosen). Therefore, potential energies derived are relative values and may be either negative or positive.

Conservation of energy

Consider a simple pendulum having a small bob of mass m oscillating between points A and C which are at height h above the bottom dead centre position B (see Fig. 8.12). When the bob is instantaneously static at A or C, there is only potential energy of the gravity force, which measured relative to B as datum is mgh. When the bob passes through B, the potential energy has reduced to zero, but the kinetic energy is $\frac{1}{2}mv^2$ where v is the instantaneous velocity of the bob. Thus, for the pendulum to keep swinging, there must be a continual interchange between potential and kinetic energy such that

$$\frac{1}{2}mv^2 = mgh$$

The formula $v = (2gh)^{1/2}$ arising from this equation is a surprising result in so far as the maximum velocity is seen, from conservation of energy, to be independent of the length of the rod and the amplitude of the angular rotation. Furthermore, it applies even for large angles of rotation where the motion is not simple harmonic.

If the pendulum falls off its rocker, its kinetic energy will initially increase as its potential energy decreases. However, even the kinetic energy will be lost eventually when it is static, having hit the ground and stopped bouncing. Even so, this does not mean that, overall, energy has been lost. It only means that it has been converted into other forms of energy such as heat and noise.

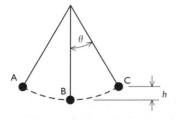

Figure 8.12 A simple pendulum.

Strain energy for a truss member

Apart from potential energy of the applied forces, the other form of potential energy of particular concern is called 'strain energy' and is due to elastic deformation. To obtain a value for the strain energy of a structure, it is necessary to identify the minimum amount of work required to achieve the particular deflected form by increasing the loads gradually from zero, such that the structure is always in equilibrium with no kinetic energy created. The total strain energy is the sum of the strain energies stored in each of the members. Hence, the simplest case to evaluate is the strain energy of a truss, because the members only carry axial load.

If, when a pin-jointed member carries an axial load p', its extension is e', the increment of work required to increase these to $p' + \mathrm{d}p'$ and $e' + \mathrm{d}e'$ is $p'\,\mathrm{d}e'$. Hence, if the final load is p with an extension e, the strain energy for the member will be $\int_{e'=0}^{e} p'\,\mathrm{d}e'$ which is the area under the graph of load against extension shown in Fig. 8.13. Usually the material behaves linearly in which case the curve OB in the figure is the diagonal of the rectangle OABC and the strain energy is $\frac{1}{2}pe$. With ℓ, E and A as the length, modulus of elasticity and cross-sectional area of the member, Eq. (6.1) may be used to write the strain energy in terms of either the axial force or extension as

$$U = \frac{p^2 \ell}{2EA} = \frac{EAe^2}{2\ell}$$

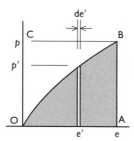

Figure 8.13 The strain energy integral for a truss member with a non-linear load–deflection diagram.

Strain energy for a beam

If a linear elastic beam carries a bending moment of M at spanwise position x and has a curvature due to bending of κ, which equals $\mathrm{d}^2 v/\mathrm{d}x^2$ where v is the deflection in the plane of bending, the strain energy stored in an incremental length dx is

$$dU = \tfrac{1}{2}M\kappa\,dx$$

and, making use of the bending formula (6.5), the strain energy due to bending of the whole beam is

$$U = \frac{1}{2}\int M\frac{d^2v}{dx^2}dx = \frac{1}{2}\int \frac{M^2}{EI}dx = \frac{1}{2}\int EI\left(\frac{d^2v}{dx^2}\right)^2 dx$$

Strain energy for complex stress systems

Where an incremental element of material of volume $d\bar{v}$ carries a 3D stress system giving rise to 3D strains, using the notation of Sections 3.5 and 6.1, the strain energy stored is

$$dU = \frac{1}{2}(\sigma_x\varepsilon_x + \sigma_y\varepsilon_y + \sigma_z\varepsilon_z + \tau_{xy}\gamma_{xy} + \tau_{yz}\gamma_{yz} + \tau_{zx}\gamma_{zx})\,d\bar{v} \qquad (8.5)$$

This equation, coupled with stress–strain relationships enables the amount of strain energy to be determined for any linear elastic structure for which the stress system or the displacement form are completely defined.

Determination of deflections for single concentrated loads

Where a linear elastic structure is subject to just a single concentrated load, it is possible to use the strain energy to compute the deflection of the load.

Ex. 8.7
Determine the amount of strain energy stored in the crane truss of Ex. 6.13 and hence calculate the deflection of the single concentrated load.

Answer

$$U = \frac{1}{2}[144.2 \times 3.47 - 216.3 \times (-5.20) - 169.7 \times (-1.60)$$

$$+ 200 \times 2.67 + 0 - 300 \times (-2.0)] = 1515\,\text{Nm}$$

NB   The contribution of every member should be non-negative.

If the truss is loaded slowly, the loss of potential energy of the applied force is $\frac{1}{2} \times 100v_E\text{kN}$, v_E being the deflection of the applied load. Equating this to the strain energy gives $v_E = 30.3\,\text{mm}$.

Ex. 8.8
Determine the amount of strain energy stored in the cantilever beam of Fig. 6.9 and calculate the deflection at the tip.

Answer
The bending moment at distance x from the root is $M = P(\ell - x)$. Hence the strain energy is

$$U = \frac{1}{2} \int \frac{M^2}{EI}\, dx = \frac{P^2}{2EI} \int_0^\ell (\ell - x)^2 \, dx$$

$$= \frac{P^2}{2EI} \left[\ell^2 x - \ell x^2 + \frac{1}{3}x^3 \right]_0^\ell = \frac{P^2 \ell^3}{6EI}$$

Equating this to the external work done when the load is applied slowly gives the end deflection as $P\ell^3/3EI$.

Ex. 8.9
a Determine an expression for the angle of twist for a length ℓ of a uniform closed hollow box of enclosed area A subject to a torque T, if the material has a shear modulus of G and the wall thickness t may vary round the section.
b Hence determine the angle of twist of a length of 30 m of the box girder of Ex. 3.10 if $G = 130\,\text{kN/mm}^2$.

Answer
a The torque T will produce a Bredt–Batho shear flow round the cross-section with a shear stress of $\tau = T/2At$ (see Section 3.6). Consider an element of the wall of length dx, perimeter distance ds and thickness t. Because only one term in Eq. (8.5) is non-zero, the strain energy for this element is $\frac{1}{2}\tau^2 \times t \times \text{d}s \times \text{d}x/G$. Thus

$$dU = \frac{T^2 \, ds \, dx}{8A^2 Gt}$$

Because the stresses do not vary in the longitudinal (x) direction and t is the only variable round the cross-section, integrating for the whole box gives

$$U = \frac{T^2 \ell}{8A^2 G} \oint \frac{ds}{t}$$

But if ϕ is the angle of twist, the work done by the torque, if applied incrementally, is $\frac{1}{2}T\phi$. Equating this to the strain energy gives

$$\frac{\phi}{\ell} = \frac{T}{4A^2 G} \oint \frac{ds}{t} \tag{8.6}$$

b $T = 14{,}000\,\text{kNm}$, $\ell = 30\,\text{m}$, $A = 52\,\text{m}^2$ and $G = 130\,\text{kN/mm}^2$

$$\oint \frac{ds}{t} = \frac{15{,}000}{25} + \frac{11{,}000}{12} + \frac{2 \times 4472}{12} = 2262$$

Hence

$$\phi = \frac{14{,}000 \times 30 \times 2262}{4 \times 52^2 \times 130 \times 10^6} = 0.68 \times 10^{-3}\,\text{rad}$$

Comment
Formula (8.6) is particularly difficult to derive without using energy concepts.

Conservative structures

When a structure has the same load–deflection curve for unloading as it has for loading, it is described as elastic. In this case any energy converted to strain energy is recoverable. Where a structure or system is conservative, the total potential energy is a useful function with which to investigate both equilibrium and stability.

The theorem of minimum total potential energy

The theorem of minimum total potential energy may be stated as 'Stable equilibrium is only possible where the total potential energy has a minimum value with respect to all possible displacements'. Figure 8.14 shows two types of potential energy function that might arise to produce stable equilibrium.

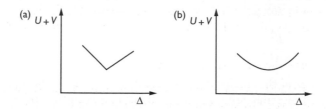

Figure 8.14 Variation of total potential energy against displacement Δ for two types of stability: (a) inelastic; (b) elastic.

Stable equilibrium of inelastic systems

An illustration of an inelastic system is the rigid block resting on a flat surface shown in Fig. 8.15(a). With no strain energy, the only potential energy is due to the gravity force arising through self-weight. If the block is rotated from rest in an anticlockwise direction, it will turn about the bottom left-hand corner, A, with the potential energy increasing until the centre of gravity is directly above A. However, if the block is rotated in a clockwise direction, it will turn about B with a similar increase in potential energy. Hence a V-shaped potential energy function corresponding to stable equilibrium derives from the sudden changeover in the position of the centre of rotation when θ passes through zero (Fig. 8.15(b)).

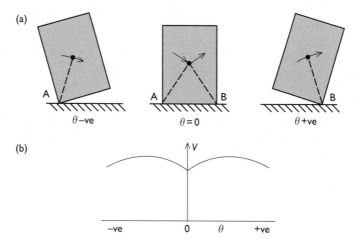

Figure 8.15 Rotation of a rigid block on a flat surface: (a) different positions of the centre of gravity and its direction of movement for θ increasing; (b) potential energy function.

It has been shown by Jennings (1986) that the potential energy function of a masonry arch consisting of voussoirs as rigid blocks has a V-shaped minimum provided that a thrust line can be drawn within the arch ring.

Stable equilibrium of elastic systems

With elastic systems, the potential energy function and its slope tend to be continuous, giving a U-shaped minimum for stable equilibrium. In this case, the theorem can be separated into two components stating that

(i) The potential energy function has a stationary value for equilibrium.
(ii) The potential energy function needs to have a positive curvature for the equilibrium to be stable.

In many situations, the stability of a structure is not in doubt and so the only concern is that of determining the equilibrium position via the stationary principle. However, the use of the equilibrium principle by itself in the case of the pendulum of Fig. 8.12 could lead to either of the equilibrium positions $\theta = 0°$ or $180°$ of which only the first is stable. It is therefore necessary to investigate the curvature of the potential energy function if there is any question about stability.

Stationary potential energy

In each of the Exs 8.7–8.9, a deflection was predicted by equating the strain energy of the structure to the work done by the external force applied gradually. This procedure was successful because there was only one applied load contributing to the external work done and it was the deflection of this load that was required. In other situations, where equilibrium conditions or deflections are required, it is necessary to invoke stationary principles in order to use energy concepts.

If the displacement of a structure by a small distance $\delta\Delta$ causes changes δU and δV to the strain energy U and the potential energy of the applied forces V, the stationary principle of the theorem of minimum total potential energy implies that, for equilibrium,

$$\delta U + \delta V = 0 \qquad (8.7)$$

Consider a truss in which the external forces are f_j and the member forces are p_i. If the displacement $\delta\Delta$ of the truss produces displacements of du_j to the external forces and extensions de_i to the members,

$$\delta U = \sum p_i \, de_i \quad \text{and} \quad \delta V = -\sum f_j \, du_j$$

Substituting these values into Eq. (8.7) gives precisely the same equation as the basic virtual work Eq. (8.1). Similarly, if beam bending is included within the strain energy formulation, Eq. (8.2) is obtained.

In view of this correspondence between the stationary principle of the total potential energy and the basic virtual work formulation, the various techniques developed in the nineteenth and twentieth centuries for determining equilibrium equations and deflections by using stationary energy principles will not be discussed here. The correspondence of these methods does, however, establish why the basic virtual work method can be used to determine equilibrium conditions. Furthermore, there are two fundamental differences between energy principles and virtual work:

a   Virtual work can be applied to infeasible displacements of a system for which the potential energy has no meaning, for instance, by moving the force system of Fig. 8.1 either into or away from the inclined plane. Thus the virtual work method is more versatile for investigating equilibrium.
b   Virtual work does not provide any information about stability.

Castigliano's theorem of least work *(Castigliano, 1919)*

Where forces act on a structure which is statically indeterminate, the internal load paths distribute in such a way that the least amount of work is done by the external forces. Furthermore, because for a conservative structure, the work done by the external forces is equal to the strain energy stored, Castigliano's theorem of least work implies also that the strain energy, when considered as a function of the redundant force variables, is a minimum.

If P is any of the redundant force variables, the strain energy U will be U-shaped with a stationary value such that $\delta U/\delta P = 0$ and also a positive curvature for which $\delta^2 U/\delta P^2 > 0$. The stationary aspect of this theorem has been the cause of much discussion in the past. It has been reckoned to be only applicable to linear elastic structures and to overlap with the virtual work method. As a result, it has fallen out of favour for the analysis of statically indeterminate structures (Matheson, 1971).

However, the minimisation aspect of the theorem is worthy of attention. Consider a uniform built-in beam which is subject to a uniformly distributed load. Figure 8.16 shows three possible symmetric solutions for the BMD which satisfy equilibrium. The first is obtained by allowing the supports to rotate freely and the last by allowing the central cross-section to rotate like a hinge. Both of these solutions, which satisfy equilibrium but not compatibility, have a total strain energy which is patently greater than the strain energy for the correct second solution (remembering that for uniform beam problems U is proportional to $\int M^2 \, dx$). Hence the theorem gives an insight into structural behaviour which could be useful in estimating approximate solutions or in assessing the validity of computed solutions. Indeed it can be

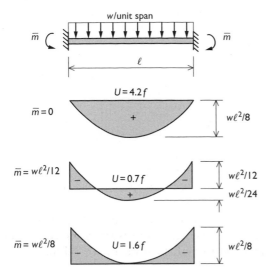

Figure 8.16 BMDs for different equilibrium solutions for a built-in beam showing the strain energy U in terms of $f = 10^{-3}\,w\ell^5/EI$.

used to deduce that load paths pass as far as possible through stiffer parts of a structure. For example, from the end views of the Gateshead Millennium Bridge shown in Fig. 6.76, it is possible to infer, knowing that the arch is stiffer in out of plane bending than the deck, that it will attract a greater proportion of the out of plane bending moments.

Q.8.8
a Determine the strain energy for the truss of Fig. 5.19(a) if all the members have $A = 880\text{ mm}^2$ and $E = 200\text{ kN/mm}^2$.
b Hence estimate the deflection of the load.

Q.8.9
a Obtain an expression for the strain energy of the third beam shown in Fig. 6.10.
b Hence justify the expression for the deflection under the load.

Q.8.10
a From Eq. (8.5), determine an expression for the elastic torsional stiffness, $T\ell/\phi$ of a hollow circular tube of length ℓ, average radius r, thickness t and shear modulus G.

b   Use this answer to devise a formula for the elastic torsional stiffness of a solid circular cylinder by considering it to be a combination of concentric hollow cylinders whose angles of twist are similar.

Q.8.11
Use the theorem of minimum total potential energy to determine the uniformly distributed lateral loading required to collapse the wall of Q.2.5 by means of the mechanism shown in Fig. 2.14.

Q.8.12
A built-in beam has an I value between the one-quarter and three-quarter points which is many times greater than that for its outer parts. If a concentrated load is applied at the centre point, sketch what the BMD and deflected shape might be expected to be like.

8.3   Energy methods for buckling

Axial shortening

To apply energy methods to buckling analysis, it is necessary to take account of the second order shortening of members due to lateral deflections. If a rigid member of length ℓ is deflected sideways by a small amount such that the slope becomes θ, the length of the member projected onto its original axis becomes $\ell \cos \theta$ (see Fig. 8.17). Because only small deflections are being considered, the series expansion for $\cos \theta$ can be truncated to give

$$\cos \theta \simeq 1 - \frac{1}{2}\theta^2$$

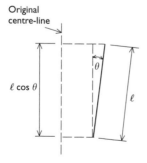

Figure 8.17 Axial projection of a laterally deflected length of column.

with the result that the axial shortening due to θ is $\frac{1}{2}\ell\theta^2$. Thus if a column consists of a set of rigid bars in series, its total shortening is given by $\frac{1}{2}\sum\ell_i\theta_i^2$ where ℓ_i and θ_i are the lengths and deflected slopes of the component members.

If a column of length ℓ bends with lateral deflection given by $v(x)$ where x is the axial coordinate measured from one end, an element of the column of length dx will have a slope of dv/dx. Hence by comparison with the previous result, the axial shortening will be

$$\frac{1}{2}\int_0^\ell \left(\frac{dv}{dx}\right)^2 dx$$

By including the reduction in potential energy due to axial shortening of a column loaded in compression, it is possible to use the theorem of minimum total potential energy to investigate both equilibrium and stability.

Ex. 8.10
Use energy principles to investigate the stability of the two-bar strut shown in Fig. 7.13(a).

Answer
The slopes, θ_{AB} and θ_{BA} of the bars AB and BC due to a lateral displacement v_B of the central joint are $\pm v_B/\ell$ (see Fig. 8.18). Hence the loss of potential energy due to axial shortening is $\frac{1}{2}P(\ell\theta_{AB}^2 + \ell\theta_{BC}^2) = Pv_B^2/\ell$. Strain energy occurs only in the restraining spring such that $U = \frac{1}{2}kv_B^2$. The total potential energy is therefore

$$U + V = \tfrac{1}{2}kv_B^2 - Pv_B^2/\ell$$

For equilibrium $\partial(U + V)/\partial v_B = 0$, thus

$$(k - 2P/\ell)v_B = 0$$

showing that the column can only be in equilibrium in a deflected position if $P = \frac{1}{2}k\ell$.
    Differentiating again

$$\frac{\partial^2(U + V)}{\partial v_B^2} = k - \frac{2P}{\ell}$$

Thus for $P < \frac{1}{2}k\ell$ the coefficient of the potential energy function is positive indicating stability, whereas for $P > \frac{1}{2}k\ell$, it is negative indicating instability.

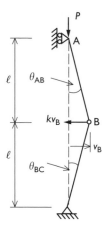

Figure 8.18 Lateral deflection of the two-bar column Fig. 7.13(a).

Ex. 8.11
Use energy principles to investigate the lateral stiffness characteristics of the three-bar strut shown in Fig. 7.14(a).

Answer
If small lateral forces f_B and f_C act at the joints B and C to induce lateral displacements as shown in Fig. 8.19, the potential energy function has terms arising from the strain energy of the springs, the axial shortening and the lateral forces. Thus

$$U + V = \frac{1}{2}k\left(v_B^2 + v_C^2\right) - P\left[\frac{v_B^2}{2\ell} + \frac{(v_C - v_B)^2}{2\ell} + \frac{v_C}{2\ell}\right]$$

$$- \left(f_B v_B + f_C v_C\right) \tag{8.8}$$

Because there are now two displacement variables, v_B and v_C, equilibrium requires the potential energy function to be stationary with respect to both, that is $\partial(U+V)/\partial v_B = 0$ and $\partial(U+V)/\partial v_C = 0$. Thus, by partially differentiating Eq. (8.8), the following stiffness equations can be determined:

$$\begin{bmatrix} (k - 2P/\ell) & P/\ell \\ P/\ell & (k - 2P/\ell) \end{bmatrix} \begin{bmatrix} v_B \\ v_C \end{bmatrix} = \begin{bmatrix} f_B \\ f_C \end{bmatrix} \tag{8.9}$$

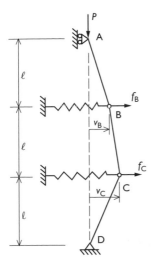

Figure 8.19 Lateral forces applied to the column of Fig. 7.14.

When $P = 0$, the deflection characteristics are just that of the two springs acting independently, that is $kv_B = f_B$ and $kv_C = f_C$. However, as P increases, the stiffness coefficients of these equations are modified, causing the column to become more flexible. If a lateral force is applied at B only, the condition $f_C = 0$ can be used to eliminate v_C from Eq. (8.9) to give

$$\left[(k - 2P/\ell)^2 - (P/\ell)^2 \right] v_B = (k - 2P/\ell) f_B$$

Although, for any value of P, the column behaves in a linear elastic manner as far as lateral loads and deflections are concerned, the response to changes in P is non-linear as shown in Fig. 8.20. The lateral response is asymptotic to the buckling load $P = k\ell/3$ where the stiffness equations:

$$\begin{bmatrix} k/3 & k/3 \\ k/3 & k/3 \end{bmatrix} \begin{bmatrix} v_B \\ v_C \end{bmatrix} = \begin{bmatrix} f_B \\ f_C \end{bmatrix}$$

become unsolvable due to singularity of the stiffness matrix.

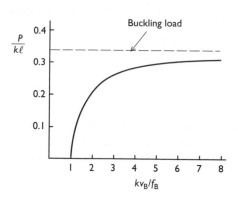

Figure 8.20 Increase in lateral flexibility as axial load is increased for the three-bar column of Fig. 8.19.

Trial modes for Euler buckling

Most buckling analyses are conducted by defining trial shapes for the buckling modes which are then employed in energy methods to investigate the buckling characteristics. It is instructive to try this out, initially, with the Euler strut because this is one of the few problems for which the correct theoretical answer is known.

Ex. 8.12

Consider an Euler strut of length ℓ and bending stiffness EI. Where v is lateral displacement at distance x from one end, the following trial deflection shapes all have zero end deflections and are symmetrical about $x = \frac{1}{2}\ell$:

a *Quadratic:* $v = a_1 x(\ell - x)$ having constant curvature.
b *Cubic:* $v = a_2 x(\ell - 4x^2/3\ell)$ for $x < \frac{1}{2}\ell$ with a mirror image for $x > \frac{1}{2}\ell$. This has zero end curvatures and zero slope at the centre.
c *Quartic:* $v = a_3 x(\ell - x)(\ell^2 + \ell x + x^2)/\ell^2$ having zero end curvatures, being a continuous function.

Answer

a $\partial v/\partial x = a_1(\ell - 2x)$ and $\partial^2 v/\partial x^2 = -2a_1$
   Hence

$$U = \tfrac{1}{2}EI(2a_1)^2 \int_0^\ell dx = 2EI\ell a_1^2$$

and

$$V = -\tfrac{1}{2}Pa_1^2 \int_0^\ell (\ell - 2x)^2 \, dx = -Pa_1^2 \ell^3 / 6$$

Thus:

$$\partial^2 (U + V)/\partial a_1^2 = 4EI\ell - P\ell^3/3$$

with a positive (stable) result when $P < 12EI/\ell^2$
This is in error by 21.6% from the known Euler load.
Similar (but more mathematically tedious) evaluations give

b $P < 10EI/\ell^2$ (1.3% error)
c $P < 168EI/17\ell^2$ (0.13% error)

Comment
As might be expected, the predicted buckling loads are more accurate if the trial mode satisfies the known end conditions and is fully continuous. However, combining trial modes in an analysis will give an even better answer than investigating them separately. For instance, an analysis combining modes 1 and 3 for the Euler strut would be initiated from

$$v = a_1 x(\ell - x) + a_3 x(\ell - x)(\ell^2 + \ell x + x^2)/\ell^2$$

where the ratio a_1/a_3 most likely to cause buckling is obtained from the ensuing eigensolution.

Over-estimation of buckling loads

A property of this procedure for buckling analysis, which can be seen in the previous results, is that predictions are always greater than or equal to the theoretical buckling loads. This is because each buckling mode investigated indicates a possible buckling behaviour consistent with the theorem of minimum total potential energy. However, it will not be the correct buckling mode if there is any other uninvestigated mode which has a lower associated buckling load.

Q.8.13
Determine, using the theorem of minimum total potential energy, the buckling load of the struts of Q.7.4 and Q.7.5 having joints which are weak in bending.

> *Q.8.14*
> A pin-ended strut of length ℓ has a second moment of area which varies according to
>
> $$I = 4I_C x(\ell - x)$$
>
> where x is the axial co-ordinate measured from one end and I_C is the I value at the centre. Estimate the buckling load by using the lateral displacement
>
> $$v = ax(\ell - x)$$
>
> as a trial deflection mode.

8.4   Stiffness and flexibility properties

The transpose relationship

In Section 6.7 a transpose relationship was detected between compatibility and equilibrium equations for linear elastic trusses. Using the notation of that section it was found that, if the compatibility relationship describing member extensions in terms of joint displacements is

$$\mathbf{e} = \mathbf{A}\mathbf{u} \tag{8.10}$$

the equilibrium relationship giving external joint forces in terms of member forces is

$$\mathbf{f} = \mathbf{A}^{\mathrm{T}}\mathbf{p} \tag{8.11}$$

In order for this relationship to hold, the member forces need to be listed in the same order as the member extensions and the joint forces need to be listed correspondingly to the joint displacements.

The product

$$\mathbf{f}^{\mathrm{T}}\mathbf{u} = [f_1 \cdots f_n] \begin{bmatrix} u_1 \\ \vdots \\ u_n \end{bmatrix} = \sum f_j u_j$$

is the work done by the external forces, $\mathbf{f}$, when subject to a disturbance $\mathbf{u}$. However, because the truss is assumed to be loaded within the linear elastic range, the virtual work Eq. (8.1) indicates that this is equal to the work,

$\sum p_i e_i$, required to deform the members. Thus,

$$f^T u = p^T e$$

and using the compatibility relationship (8.10)

$$f^T u = p^T A u$$

The only way this relationship can be true, whatever virtual displacement is represented by u, is for

$$f^T = p^T A$$

In order to transpose a matrix equation involving products, it is necessary to transpose the individual matrices and reverse the order of any products. Thus transposing this last equation proves Eq. (8.11) to be true.

The transpose relationship applies to structures more generally by defining forces as including also moments, with the corresponding displacements being their rotations.

Ex. 8.13
A uniform cantilever beam AB of length ℓ carries a force P_B and a moment M_B at the free end as shown in Fig. 8.21(a). By comparing with the simply-supported beam subject to end moments M_{AB} and M_{BA} shown in Fig. 8.21(b), develop equilibrium relationships which specify P_B and M_B in terms of M_{AB} and M_{BA} such that the BMDs will be the same in both cases. Hence use the transpose relationship to deduce the corresponding compatibility equations.

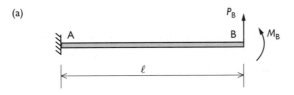

Figure 8.21 A cantilever beam: (a) loading; (b) loading on an equivalent simply-supported beam.

Answer

In both cases the BMD will be linear, so it is only necessary to ensure the same end moments to guarantee similar BMDs.

But $M_{BA} = M_B$ and from equilibrium $M_{AB} = P_B\ell + M_B$.

Hence the required equilibrium equations are

$$\begin{bmatrix} P_B \\ M_B \end{bmatrix} = \begin{bmatrix} 1/\ell & -1/\ell \\ 0 & 1 \end{bmatrix} \begin{bmatrix} M_{AB} \\ M_{BA} \end{bmatrix}$$

From the transpose relationship, the corresponding compatibility equations are

$$\begin{bmatrix} \theta_{AB} \\ \theta_{BA} \end{bmatrix} = \begin{bmatrix} 1/\ell & 0 \\ -1/\ell & 1 \end{bmatrix} \begin{bmatrix} v_B \\ \theta_B \end{bmatrix}$$

Stiffness matrix properties

Symmetry

In Section 6.7 it was found that the overall stiffness matrix for a truss can be constructed according to

$$\mathbf{K} = \mathbf{A}^T \overline{\mathbf{K}} \mathbf{A}$$

where $\mathbf{A}$ is the compatibility matrix and $\overline{\mathbf{K}}$ is the diagonal matrix of member stiffnesses. Since $\overline{\mathbf{K}}^T = \overline{\mathbf{K}}$, the reversal rule for transpose products gives

$$\mathbf{K}^T = (\mathbf{A}^T \overline{\mathbf{K}} \mathbf{A})^T = \mathbf{A}^T \overline{\mathbf{K}}^T \mathbf{A} = \mathbf{A}^T \overline{\mathbf{K}} \mathbf{A} = \mathbf{K}$$

Hence the overall stiffness matrix must be its own transpose and (as noted in Section 6.7) will be symmetric. This property is not restricted to trusses. It applies universally to the stiffness matrices of all linear elastic structures.

The positive definite property

The strain energy U of an elastic structure is equal to the work done by the external forces, if applied gradually. Thus, if a linear elastic structure is subject to forces $\mathbf{f}$ which produce displacements $\mathbf{u}$, the strain energy is $U = \frac{1}{2}\mathbf{f}^T\mathbf{u}$. Transposing the stiffness equations $\mathbf{f} = \mathbf{Ku}$ and making use of symmetry of the stiffness matrix gives

$$\mathbf{f}^T = \mathbf{u}^T\mathbf{K}^T = \mathbf{u}^T\mathbf{K}$$

Hence the strain energy is given by

$$U = \frac{1}{2}\mathbf{u}^T\mathbf{Ku}$$

For a structure to be stable it is necessary for the stiffness matrix $\mathbf{K}$ to satisfy the condition

$$\mathbf{u}^T \mathbf{K} \mathbf{u} > 0$$

for any possible direction of movement described by a non-null vector $\mathbf{u}$. A matrix satisfying this condition is said to be 'positive definite'.

From matrix theory it can be shown that a symmetric positive definite matrix has the following properties (Jennings and McKeown, 1993):

a   All of its eigenvalues are real and positive.
b   The leading diagonal elements (those with $i = j$) are all positive.
c   The determinant (being the product of the eigenvalues) is positive.
d   If corresponding rows and columns are omitted, the determinant of the remaining matrix is positive.

Whereas it is difficult to check by hand that a matrix larger than 2×2 is positive definite, gross errors are easily spotted. For instance spot checks conducted on the symmetric stiffness matrix (6.22) indicate that all leading diagonal elements are positive (i.e. $k_{ii} > 0$) and also all determinants of the form $k_{ii}k_{jj} - k_{ij}^2$ are also positive. If there was a gross error, it is likely that at least one of these checks would have produced a negative result.

Ex. 8.14
A uniform elastic cantilever beam of length ℓ, cross-sectional area A and second moment of area I has displacements u, v and θ imposed at the free end by means of external forces P, S and M, as shown in Fig. 8.22 (here the rotation, θ, is grouped with the displacements and the moment, M, with the forces). The stiffness equations have been evaluated as

$$\begin{bmatrix} P \\ S \\ M \end{bmatrix} = \begin{bmatrix} EA/\ell & 0 & 0 \\ 0 & 12\,EI/\ell^3 & -6\,EI/\ell \\ 0 & -6\,EI/\ell & 2\,EI/\ell \end{bmatrix} \begin{bmatrix} u \\ v \\ \theta \end{bmatrix}$$

Examine the stiffness matrix to see if you can detect any errors.

Answer
The diagonal elements are all positive. However, if row and column one are deleted, the determinant of the remainder is $(24E^2I^2)/\ell^4 - (36E^2I^2/\ell^2)$. First the dimensions of these two terms

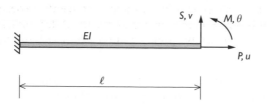

Figure 8.22 Another cantilever beam.

do not match. The off-diagonal elements should be in the units of moment/unit displacement or force/unit rotation (e.g. kN units) whereas $6EI/\ell$ has typical units kNmm. One error is that they should read $6EI/\ell^2$. However, when this is corrected, the 2×2 determinant is negative indicating a further error. The last coefficient (which is the end rotational stiffness should be $4EI/\ell$ instead of $2EI/\ell$). The correct stiffness equations are

$$
\begin{bmatrix} P \\ S \\ M \end{bmatrix} = \begin{bmatrix} EA/\ell & 0 & 0 \\ 0 & 12\,EI/\ell^3 & -6\,EI/\ell^2 \\ 0 & -6\,EI/\ell^2 & 4\,EI/\ell \end{bmatrix} \begin{bmatrix} u \\ v \\ \theta \end{bmatrix} \tag{8.12}
$$

The flexibility matrix

The flexibility matrix, **F**, of a linear elastic structure relates displacements to forces according to

$$
\mathbf{u} = \mathbf{Ff}
$$

If **K** is the stiffness matrix relating the same displacements and forces such that $\mathbf{f} = \mathbf{Ku}$, it can be shown that **F** is the inverse of **K**. Furthermore, if **K** is symmetric positive definite, **F** will be also.

For instance the flexibility equations for the cantilever beam of Fig. 8.22, which can be derived by inverting Eq. (8.12) are

$$
\begin{bmatrix} u \\ v \\ \theta \end{bmatrix} = \begin{bmatrix} \ell/EA & 0 & 0 \\ 0 & \ell^3/3EI & \ell^2/2EI \\ 0 & \ell^2/2EI & \ell/EI \end{bmatrix} \begin{bmatrix} P \\ S \\ M \end{bmatrix} \tag{8.13}
$$

which can be checked for symmetry and positive definiteness.

Flexibility matrices are generally easier to understand and interpret than stiffness matrices. Also it is easier to obtain their coefficients experimentally. Thus the second and third columns of the above flexibility matrix are obtained from the deflections caused by applying unit loads $S = 1$ and $M = 1$ as shown in Fig. 8.23.

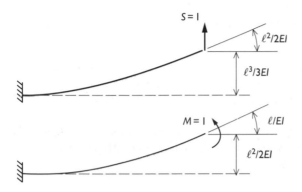

Figure 8.23 Significance of the flexibility matrix coefficients for the cantilever beam.

Maxwell's reciprocal theorem

Maxwell's reciprocal theorem states that the deflection at point A (of a linear elastic structure) due to a unit load at point B, designated Δ_{AB}, is equal to the deflection at point B due to a unit load at point A, designated Δ_{BA}. The traditional justification is as follows.

Consider the strain energy of a structure due to unit loads applied at both points A and B. If the unit load at A were applied first gradually, the work done would be $\frac{1}{2}\Delta_{AA}$ (where Δ_{AA} represents the deflection at A due to unit load at A). If then the unit load at B is applied gradually, an additional amount of work $\Delta_{AB} + \frac{1}{2}\Delta_{BB}$ is done. The Δ_{AB} term does not have a factor of $\frac{1}{2}$ because the unit load at A is present the whole time that the load at B is being applied. The total strain energy is therefore

$$\tfrac{1}{2}\Delta_{AA} + \Delta_{AB} + \tfrac{1}{2}\Delta_{BB}$$

However, if the order of application of the loads is reversed, the total strain energy is

$$\tfrac{1}{2}\Delta_{AA} + \Delta_{BA} + \tfrac{1}{2}\Delta_{BB}$$

Since the final situations are the same and energy is conserved, $\Delta_{AB} = \Delta_{BA}$.

Maxwell's reciprocal theorem could, alternatively, have been justified via the flexibility matrix, because Δ_{AB} and Δ_{BA} represent corresponding coefficients on either side of the diagonal of the flexibility matrix equal through symmetry.

An example of how Maxwell's reciprocal theorem is useful in understanding structural behaviour relates to the shear centre of a beam (considered in Section 4.9). Because a load applied in any lateral direction through the shear centre of a beam produces no twist, Maxwell's theorem indicates that a torsional moment applied to the beam produces no movement of the shear centre in any lateral direction.

Geometric non-linearity

Where members are slender, it was found in Chapter 7 that the axial load they carry can influence their lateral stiffness. Consider, for instance, a uniform pin-ended member of length ℓ, with axial force P and extension e, which deflects laterally by v so that the inclination θ (assumed to be small) is $\theta = v/\ell$ as shown in Fig. 8.24. The lateral force S required for equilibrium is $S = Pv/\ell$. Hence the member stiffness equations including lateral as well as axial displacements are

$$\begin{bmatrix} P \\ S \end{bmatrix} = \begin{bmatrix} EA/\ell & 0 \\ 0 & P/\ell \end{bmatrix} \begin{bmatrix} e \\ v \end{bmatrix}$$

Where P is positive, the P/ℓ term provides the pendulum stiffness which aids the stability of cable-supported structures. When P is negative, this term provides the destabilising effect of the compressive force. However its inclusion makes the stiffness matrix force dependent. The response will no longer be linear, with the structure becoming stiffer or more flexible as load is added (Figures 7.2 and 8.20 illustrate the increasing flexibility of a structure as the buckling load is approached).

To include such terms in an analysis by the stiffness method, it is first necessary to guess the axial loads so that stiffness equations can be specified and solved. Once solved, the axial forces can be revised and the solution repeated iteratively until no further adjustments are necessary. Extension of this technique to more complex structures (e.g. involving bending) requires the development of the appropriate member stiffness equations.

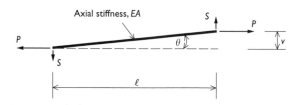

Figure 8.24 Equilibrium of a pin-ended member subject to sway displacement.

The following are some features of non-linear structural behaviour due to change of geometry which apply generally.

a   The stiffness matrix should be considered as defining incremental changes in the applied forces produced by incremental changes in the displacements.

b   The modifying terms to the stiffness matrix, which are dependent on internal forces, constitute a 'geometric' stiffness matrix which adds to the standard stiffness matrix.

c   The incremental stiffness and flexibility matrices are both symmetric and, whilst the structure is stable, positive definitive.

d   Hence Maxwell's reciprocal theorem still applies, but only for incrementally small 'unit' forces.

Q.8.15
Which of the following are clearly not valid symmetric positive definite stiffness or flexibility matrices and why?

$$
\text{(i)}\begin{bmatrix} 4 & 18 & 0 \\ 18 & 192 & 4 \\ 0 & 4 & 1 \end{bmatrix}
\qquad
\text{(ii)}\begin{bmatrix} 6 & 4 & 0 \\ 4 & 6 & 0 \\ 0 & 0 & -20 \end{bmatrix}
\qquad
\text{(iii)}\begin{bmatrix} 10 & -8 & -2 \\ -8 & 12 & -4 \\ -2 & -4 & 4 \end{bmatrix}
$$

$$
\text{(iv)}\begin{bmatrix} 12 & -8 & -2 \\ -8 & 4 & -4 \\ -2 & -4 & 12 \end{bmatrix}
\qquad
\text{(v)}\begin{bmatrix} 10 & 4 & -1 \\ -4 & 6 & 2 \\ 1 & -2 & 3 \end{bmatrix}
\qquad
\text{(vi)}\begin{bmatrix} 1 & -1 & \\ -1 & 2 & -1 \\ & -1 & 1 \end{bmatrix}
$$

Q.8.16
Given the stiffness equation for the simply-supported beam shown in Fig. 8.21(b):

$$
\begin{bmatrix} M_{AB} \\ M_{BA} \end{bmatrix} = \frac{2EI}{\ell}\begin{bmatrix} 2 & -1 \\ -1 & 2 \end{bmatrix}\begin{bmatrix} \theta_{AB} \\ \theta_{BA} \end{bmatrix}
$$

derive, using the results of Ex. 8.13, the flexibility matrix for the cantilever beam.

8.5 Influence lines

What use is an influence line?

An influence line (IL) is a plot of a particular parameter, such as an internal force, as a function of the position of the load (it is convenient to use a unit load for this purpose). For instance the bending moment at the root of a

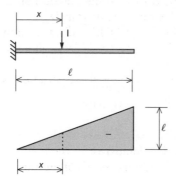

Figure 8.25 Influence line for the root bending moment of a cantilever beam.

cantilever beam of length ℓ due to a unit load at distance x from the root is $-x$ giving the result shown in Fig. 8.25.

The main use of such a graph is to indicate where movable loads need to be placed in order to generate the most severe effects. From Fig. 8.25 it may be deduced that the worst root bending moment for the cantilever is obtained as follows:

a   If there is a single concentrated load, it needs to be placed at the tip (e.g. a single person on a diving board).
b   If there is a uniformly distributed load it needs to be placed over the whole beam (e.g. if it is a balcony support on which people might stand anywhere).
c   If there is a train of loads, arising for example, from the axles of a locomotive or heavy goods vehicle, one load (possibly the heaviest) needs to be placed at the tip. This is because moving a train of loads to the right will always increase the root bending moment, that is until one of the loads reaches the tip.

With more complicated structures even a qualitative knowledge of the shape of the relevant ILs will give a designer better insight into which are the most critical loading cases to consider, thus enabling the number of computer analyses to be reduced.

Ex. 8.15
Figure 8.26(a) shows a girder bridge ABCDEF, simply-supported at C and E.

a   With the dimensions shown, draw the ILs for (i) the reaction at C,
    (ii) the shear force at D and (iii) the bending moment at D.
b   If a vehicle with axle loads of 120 kN and 80 kN, distance 3 m
    apart, slowly crosses the bridge in either direction, determine the
    maximum values of each of these parameters.
c   Determine the maximum possible value for each of these parameters
    due to pedestrian loading at 16 kN/m.

Answer
a   (i)   When the unit load is distance $\bar{x}$ to the left of E, the reac-
          tion at C is given by $R_C = \bar{x}/16$ giving the IL shown in
          Fig. 8.26(b).
    (ii) When the unit load is distance x to the right of C, the
          reaction at E is given by $R_E = x/16$. Hence, considering
          equilibrium of portion DF of the beam, $S_D = x/16 - (1)$
          where the term in brackets is present only when the unit
          load is to the right of D. This gives the IL for S_D shown in
          Fig. 8.26(c).

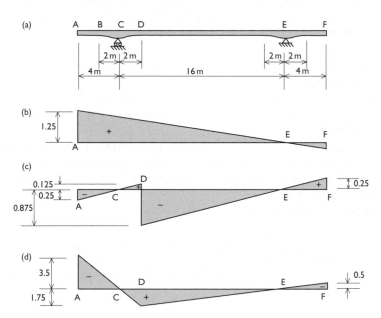

Figure 8.26 Some influence lines for a statically determinate beam: (a) configuration; (b) IL for (upward) reaction at C; (c) IL for shear force at D;
(d) IL for bending moment at D in m units.

(iii)   When the unit load is to the left of D, $M_D = 14R_E$ and when it is to the right of D, $M_D = 2R_C$. Hence the IL for M_D is as shown in Fig. 8.26(d).

b   (i)   With the 120 kN load at A and the other load to the right,

$$R_B = 1.25 \times 120 + 1.25 \times (17/20) \times 80 = 235 \, \text{kN}$$

(ii)   With the 120 kN load just to the right of D and the other load to the right,

$$S_D = -0.875 \times 120 - 0.875 \times (11/14) \times 80$$
$$= -160 \text{kN}$$

(iii)   With the 120 kN load at A and the other load to the right,

$$M_D = -3.5 \times 120 - 3.5 \times (1/4) \times 80$$
$$= -490 \, \text{kN (i.e. hogging)}$$

(The maximum sagging bending moment occurs with the 120 kN load at D and is 320 kN.)

c   (i)   With the distributed load acting from A to E, the reaction is 16 kN/m multiplied by the area of the IL between A and E. This gives

$$R_B = 16 \times 1.25 \times 20/2 = 200 \text{ kN} \tag{8.14}$$

(ii)   With the distributed load acting from A to C and D to E,

$$S_C = 16(0.25 \times 4/2 + 0.875 \times 14/2) = 106 \, \text{kN}$$

(iii)   The maximum sagging bending moment is with the distributed load acting from C to E. This gives

$$M_D = 16 \times 1.75 \times 16/2 = 224 \, \text{kNm}$$

(The maximum hogging bending moment is with the load acting on both the cantilever ends and is 128 kNm.)

Müller–Breslau's principle

At first sight ILs may look peculiar. However Müller–Breslau, using energy concepts, discovered the logic which is the key to understanding them. His principle may be stated as follows: 'The deflected shape of a structure* due to a localised unit distortion is the IL for the internal force corresponding to that distortion.'

*NB   If the structure is statically indeterminate, it also needs to be linear elastic.

Application to statically determinate structures

To obtain ILs for shear force and bending moment at a particular cross-section of a beam, introduce corresponding unit distortions there, as shown in Fig. 8.27, and then place the distorted beams onto their supports as shown in Fig. 8.28. To obtain the IL for the axial force in a member, a unit contraction should be given to it. Furthermore, if imaginary pin-ended members are used to simulate supports (as in Fig. 6.40), it follows that the IL for any reaction can be obtained by moving the reaction point a unit distance. Thus

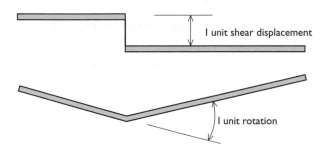

Figure 8.27 Localised distortions for use with Müller–Breslau's principle.

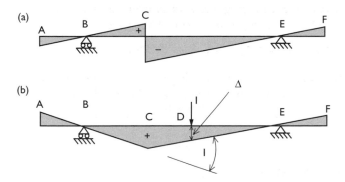

Figure 8.28 Use of virtual displacements to identify ILs for a simply-supported beam with overhangs: (a) IL for shear force at C; (b) IL for bending moment at C.

the shape of all three ILs derived for Ex. 8.15 could have been predicted using Müller–Breslau's principle.

To justify Müller–Breslau's principle for statically determinate structures, consider a beam with a unit force applied at some point D. If a virtual displacement corresponding to the required IL takes place (as illustrated in Fig. 8.28(b)) in which the unit force displaces by $\Delta(x)$, the external work done is $1 \times \Delta(x)$ and the internal work done is $M \times 1$. Hence equating these gives $M = \Delta(x)$, proving that $\Delta(x)$ describes the required bending moment as a function of x.

When considering statically determinate structures in this way, Müller–Breslau's principle is describing equilibrium relationships and hence the ILs derived do not depend on the structure being linear elastic. The lines are, in the main part, linear.

Application to statically indeterminate structures

The same principle applies to statically indeterminate structures which are linear elastic. If, for instance, a beam is continuous over three supports, the ILs for shear force and bending at a cross-section can be obtained by forcing distorted beams of the form shown in Fig. 8.27 to fit the three support conditions, as shown in Fig. 8.29. Similarly the IL for a reaction can be obtained by providing the beam with a displacement there.

The justification for Müller–Breslau's principle in this case is through Maxwell's reciprocal theorem. For instance, if the IL for the reaction R at the left-hand support of the continuous beam is required, consider the response of the beam to a unit load with this reaction set to zero. Figure 8.30(a)

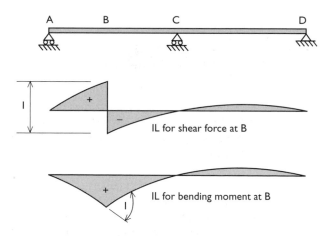

Figure 8.29 Sketched ILs for a continuous beam.

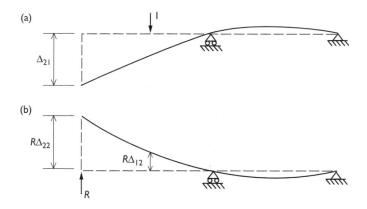

Figure 8.30 Use of Maxwell's reciprocal theorem to justify Müller–Breslau's principle: (a) removal of support; (b) reinstatement of support reaction.

shows the resulting distorted beam in which the removed support point deflects by Δ_{21}. Figure 8.30(b) shows the response of the beam to a reaction force R which includes a displacement $R\Delta_{22}$ at the removed support. In order to bring the displacement at the support to zero it is necessary that $R = \Delta_{21}/\Delta_{22}$. However, from Maxwell's reciprocal theorem this equals Δ_{12}/Δ_{22}. Furthermore, because Δ_{12} and Δ_{22} are both caused by loading at point 2 (the support), it is possible to scale them in proportion until $\Delta_{22} = 1$, in which case $R = \Delta_{12}$. Thus the shape of the beam when deflected by a unit amount at the support is the IL for the support reaction. The principle can be justified in a similar way for bending moments, for instance, by introducing an artificial hinge at the position of the required bending moment and then determining the bending moment across the hinge required to reduce its rotation to zero.

Use of sketched influence lines

In many cases sketched ILs may be sufficient to identify where movable loads need to be placed in order to obtain the most critical design cases.

Ex. 8.16
The two span continuous beam shown in Fig. 8.29 may be subject to downward acting distributed live loading acting anywhere across the two spans. Indicate which loading cases would need to be examined

to determine critical values of

a   Shear force at B
b   Bending moment at B
c   Reaction at A.

Answer
The following results are obtained by examining where the ILs are positive and where they are negative:

a   For negative shear force at B, load BC only.
    For positive shear force at B, load AB and CD.
b   For sagging bending moment at B, load span AC only.
    For hogging bending moment at B, load span CD only.
c   For upwards reaction at A, load span AC only.
    For downwards reaction at A, load span CD only.

Influence lines for trusses

Influence lines are particularly useful for considering vehicle loads on bridges, many of which are trusses. In truss analysis, however, it is assumed that all loading occurs at the joints. This means that it is only possible to directly compute certain key points on any IL, being those points where the unit load lies directly over a deck joint. The shape of ILs between these key points depends on the deck structure itself. A convenient assumption is that the deck structure comprises a series of simply-supported beams spanning between successive joints in the deck chord as shown in Fig. 8.31. If this is the case, the loads transmitted to the joints of the truss vary linearly as the unit load traverses each anxiliary beam. Hence ILs for the truss will always be linear between the key points.

From Müller–Breslau's principle it follows that, where the deck chord of a bridge is horizontal, the IL for the axial force in a member corresponds to the deflection of the deck chord when the member concerned is given a unit

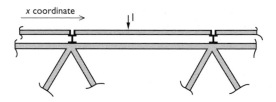

Figure 8.31 Use of auxiliary beams to load a truss only at the joints.

axial contraction. Hence the shape of ILs for trusses can be easily sketched as shown in Fig. 8.32.

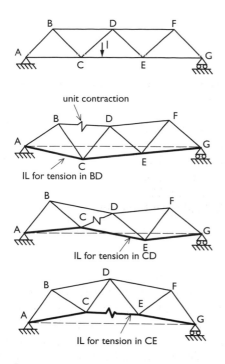

Figure 8.32 Influence lines for a statically determinate truss (shown as bold lines).

Q.8.17
For the girder bridge shown in Fig. 8.33 having three simple supports at A, B and D and a hinge at C, determine from first principles

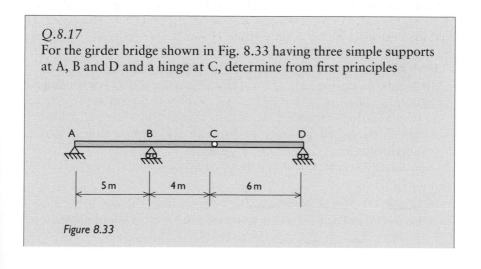

Figure 8.33

(i)   The IL for reaction at A
(ii) The IL for reaction at B
(iii) The IL for shear force just to the left of B
(iv) The IL for shear force just to the right of B
(v)   The IL for bending moment at B.

Hence determine the most critical reactions at A and B, shear forces near to B and bending moments at B when

a   two concentrated loads of 120 kN and 80 kN, 3 m apart, pass across the bridge in either direction.
b   a pedestrian loading is present with a maximum value of 16 kN/m.

Q.8.18
Use Müller–Breslau's principle to sketch the shape of the influence lines for Q.8.17. What can you say about the sign of the local distortion required to generate an IL when using the principle?

Q.8.19
For the continuous beam shown in Fig. 8.34, in which B and D are the mid-points of spans AC and CE respectively, sketch ILs for the following:

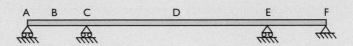

Figure 8.34

(i) Reaction at A, (ii) Reaction at C, (iii) Bending moment at B, (iv) Shear force to the right of C, (v) Bending moment at C, (vi) Bending moment at D.

   Hence identify, in each case, where a uniformly distributed load needs to be placed in order to produce the most critical design requirements.

Q.8.20
Sketch the ILs for axial force in the members CE, CD and BD of the truss shown in Fig. 8.35 if it is loaded via the bottom chord joints.

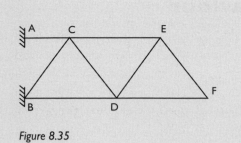

Figure 8.35

Q.8.21
Is it possible to use energy methods to determine ILs for deflection?

Chapter 9

Ultimate behaviour

Whereas it might be hoped that, under normal working conditions, all the material within structures may remain within the elastic range, it is important to know what eventually will happen if they are loaded to failure. Furthermore, it seems sensible to try to ensure that structures have similar load factors against ultimate collapse, if that is the main consideration in their design.

Sometimes structures are subject to abnormal conditions in which the principal considerations are survival of people and survival of the structure in that order. Abnormal conditions arise, for instance, in earthquakes or fires, when structures are hit by outside bodies such as road vehicles, ships or aircraft or when they are subject to blast loading. In such cases it is reasonable to expect that damage will be as localised as possible so that danger to people is minimised.

On completion of this chapter you should be able to do the following:

- Appreciate the difference between brittle and ductile behaviour of materials.
- Recognise the influence that this difference has on good design practice.
- Calculate plastic moments for beams and plastic collapse loads for beams and frames.
- Calculate ultimate loads for reinforced concrete beams.
- Understand some ways in which the worst effects of earthquakes, impact and fire might be alleviated through good design.

9.1 Ductility and brittleness

Ductility of mild steel

Mild steel behaves elastically up to its yield stress, at which stage a pronounced strain (as much as 1–2%) may take place before it will carry any further increase in stress (Fig. 9.1(a)). Yielding is caused by slip planes developing within individual crystals (Illston and Domone, 2001). When this happens 'Lüder lines', corresponding to the directions of high shear strain, may be detected on the surface. Whereas stress–strain curves of metals are normally taken from the performance of bars loaded in unidirectional

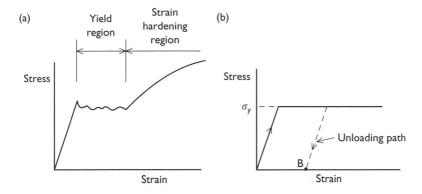

Figure 9.1 Load–deflection characteristics of mild steel: (a) typical; (b) idealised.

tension, it is the maximum shear stress (corresponding to half of the maximum difference in principal stresses) which determines when yield will occur.

Beyond the region of yielding there is a strain hardening region which is not of much direct use structurally because, at such large strains, the structure would be too distorted to be effective. However, the existence of strain hardening is important in so far as it ensures that it is safe to allow yielding to take place without the risk of the material failing completely. When yielding occurs, the strain is part elastic and part plastic. If unloading subsequently takes place, only the elastic strain is recoverable as seen by the typical return path in the idealised elastic–plastic behaviour shown in Fig. 9.1(b).

If mild steel has been stressed into the yield zone and then unloaded, any re-loading starts off from the new unloaded position, B, on Fig. 9.1(b). Hence the Lüder zone will be reduced or even eliminated depending on how much previous plastic deformation took place. This is the principle of work hardening as a method of increasing the performance of metals.

General ductile behaviour

Other metals, including high tensile steels, have yield characteristics, but there is usually a less pronounced discontinuity in behaviour, typical curves being shown in Fig. 9.2. When there is no definite yield point (e.g. with aluminium alloys), the 0.2% proof stress is used as a benchmark to indicate breakdown of elastic behaviour. The 0.2% proof stress is the stress at which there is a 0.2% permanent strain if the load is removed.

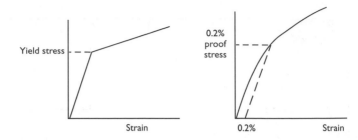

Figure 9.2 Ductile behaviour of metals.

The benefit of ductility

The important benefit of ductility is that the development of plastic strain in any part of a structure allows additional load to be carried by alternative load paths. Thus failure only occurs when all the possible load paths are at maximum capacity. This benefit will be greatest when the structure is highly redundant with many alternative load paths.

Ex. 9.1

It is required to compare the load carrying capacity of the two trusses shown in Fig. 9.3 in which members AB, BC and BD have cross-sectional areas of A, A and $\frac{1}{2}A$ respectively. They are to be made from round bars of steel which have a flat yield characteristic with a yield stress of σ_Y as shown in Fig. 9.1(b). Assuming that the joints are strong enough to carry all the applied loads, evaluate the percentage gain in load carrying capacity from adding the central member if

a   failure is deemed to take place when the yield stress is first reached;
b   yielding is allowed in the members.

Answer
Because of symmetry $P_{BC} = P_{AB}$ throughout.

a   For the two-bar truss $P = \sqrt{2}P_{AB} = \sqrt{2}A\sigma_{AB}$. Hence $P_{max} = 1.414A\sigma_Y$.
   But for the three-bar truss, if e is the vertical displacement of B, $e_{AB} = e/\sqrt{2}$ and $\ell_{AB} = \sqrt{2}h$. Hence $\sigma_{AB} = Ee/2h$ whereas $\sigma_{BD} = Ee/h$.

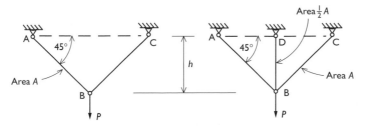

Figure 9.3 Simple trusses without and with redundancy.

Member BD is the first to reach yield when $Ee/h = \sigma_Y$ at which stage $\sigma_{AB} = \frac{1}{2}\sigma_Y$.

Thus from equilibrium

$$P_{max} = \frac{2P_{AB}}{2\sqrt{2}} + P_{BD} = \frac{2A\sigma_Y}{2\sqrt{2}} + \frac{A\sigma_Y}{2} = 1.207\sigma_Y$$

NB   This indicates that the extra member gives a reduction (rather than an increase) in load carrying capacity of 15%.

b   The maximum load that can be taken by the two-bar truss is when the yield stress is reached. As before $P_{max} = 1.414A\sigma_Y$.

However for the three-bar truss, once the yield stress is reached in member BD it can develop plastic strain whilst maintaining its load at $\frac{1}{2}A\sigma_Y$. As it stretches, the stresses in members AB and BC will increase, with failure only taking place when these are also at yield.

Thus

$$P_{max} = \frac{2A\sigma_Y}{\sqrt{2}} + \frac{A\sigma_Y}{2} = 1.914A\sigma_Y$$

This is a 35% increase in load capacity over the two-bar truss.

A particular advantage of ductility in statically indeterminate structures is that engineers do not need to worry so much about sources of stress which produce self-equilibrating systems such as due to lack of fit, movement of supports, and temperature strains. Suppose, for instance, that member BD of the three-bar truss was forced into place because its length did not exactly match the distance between B and D. During initial loading the elastic forces would be different to those calculated in the answer to part (a), resulting in yield being first reached at a different load. However, eventual collapse would still only occur when all three bars reach their yield stress. Hence the presence of the initial stresses would have made no difference to the plastic collapse load.

Modification of stiffness

Once parts of a structure have started to yield, the stiffness properties are modified. In the case of the three-bar truss, the stiffness (as measured by the slope of the curve of P/e) reduces to 59% of its previous value once member BD is yielding. In many cases this loss of stiffness is of no concern. However it is very important when a material, whose stiffness is required to prevent compression buckling, yields.

An example of this is a pin-ended strut carrying a large axial compressive load which is subject also to some lateral loads as shown in Fig. 9.4. If the yield stress is reached at A due to the combination of bending moment and axial compression, the bending stiffness at the centre section will start to reduce. This will result in a larger curvature there, more lateral deflection and hence a further increase in the bending moment.

Thus it requires only a small yield zone to develop at A before the strut will buckle, despite being well below the Euler load. This is why, in the construction of the first Quebec Bridge, raising permissible stress levels closer than usual to the elastic limit in compression members was particularly dangerous.

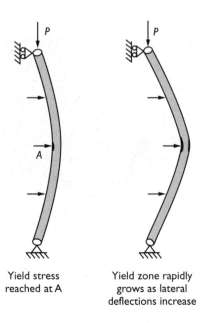

Yield stress   Yield zone rapidly
reached at A   grows as lateral
                     deflections increase

Figure 9.4 Collapse of a beam-column.

Brittle behaviour

Cast iron, used extensively in the nineteenth century was well known for being unreliable in tension because of the possibility of fracture. The important consideration with that and other brittle materials is the stress levels. Fracture is most likely to occur at stress concentrations (e.g. near rivet holes, bolt holes or scratches) or at imperfections, particularly where there have been flaws in casting.

Because it is the magnitudes of the stresses which are important, it would not be a good idea to add the third member BD in the trusses of Fig. 9.3 if the material is brittle. This member, being more highly stressed than the other two, is likely to fail early in the loading process. Indeed, it could be classified as a form of stress concentration.

Glass, another material which is well known for being brittle, is being increasingly used as the main cladding material in facades and attria. In view of its brittle nature, the following rules of design practice tend to be observed:

- A fail-safe approach is required where, if a component breaks, the rest of the structure will remain intact. This usually means that glass panels are hung from or supported by the main structure, but there are exceptions (Fig. 9.5).
- Replacement of broken panels should be possible.

Figure 9.5 Yarakucho glass canopy, Tokyo which received a Structural Achievement Award from the Institution of Structural Engineers in 1997 (Stansfield, 1999).

- Toughened glass which breaks into small pieces should be used where a breakage would cause a danger to the public.
- Methods for cleansing and maintenance need to be devised at the design stage.

Some types of brittle materials are fibrous in nature. These include carbon, glass and polyester fibres which are usually set in a matrix of softer material. Creep stress rupture is a danger in some of these materials, where there is a reduction in strength if they are subjected continuously to a high level of tensile loading.

Timber is a natural fibrous material which exhibits brittle behaviour. It needs special attention because of the differing strengths across and in alignment with the fibre direction and also because of knots and other irregularities. Where strength is important, careful visual inspection and grading is required. Laminating timber (i.e. gluing layers together) is a way of improving performance by reducing the effects of irregularities, avoiding splitting (common in bulk timber) and allowing different shapes to be manufactured with the most suitable fibre directions throughout (see Fig. 9.6). Plywood is a common form of lamination which gives good 2D bending characteristics to flat boards.

Figure 9.6 Laminated timber portal frames, Göteborg, 1986.

Concrete in compression

Concrete is a material whose internal structure is complex because of the aggregate content and possible existence of void spaces. When loaded in

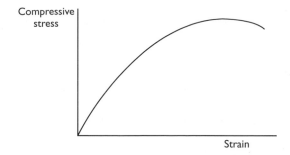

Figure 9.7 A typical stress–strain curve for concrete.

compression it has a non-linear load–deflection graph of the form shown in Fig. 9.7. The stiffness (indicated by the slope of the curve) decreases as the most highly stressed internal load paths give way. When the peak stress is reached, the concrete fails by crushing. Thus concrete behaves, to begin with, like a ductile material, but later exhibits brittle characteristics.

Concrete is also affected by shrinkage and creep. Shrinkage occurs in the initial curing process and when drying out. If shrinkage is prevented from taking place, tensile forces will develop which may cause cracking. Creep occurs when a sustained load is applied. The initial elastic deformation is supplemented by extra creep deformation whose rate decreases with time and is only partly recovered if the load is removed. These factors need to be taken into account in the design of concrete structures and in the assessment of their deflections.

Crack growth

Most structural materials contain microcracks which have the potential of extending when they are stressed. With a brittle material these cracks grow rapidly once the stress is high enough and this can lead to sudden fracture. Also the impact resistance of the material will be low. In contrast a ductile material absorbs a large amount of energy by plastic deformation as the cracks extend. Whereas it will have a high degree of resistance to a single load application, it may be vulnerable to fatigue failure due to repeated cycles of tensile loading (see Sections 3.7 and 5.11).

It is also important to note that the brittleness and ductility of materials can be influenced by several factors including chemical composition and temperature. For instance, not all forms of cast iron are brittle. On the other hand, all metals have a temperature below which they become brittle. Hence brittle fracture is a special hazard in very cold weather.

Q.9.1

If the trusses of Ex. 9.1 have the following properties: $E = 200\,\text{kN/mm}^2$, $A = 420\,\text{mm}^2$, $h = 5\,\text{m}$ and the material has a stress–strain characteristic idealised as shown in Fig. 9.1, with a yield stress of $240\,\text{N/mm}^2$, determine their load–deflection diagrams.

9.2 Ultimate strength in bending

Development of a plastic hinge

Consider a rectangular beam of breadth b and depth d made from elastic–plastic material having stress–strain diagrams for both tension and compression idealised as shown in Fig. 9.1. If it is assumed that, an increasing bending moment is applied to a cross-section, the strain is proportional to the distance from the neutral axis, there will be two distinct phases in the response as follows:

a   Until the outer fibres reach $\pm\sigma_Y$, the cross-section will remain elastic with a linear diagram of stress.

b   Above this value zones of constant yield stress will develop at the outer fibres, which will increase in depth as the bending moment increases.

If ε_Y is the strain at which yield is first reached such that $E\varepsilon_Y = \sigma_Y$, Fig. 9.8 shows how the bending moment increases with the outer fibre strain. The two key bending moments are the 'yield moment' M_Y (the bending moment at which yield first occurs, and the 'plastic moment' M_P, the maximum bending moment which develops at very large strains and to which the graph is asymptotic. The plastic moment is obtained by assuming that all the material has reached the yield stress with the sign of the stress reversing at the neutral axis (Fig. 9.9). From this figure $M_P = bd^2\sigma_Y/4$ and there is a 'plastic modulus' given by

$$Z_P = M_P/\sigma_Y = bd^2/4$$

which may be contrasted with the elastic modulus $Z = M_Y/\sigma_Y = bd^2/6$.

When the plastic moment is approached closely, the outer fibre strains, and also the curvature, increase without any significant change in the bending moment. This condition is called a 'plastic hinge'. If a statically determinate structure, which is being loaded proportionally, develops a plastic hinge, it must at that stage collapse.

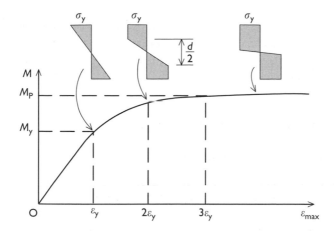

Figure 9.8 Relationship between bending moment and outer fibre strain (which is proportional to curvature) for a rectangular elastic–plastic beam. Changes in the stress diagram are shown.

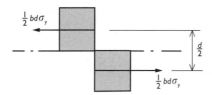

Figure 9.9 Stress diagram for a rectangular cross-section with a fully developed plastic moment.

Shape factor

The ratio $M_P/M_Y = Z_P/Z$, which for the rectangular beam in Fig. 9.9 is 1.5, indicates the reserve of strength still available when yield is first reached and is known as the 'shape factor'. Figure 9.10 shows a progression of symmetric cross-sections having reducing shape factor, the properties of which are shown in Table 9.1. Cross-sections which have most of their area concentrated into two flanges, making them efficient in bending, have a shape factor close to unity. This is because, at first yield, there is little reserve of strength left in the material. However, in comparing different cross-sections, it is important that cross-sections have sufficient wall thicknesses to prevent local buckling occurring in advance of the full plastic hinge forming.

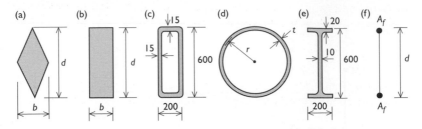

Figure 9.10 A progression of cross-sectional shapes.

Table 9.1 Shape factors for the cross-section shown in Fig. 9.10

Property	(a) Diamond	(b) Rectangle	(c) Hollow rectangle	(d) Hollow circular tube $(t \ll r)$	(e) I	(f) Elementary beam
$Z = M_Y/\sigma_Y$	$bd^2/48$	$bd^2/6$	3.25×10^6	$\pi r^2 t$	2.73×10^6	$A_f d$
$Z_P = M_P/\sigma_Y$	$bd^2/12$	$bd^2/4$	4.19×10^6	$4r^2 t$	3.10×10^6	$A_f d$
Shape factor	4	1.5	1.29	1.27	1.13	1

> *Ex. 9.2*
> Determine the shape factor for the I section which is case (e) of
> Fig. 9.10.
>
> *Answer* (using mm units)
> In the fully plastic state, the top flange carries a force of $4000\sigma_Y$ at
> distance 290 above the neutral axis. The portion of web above the
> neutral axis carries a force of $2800\sigma_Y$ with its resultant at distance
> 140 above the neutral axis.
> Hence, taking account of symmetry,
>
> $$M_P = 2(290 \times 4000\sigma_Y + 140 \times 2800\sigma_Y) = 3.10\sigma_Y \times 10^6$$
>
> However $M_Y = 2.73\sigma_Y \times 10^6$.
> Hence the shape factor is $3.10/2.73 = 1.13$.

Particular care needs to be taken with unsymmetric cross-sections
(Types III and IV of Fig. 4.19) because the neutral axis is no longer likely
to coincide with the centroid of the cross-section once yield zones start to
develop. The correct definition for the neutral axis is the locus of positions
of zero stress necessary to produce pure bending (i.e. with no axial force).

When the plastic moment is reached, since all the material is at the yield stress, the condition of no resultant axial load requires the cross-sectional areas above and below the neutral axis to be equal.

Ex. 9.3
Determine the plastic moment of the T section shown in Fig. 9.11 if its yield stress is 240 N/mm². Hence also determine the shape factor.

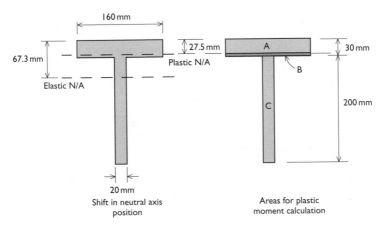

Figure 9.11 A T beam.

Answer
The cross-sectional areas of the flange and web are 4800 mm² and 4000 mm². Hence 4400 mm² of the cross-section must be above the neutral axis placing it in the flange at a depth of 27.5 mm (note that this is quite different from the elastic neutral axis position which is in the web at a depth of 67.3 mm). Splitting the cross-section into three regions, as shown in Fig. 9.11, provides a basis for calculating the plastic moment as shown in Table 9.2, giving an answer of 113.0 kNm.

Table 9.2 Calculation of the plastic moment of the T section of Fig. 9.11

Region	Area mm²	Force kN	Arm mm	Moment kNm
A	4400	1056	13.75	14.5
B	400	−96	−1.25	0.1
C	4000	−960	102.50	98.4
Total				113.0

> The I value for the section is $42.5 \times 10^6 \, \text{mm}^2$. Hence, with the maximum fibre distance from the centroid as $162.7 \, \text{mm}$, the elastic modulus is $261.5 \times 10^3 \, \text{mm}^3$. Dividing the plastic moment by the yield stress gives the plastic modulus as $471 \times 10^3 \, \text{mm}^3$. Hence the shape factor is 1.8.

Rigid-plastic theory

In rigid-plastic theory the elastic phase of the stress–strain diagram is ignored as shown in Fig. 9.12. A structure made from with this hypothetical material would not deflect until sufficient yield zones exist to produce a mechanism. This concept provides a simple means of directly predicting plastic collapse, bypassing any need to determine deflection characteristics. Thus, if the three-bar truss of Fig. 9.3 were made of rigid-plastic material, no movement would take place until all the bars reached their yield stress. Similarly the simply-supported beam ABC of length ℓ with a uniformly distributed load of w, shown in Fig. 9.13, would remain rigid until the central bending moment M_B reached its full plastic moment. Hence collapse takes place when $M_P = w\ell^2/8$ with the collapse mechanism shown in the figure, in which lengths AB and BC remain rigid while the cross-section at B acts as a plastic hinge.

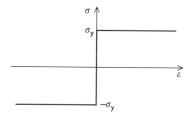

Figure 9.12 Rigid-plastic stress–strain diagram.

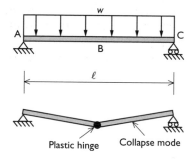

Figure 9.13 Collapse of a simply-supported beam according to rigid-plastic theory.

Ultimate strength of reinforced concrete beams

Because of the non-linear nature of the stress–strain curve for concrete in compression and also because of its variability, methods of determining ultimate bending strength of reinforced concrete beams use simple idealised stress–strain curves such as the one shown in Fig. 9.14. This curve differs from the rigid-plastic assumption for steel and has the following properties:

(i) The magnitude of the stress block is assumed to be two-thirds of the strength of the concrete as determined from a standard compression test.
(ii) For strains of less than 0.00035, the stress is assumed to be zero.
(iii) Beyond a strain of 0.0035, the concrete is assumed to be crushed and incapable of carrying any load. Hence at the point of failure the outer fibre strain of the concrete in compression is assumed to be 0.0035.

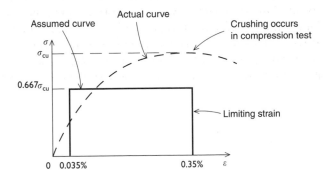

Figure 9.14 Approximation of a stress–strain curve for concrete in compression.

Ex. 9.4

A rectangular concrete beam of width 250 mm has three reinforcing bars of diameter 28 mm set with their centrelines at depth 360 mm from the top of the section as shown in Fig. 9.15. If the yield stress in the reinforcement is $\sigma_Y = 400\,\text{N/mm}^2$ and the crushing strength of the concrete obtained from a cube test is $\sigma_{cu} = 40\,\text{N/mm}^2$, estimate the ultimate bending moment that it can carry, checking whether it is under- or over-reinforced.

Answer

Area of steel $= 3 \times \pi \times 14^2 = 1847\,\text{mm}^2$

Hence the steel force at yield is $0.400 \times 1847 = 739\,\text{kN}$

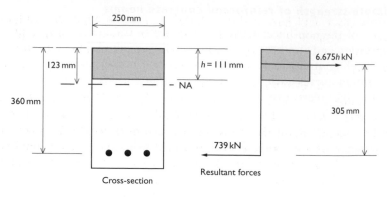

250 mm

123 mm

360 mm

$h = 111$ mm

NA

6.675h kN

305 mm

739 kN

Resultant forces

Cross-section

Figure 9.15 Analysis for ultimate load capacity of a reinforced concrete beam.

The concrete stress block is assumed to have a stress of $0.667\sigma_{cu} = 26.7\,\text{N/mm}$. If its depth is h, the concrete force is $26.7 \times 250h = 6675h\,\text{N}$. The forces are therefore as shown in Fig. 9.15. Equating these two forces gives $h = 111\,\text{mm}$.

If the strains at the top and bottom of the concrete stress block are respectively 0.0035 and 0.00035, the neutral axis must be at depth $111/0.9 = 123.0\,\text{mm}$. Because this is less than one-half of 360 mm the beam is under-reinforced. At failure the lever arm will be $360 - \frac{1}{2} \times 111 = 305\,\text{mm}$ giving the ultimate bending moment as

$$M_{ult} = 739 \times 0.305 = 225\,\text{kNm}$$

Ex. 9.5
Using concrete having a compressive strength $\sigma_{cu} = 36\,\text{N/mm}^2$ and steel having a yield strength $\sigma_Y = 500\,\text{N/mm}^2$, design a rectangular reinforced concrete beam to have an ultimate bending moment capacity of no less than 440 kNm, making the width approximately three-fourths of the depth to the reinforcement. Reinforcement bar diameters should be no more than 34 mm with at least 30 mm of concrete surrounding each bar.

Answer
Let the width and depth to reinforcement of the beam be b and d, respectively. Place the neutral axis at half the depth (i.e. as low as possible without making the beam over-reinforced). The concrete

stress block will have a depth $0.45d$. Hence the concrete resultant force is $0.667 \times 36 \times 0.45bd = 10.8bd$. But the lever arm will be $0.775d$. Hence the ultimate bending moment will be $10.8 \times 0.775bd^2 = 8.37bd^2$. Assuming that $b \simeq 0.75d$ and equating this bending moment to the required bending moment of 440×10^6 Nmm gives $d \simeq 412$ mm.

Assume $d = 415$ mm. The lever arm is 322 mm. Hence the steel and concrete forces are 1366 kN. Equating the concrete force to $10.8bd$ gives $b = 305$ mm. Also equating the steel force to $500A_{st}$ gives the required steel area as 2732 m^2. Sufficient steel area is provided by four bars at 30 mm diameter, which may be placed in a single row with five gaps of 37 mm each.

Q.9.2
A hollow beam is to have the cross-section shown in Fig. 9.16. Determine the elastic modulus, the plastic modulus and hence the shape factor.

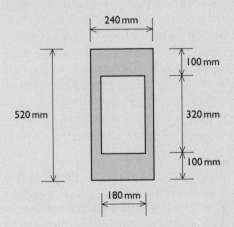

Figure 9.16 A rectangular hollow section.

Q.9.3
Two rectangular members of cross-section 20 mm $\times$ 80 mm and 10 mm $\times$ 80 mm act as the flanges of a beam, being linked by shear bracing such that the overall depth is 100 mm as shown in Fig. 9.17. If

the flanges both have a yield stress of 45 N/mm^2 and the bracing does not contribute to the bending strength, determine the plastic moment of the beam.

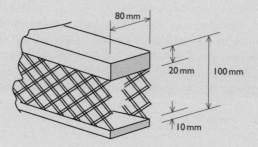

Figure 9.17 A hollow open-webbed beam.

Also determine what plastic moment can be achieved, without increasing the volume of material, by making both flanges of depth 15 mm.

Q.9.4
A reinforced concrete beam having width 300 mm and depth 460 mm has 3 × 34 mm diameter steel reinforcing bars. If the crushing strength of the concrete is $\sigma_{cu} = 32$ N/mm^2 and the yield strength of the steel is $\sigma_Y = 440$ N/mm^2, determine the maximum span, simply-supported at the ends, which can safely support a maximum (already factored) distributed load of 5 kN/m.

Q.9.5
For the solution to Ex. 9.5, check that the steel is yielded in the assumed ultimate condition, if $E_{st} = 200$ kN/mm^2.

9.3 Ductile failure of redundant structures

Equilibrium requirements for collapse

When considering the three-bar truss shown in Fig. 9.3, it was discovered that the onset of yield does not necessarily result in collapse. Similarly, where a structure carries bending moment, the formation of a plastic hinge does not necessarily result in collapse either.

A propped cantilever with a concentrated load

Consider a uniform cantilever ABC of length ℓ built in at A and propped at C, carrying a concentrated load, P, at its centre B. The elastic solution for the BMD (evaluated in Q.6.10) is shown in Fig. 9.18. If the load P is increased, a plastic hinge will form first at A when $P = \hat{P} = 5.32M_P/\ell$. With joint A acting as a hinge, the beam will carry additional load as a simply-supported beam. The maximum amount of additional load is $\bar{P} = 0.68M_P/\ell$ at which stage a second plastic hinge forms at B. Once this hinge forms, collapse is by the mechanism shown in Fig. 9.18.

If the collapse condition were to be appreciated from the start, this could have been analysed directly using equilibrium as follows. With M_A and M_B as the bending moments at A and B (sagging positive), the reaction at C as $P/2+M_A/\ell$, the bending moment at B is $M_B = P\ell/4+M_A/2$. By substituting $M_B = M_P$ and $M_A = -M_P$, this equation yields

$$P = 6M_P/\ell$$

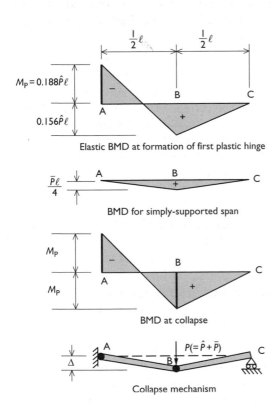

Figure 9.18 Loading stages of a propped cantilever carrying a concentrated load.

which is the collapse load. Note that, in this case, two plastic hinges are required to form a mechanism which is one more than the number of redundancies in the structure.

A propped cantilever with distributed load

Performing a similar analysis for the case of a uniformly distributed load w/unit span applied to the same beam, the elastic solution gives $M_A = -0.125w\ell^2$ and a maximum sagging bending moment of $0.0703w\ell^2$ (see the solution to Q.6.9). Hence the first plastic hinge forms at A with $w = \hat{w} = 8M_P/\ell$ at which time the maximum sagging bending moment is $0.5624M_P$. Adding further distributed load $\overline{w}$ with the beam simply-supported gives a sagging bending moment at the centre of $\overline{w}\ell^2/8$. Hence it is safe to add $\overline{w} = 0.4376M_P \times 8/\ell^2 = 3.5M_P/\ell^2$ making the total load $w = \hat{w} + \overline{w} = 11.5M_P/\ell^2$. However, because the maximum sagging bending moments for the elastic and the supplementary loading cases occur at different spanwise positions, nowhere will the sagging bending moment reach precisely M_P. Hence $11.5M_P/\ell^2$ will slightly underestimate the collapse load.

The static theorem for plastic collapse

These results are illustrations of the following theorem which is generally applicable to beams and frames fabricated from ductile material: 'If a set of internal forces in equilibrium with the applied loads has $|M| \leq M_P$ throughout, the structure is at or below the collapse condition'. One of the names for this is the static theorem (alternative names are the lower bound theorem or the safe theorem).

Mechanism requirements for collapse

In order for a plastic collapse mechanism to be activated, it is necessary for the work done by the external forces in any movement to overcome the energy absorbed by the plastic hinges.

The propped cantilever with concentrated load

Consider a movement of the mechanism shown in Fig. 9.18 in which the central deflection is Δ. The rotations of the plastic hinges are $\theta_A = 2\Delta/\ell$ and $\theta_B = 4\Delta/\ell$ (note that the signs of the rotations have been ignored). Hence the energy absorbed by the two plastic hinges is

$$M_P(|\theta_A| + |\theta_B|) = 6M_P\Delta/\ell$$

(Because energy can only be absorbed at each plastic hinge, all of the contributing terms must be positive.)

However, with the external work done equal to $P\Delta$, the mechanism will be on the point of moving when $P = 6M_P/\ell$, which is the same result as before.

The propped cantilever with distributed load

When the load is distributed the maximum sagging bending moment will occur closer to C than to A. If it is assumed to occur at distance 0.55ℓ from A, with Δ as the displacement there, the rotations of the plastic hinges at A and B are $\Delta/0.55\ell$ and $(\Delta/0.55\ell + \Delta/0.45\ell)$ respectively (see Fig. 9.19), giving the energy absorption as $5.86M_P\Delta/\ell$. However, the distributed load drops on average a distance $\frac{1}{2}\Delta$ and hence the external work done is $\frac{1}{2}w\Delta\ell$. Equating the external work done to the energy absorbed gives

$$w = 11.72M_P/\ell^2$$

There will be different possible collapse loads predicted from mechanisms with different positions for the sagging plastic hinge. Of these the one reached first, as the load is increased, must be the theoretical collapse load. Hence the prediction $w = 11.72M_P/\ell^2$, if not correct, must be an over-estimate of the collapse load.

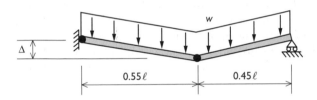

Figure 9.19 Assumed mechanism for the propped cantilever carrying a distributed load.

The kinematic theorem for plastic collapse

These results are illustrations of the following second theorem which is also generally applicable to beams and frames fabricated from ductile material: 'The load required to activate a rigid-plastic mechanism must be greater than or equal to the collapse load'. One of the names for this is the kinematic theorem (alternative names are the upper bound theorem or the unsafe theorem).

Use of bounds on the plastic collapse load

If both methods of analysis are applied to the same problem, the critical load will be bracketed between two bounds. For the propped cantilever subject to the distributed load, for instance, it has been ascertained that

$$11.5 M_P/\ell^2 < w_{cr} \le 11.72 M_P/\ell^2$$

(the accurate theoretical solution is $w_{cr} = 11.657 M_P/\ell^2$).

However, it is more common to use just one of these methods. For the propped cantilever with the single concentrated load considered earlier, either method could have been used on its own with confidence that the answer was correct. Ways of using one or other of the techniques to analyse some more complex problems will be discussed below.

Mechanisms for collapse of beams and frames

Fundamental to using the kinematic theorem safely is the need to investigate the collapse of sufficient mechanisms that the critical one is bound to be amongst them. Furthermore, to avoid unnecessary work, it is helpful if mechanisms which have no hope of being critical can be discarded quickly.

If a member of a frame is uniform and is subject only to concentrated lateral loads, the only positions which need to be considered for plastic hinges are the ends of the member and the positions at which the external loads are applied (there cannot be a unique maximum bending moment anywhere else because the bending moment diagram will be linear between these points).

Figure 9.20 shows the feasible mechanisms for five statically indeterminate beams. These have been designated taking account of the following:

- A plastic hinge at a load point can only rotate in sympathy with the direction of the load. Thus the third mechanism for beam (a) is not feasible for beam (b).
- The number of plastic hinges may be less than the number of redundancies plus one in cases where there is only a partial collapse of the structure, as with beam (c).
- Any redundancy associated with axial movement of a beam should be discounted when considering the number of plastic hinges. For instance beam (d) has only two redundancies as far as this investigation is concerned.
- Mechanisms marked with a cross are clearly not critical. For instance, the second mechanisms for beams (a) and (b) have less external work than the first mechanisms but the energy required to activate their plastic hinges is similar. In beam (c) the external work is similar between the first two mechanisms but the second requires more energy to activate it.

Figure 9.21 shows feasible mechanisms for three loaded frames having uniform members. In each case the plastic moment for the beam or the

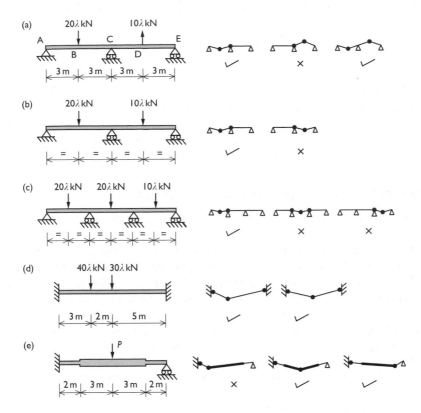

Figure 9.20 Some loaded beams, showing alongside feasible mechanisms. All are uniform except the last which has an enhanced plastic moment over the centre span. Those mechanisms which can be discounted as being critical are marked with a cross.

rafter members is greater than that for the columns. The following should be noted:

- Movement of a mechanism should not change the length of any member. Hence the position of the knee joint in frame (a) is fixed by the two members and cannot displace in any of the mechanisms.
- Where a plastic hinge forms at an eaves joint, it will be in the column because it is the weaker in bending of the two connected members.
- In (c) there are two possible mechanisms each with two plastic hinges. One is the first mechanism shown and the other its mirror image. These will require the same load to activate them and can be combined to give the symmetric mode with three plastic hinges having the same associated load. This symmetric form is likely to be preferred for analysis.

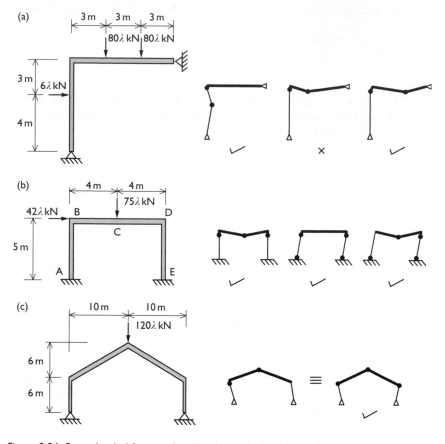

Figure 9.21 Some loaded frames, showing alongside feasible mechanisms.

Use of the kinematic theorem for analysis

Ex. 9.6
Estimate the load factor λ at which collapse of beam (a) of Fig. 9.20 will take place if its plastic moment is 35 kNm.

Answer (using kN and m units)
Consider the first of the mechanisms shown with B deflecting a distance Δ. The external work done is $20\lambda\,\Delta$. The plastic hinges at B and C

rotate through angles of $2\Delta/3$ and $\Delta/3$ respectively. Hence to activate the mechanism;

$$20\lambda\Delta = 35(2\Delta/3 + \Delta/3)$$

which gives $\lambda = 1.75$.

The second mechanism shown does not need to be analysed. For the third mechanism B and D will deflect by the same distance, say Δ. Evaluation of the work equation gives

$$30\lambda\Delta = 35(2\Delta/3 + 2\Delta/3)$$

which gives $\lambda = 1.55$. Because the third mechanism has the lowest activating load, it will cause collapse and $\lambda = 1.55$ is the collapse load factor.

Ex. 9.7 (using kN and m units)
Estimate the load factor λ at which collapse of frame (b) of Fig. 9.21 takes place, if the plastic moments of the beam and columns are 198 kNm and 126 kNm respectively.

Answer
For the first mechanism let Δ be the deflection of C. The rotations of the plastic hinges at B, C and D are $\Delta/4$, $\Delta/2$ and $\Delta/4$ respectively. Hence the work equation is

$$75\lambda\Delta = 126 \times \frac{\Delta}{4} + 198 \times \frac{\Delta}{2} + 126 \times \frac{\Delta}{4}$$

Hence $\lambda = 2.16$.

For the second mechanism let Δ be the horizontal deflection of the beam. The plastic hinges at A, B, D and E all rotate by $\Delta/5$. Hence the work equation is

$$42\lambda\Delta = 126 \times \frac{4\Delta}{5}$$

which gives $\lambda = 2.4$.

If, for the third mechanism, the horizontal deflection of the beam is Δ, C will drop by $4\Delta/5$. The rotations of the plastic hinges at A, C, D and E are $\Delta/5$, $2\Delta/5$, $2\Delta/5$ and $\Delta/5$ respectively. Hence the work equation is

$$42\lambda\Delta + 75\lambda \times \frac{4\Delta}{5} = \frac{\Delta}{5} \times 126 + \frac{2\Delta}{5} \times 198$$
$$+ \frac{2\Delta}{5} \times 126 + \frac{\Delta}{5} \times 126$$

which gives $\lambda = 1.76$. Hence plastic collapse is at a load factor of 1.76 initiated by the third mechanism shown in Fig. 9.21.

Use of the static theorem for design

The static method is particularly suitable for design in the situation where the plastic moments of members require to be chosen such that one or more sets of forces do not produce collapse. A quick and safe method which is not necessarily economical is to find any equilibrium solution and choose the plastic moments of members to be no less than the largest bending moments acting on them. However, it is also possible to refine the technique by varying the magnitude of redundancies or by varying the plastic moments in order either to produce more economical solutions or to satisfy other design requirements.

Ex. 9.8
Suggest possible safe designs for beam (a) of Fig. 9.20 based on the criterion that collapse is not to occur below $\lambda = 1$ for the given loading.

Answer
First trial: A quick answer is obtained by assuming a value for the bending moment at C, say $M_C = 0$. This gives the first BMD shown in Fig. 9.22. If a beam is chosen with a plastic moment of 30 kNm, the safety criterion is satisfied.
Second trial: In this design, the full plastic moment is only needed in the region of B. The plastic moment for the span CE could be reduced to 15 kNm. Thus the bending moment capacity envelope is reduced as shown in the second BMD of Fig. 9.22.
Third trial: It is possible to see if improvements can be made by varying the single redundancy. If M_C is the bending moment, sagging positive,

at C, equilibrium of spans AC and CE give

$$M_B = 30 + \tfrac{1}{2}M_C$$

$$M_D = -15 + \tfrac{1}{2}M_C$$

Reducing M_C in order to reduce M_B will also increase the modulus of M_D. Thus benefit is gained, but only until $M_B = -M_D$. Solving these three equations gives $M_C = -15\,\text{kNm}$, $M_B = 22.5\,\text{kNm}$ and $M_D = -22.5\,\text{kNm}$. Thus a uniform beam can be used safely having a plastic moment of 22.5 kNm (the third BMD in Fig. 9.22).

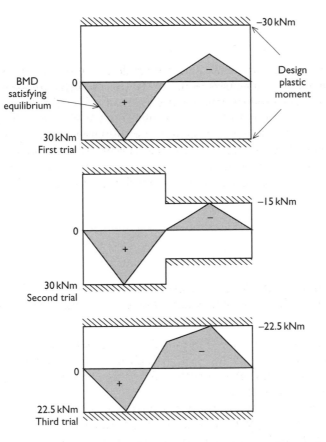

Figure 9.22 Use of the static theorem to produce different designs for a continuous beam.

A note on terminology

With small problems using hand solution or with larger problems using linear programming methods on a computer, the correct theoretical collapse load can usually be obtained when using the kinematic method. The answer will be the same as for the static method if that is optimised. Hence the term 'unsafe' and 'safe' for the theorems can often be misleading. Whereas the names upper and lower bound are appropriate for analysis problems, the bounds are reversed when considering design. Thus the kinematic method gives a lower bound to the required plastic moment and the static method gives an upper bound. It is for these reasons that the less familiar terminology 'static' and 'kinematic' have been used in this text.

Applicability of plastic methods

Care has to be taken when using plastic design, not only that the material is sufficiently ductile, but also that loss of stiffness at the onset of yield does not cause premature buckling. Thus thin-walled sections are considered unsuitable for plastic design because of the likelihood of local buckling. Care also needs to be taken where large alternating forces are present, in so far as it is important that plastic hinges should not be continually worked backwards and forwards. This danger is illustrated by bending a paper clip backwards and forwards. It fractures after only a few cycles. Whereas plastic deformation may take place, structures should 'shakedown' to a situation where subsequent alternations in load are carried by entirely elastic deformation. It should also be noted that where a member is carrying significant axial load, this will reduce the magnitude of the bending moment required to produce a plastic hinge. Horne and Morris (1981) may be consulted for more detailed aspects of the application of plastic theory to low-rise frames.

Plastic theory can be used to predict the ultimate collapse of plates via yield-line theory. Bolt groups and plates as elements of joints are also amenable to plastic analysis (Fig. 9.23). The extent to which plastic theory

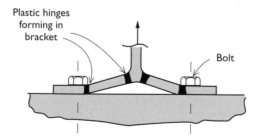

Figure 9.23 Plastic failure of a bolted joint subject to tensile force.

is applicable to reinforced concrete structures depends on the amount of plastic deformation required to reach the plastic collapse condition and the degree to which cross-sections are under-reinforced. This is because the more under-reinforced a cross-section is, the higher will be the permissible plastic curvature before crushing of the concrete takes place.

Q.9.6
Figure 9.24 shows a uniform propped cantilever subject to three imposed loads.

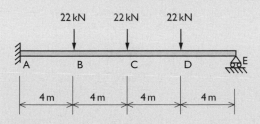

Figure 9.24

a By assuming the bending moment at the root to be $M_A = -100\,\text{kNm}$, obtain a value for a suitable plastic moment for the beam using the static theorem of plastic collapse.
b By considering variations in the value of M_A, optimise the required plastic moment.
c Investigate the collapse load of all possible mechanisms using the kinematic theorem and show that this gives the same result as the static theorem for the optimum design.

Q.9.7
A uniform cantilever AC of length ℓ, propped at C, could be subject to a single concentrated load of magnitude P acting at some point B distance x from the root A. Determine the distance x which requires the largest value of M_P to prevent plastic collapse and obtain an expression for the required M_P in terms of P and ℓ.

Q.9.8
Identify all the possible mechanisms which would need to be analysed using the kinematic theorem in order to determine the plastic collapse

load for the pitched roof portal frame shown in Fig. 9.25, assuming that all members have the same plastic moment.

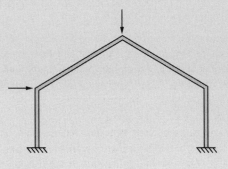

Figure 9.25

Q.9.9
In the frame of Fig. 9.26, beams AC and DF have a plastic moment of 50 kNm and columns CF and FG have a plastic moment of 20 kNm. With joints A, D and G assumed to be pinned and the dimensions as shown, determine the magnitudes of the loads P_B and P_E, applied to the centre points of the beams, which will just cause collapse.

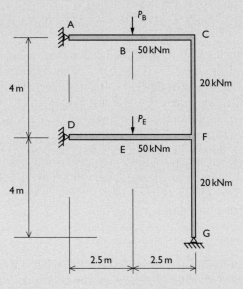

Figure 9.26

9.4   Earthquake resistance

Earthquakes as a hazard

Shah (1983) recorded an average of 13,000 deaths per year from 180 earthquakes which took place from 1947 to 1980. The damage to buildings and facilities is also a very awesome statistic. Major earthquakes occur where different continental plates (parts of the earth's crust) come into contact with each other. Fault lines may lie dormant for years but stresses can be gradually building up across them. When slippage takes place it is by a repeated slip-stick motion producing an irregular oscillating ground disturbance which carries on until the local stresses have been relieved.

Motion has both horizontal and vertical components. Although fault lines do not always appear on the surface, in 1906 the San Andreas fault produced a 6.4 m horizontal shift on the surface at San Francisco. In 1964 the Great Alaska Earthquake caused a 1.5 m vertical movement. Trees killed by sea water inundation on the downward slipping side of this fault, can still be seen after 40 further years. Major earthquakes tend to be followed by aftershocks caused by disturbances to neighbouring parts of the fault line. Unfortunately fore-shocks, which would provide a warning, rarely occur.

Earthquakes can cause damage, death and injury in several ways by

- severing services of water, gas and electricity with the possibility of fires being started;
- causing ground failure (landslips, soil liquefaction, etc.);
- generating tsunamis (water waves) resulting from sea bed or lake bed movements;
- causing building collapses.

Soil liquefaction tends to occur where there is soft soil or reclaimed land. Pore pressure builds up to such an extent that the soil behaves like a heavy fluid causing bearing failure of foundations or, where the ground is sloping, imposes large lateral forces. As a result buildings may sink, topple or start to break up. However, a very large percentage of damage and deaths arises from the collapse of buildings due to the shaking motion. It is not much of an exaggeration to say that it is not earthquakes that kill people but falling structures.

A slow learning curve

In the late 1960s, the plate tectonic theory was put forward, explaining the reason, via continental drift, why some areas of the world have a higher earthquake risk than others. In the 1970s, earthquake engineering was well advanced in the USA (Newmark and Rosenblueth, 1971). However, there are various reasons why technical advances do not produce immediate world-wide impact in stopping the devastation and loss of life.

Building methods are changed through codes of practice. However, provision of better earthquake resistance requires more care in design and construction. It may also involve additional cost (although not necessarily a lot more). Furthermore, clients may be inconvenienced (e.g. by requirements to include shear walls or bracing members where open access is desired). Particularly where clients, architects, engineers and builders have not themselves experienced severe earthquakes, there is likely to be a reluctance to change previous practices.

Altering building practices does not, however, reduce vulnerability of existing buildings unless strengthening measures (termed 'retrofits') are undertaken. Often, when earthquakes strike, structures which suffer most damage are those where owners have lacked the finance to undertake retrofit measures. Furthermore, as new forms of construction are developed, it is only when any of them are subjected to a major earthquake that theory can be supplemented by hard experience. As populations increase in vulnerable regions, risks also increase unless positive steps are taken to reduce them.

Different countries have different building materials and styles of construction. Hence lessons in one country do not necessarily apply directly to other countries. Unlike the USA and Japan, countries which have weak economies tend not to have the resources to carry out meaningful research programmes. It is in those countries that the largest death tolls usually occur, areas of poor housing that have had no planning input or control being particularly vulnerable.

Theoretical considerations

A low-rise building which is strong and stiff and has plenty of damping in the form of construction will tend to move with the ground motion. The mass of the structure and anything it is supporting will generate inertia forces which are approximately proportional to the ground acceleration. The Hyogo-Ken Nanbu earthquake which struck Kobe on 17 January, 1995 was recorded as having accelerations on rock of more than $0.8g$ horizontally and $0.3g$ vertically, with associated maximum displacements of the order of 200 mm and 100 mm respectively. However, at the surface there was amplification due to soft ground conditions resulting in accelerations of up to $1.5g$ horizontally and $1.2g$ vertically (Brebbia, 1996). Hence forces to be resisted in the most violent earthquakes may be greater than the self-weight and may act in any direction.

The weight of the structure itself is important as this forms a major component of the inertia forces to be resisted. For a structure to be strong, yet light, material preference may be in the order timber, steel, reinforced concrete and (last) masonry. Masonry (if not prestressed) has the problem of being unable to carry tension if cracked, and so is particularly susceptible to damage in earthquakes.

If a building is not well damped, its elastic properties will cause any induced energy to be retained. Oscillations of the building could overswing severely, particularly where a natural frequency of vibration of the building is close to the frequency of the earthquake motion. Energy dissipation benefiting the damping characteristics arise in a number of ways including hysteresis in the stress–strain curve of the material and friction in joints. Where movements are severe, energy dissipation can occur, for example, through friction in masonry cracks or ductility when steel frames or the bars of reinforced concrete yield.

The response of a multi-storey building to an earthquake depends on the mass, strength, stiffness and damping properties of each of the storeys. In view of these many factors care has to be taken making generalisations.

Newmark and Rosenblueth (1971) pay particular attention to the need to eliminate sway modes by providing bracing or shear resisting panels in multi-storey buildings. They also give a warning about 'inverted pendulums' (e.g. Fig. 9.27) where the base of the column needs to be sufficiently robust to carry all the imposed shear and bending forces. They also highlight the particular vulnerability of hydraulic structures such as dams and storage tanks where large forces arise from the inertia of the liquid.

With very tall buildings, the principal natural frequencies of horizontal motion will be significantly slower than the frequencies of earthquakes. The large inertia of the main part of such buildings, particularly if well damped, will cause them to remain mainly static, hence the techniques for earthquake-resistant design concentrate on their lower storeys.

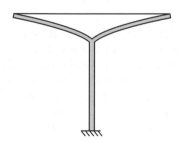

Figure 9.27 Cross-section of a canopy as an example of an inverted pendulum.

Failure patterns in buildings

Traditional structures of masonry do not perform well in earthquakes. Where horizontal earth movements align with walls, diagonal cracks tend to arise where walls have been unable to resist the tensile component of the (reversing) shear forces. Openings for windows and doors provide a

Figure 9.28 Diagonal cracks in an adobe brick building construction at Erzincan, 1992.

Figure 9.29 Collapse of a stone masonry end wall bringing down the roof of a single storey school building near Erzincan, 1992.

particular focus for these cracks. Figure 9.28 shows some damage in the 1992 Erzincan earthquake in Turkey, in which the maximum horizontal acceleration was 0.5g, where traditional buildings are of adobe (a sun-baked clay/straw brick). When earthquake movements occur at right angles to the alignment of walls, they may topple sideways, bringing down any floor beams or roof structure that they support, as shown in Fig. 9.29.

Two ways of strengthening masonry structures against earthquake damage are by including reinforced concrete ring beams above the level of openings and by using a cement rendering in which a wire mesh is placed (Coburn and Spence, 1992). Both buildings shown in Figs 9.28 and 9.29 had ring beams in place. The ring beam which can be seen in Fig. 9.28 may have prevented complete collapse, although the one in the school building, Fig. 9.29 was not able to prevent the end wall collapsing sideways.

An earlier earthquake in the Erzincan region in 1939, with a death toll of 33,000, had destroyed 140,000 homes. In 1992 the death toll was 400 with over 700 serious injuries. The lower death toll was partly due to the small number of traditional buildings present (Williams, 1992). Many of the buildings suffering damage or collapsing were of reinforced concrete construction of up to six storeys with masonry infill panels which were built *in situ*. The death toll would have been higher if the earthquake had occurred when office buildings were occupied. Serious collapses occurred with reinforced concrete structures which did not conform to the 1975 Turkish code of practice. Problems arose through the presence of weak storeys (having less resistance to lateral forces), poor design of reinforcement, weak concrete, poor design of infill panels and asymmetric construction. Figure 9.30 shows two structures in which complete storeys have collapsed. It is with pancake collapses such as these that most fatalities occur. In the first of these, asymmetry in the structure or mass distribution could have caused torsional motion giving rise to additional forces.

In the Armenia earthquake of 1988 estimates of the death toll were between 25,000 and 100,000. Here there were many pancake collapses of multi-storey framed buildings. Collapses have been attributed, to a large extent, to inadequate splices between beams, columns and shear walls in the newer prefabricated forms of construction and inadequate ties allowing end walls to fall out in the older masonry structures (Alexander, 1993). Pancake collapse of multi-storey framed buildings in Bhuj have also been a major factor in the five-figure death toll in the 2001 Gujarat earthquake, India (Arnold, 2001).

Kobe had not had a major earthquake since 1596 and most of the houses were constructed prior to the introduction of earthquake resistant forms of construction in the 1970s. Traditional forms of Japanese construction did not fare well here. Being a warmer climate than in the north of Japan where they use plywood, the timber houses had walls of mud-reinforced bamboo lattice and were divided by paper partitions. However, these light structures supported heavy clay or concrete tile roofs designed to resist typhoons. Collapse of these types of houses and fires caused by severed gas mains and overturned oil stoves cost 4000 lives. A total of 100,000 buildings completely collapsed and almost 300,000 more were damaged (Chandler and Pomonis, 1997). These included many multi-storey buildings almost

(a)

(b)

Figure 9.30 Collapse of two four-storey reinforced concrete framed buildings at Erzincan, 1992: (a) here the lower two floors have collapsed; (b) the left-hand side of this building, only a part of which is showing on this photograph, has suffered a complete pancake collapse into the double basement.

all of which were of reinforced concrete built before 1981. In 1981, a new code of practice was introduced requiring the shear ties and links to be more closely spaced in order to hold any disintegrated concrete in place.

Multi-storey buildings

Experience has shown that a particularly important factor in the performance in earthquakes of multi-storey buildings is the degree of continuity in the structural properties between storeys. It is common for buildings with infilled panels in upper storeys to have a 'soft' storey at ground level due to openings to accommodate entrances, shop windows or car ports. The stiffness of the upper floors to sway movement, as compared with the more flexible soft storey, causes most of the sway movement to take place at the soft storey putting it at high risk of initiating a pancake collapse. Thus the increased sway stiffness of the upper storeys may actually contribute to, rather than reduce, the risk of failure. Similarly, soft storeys higher up in a building or marked changes in structural properties between storeys can create hot spots for dynamic stresses which can also put buildings at risk.

Bridges and elevated roadways

Bridges and viaducts have often suffered damage from earthquakes. In the 1971 Sylmar earthquake, USA, 66 bridges were damaged on the Golden State Freeway. One of the problems encountered (not for the first time, see Newmark and Rosenblueth, 1971), was that simply-supported spans were thrown off their supports. Figure 9.31 shows how restraints were introduced in retrofit programmes to prevent both lift off and lateral movements at the supports. However, in the 1989 San Francisco earthquake, collapse of spans of the Cyprus Street Viaduct occurred despite these type of restraints being present. In this case the collapse was due to failure of reinforced concrete columns supporting, on either side, the upper level of a two-tier elevated roadway. Reinforcement round the outside of the columns burst outwards due to the alternating tensile and compressive strains, and the neighbouring concrete disintegrated. To combat such column failures, some retrofit programmes now involve enclosing columns in a steel jacket and filling any internal space between the jacket and the original column with grout (see Fig. 9.31). Carbon fibre or reinforced concrete jackets may also be used.

   At Kobe, long sections of the Hanshin Expressway supported on only a single line of columns toppled sideways. Figure 9.32 shows how the bottom of the columns disintegrated with much of their main reinforcement bursting out. There is a noticeable similarity in this design with the inverted pendulum

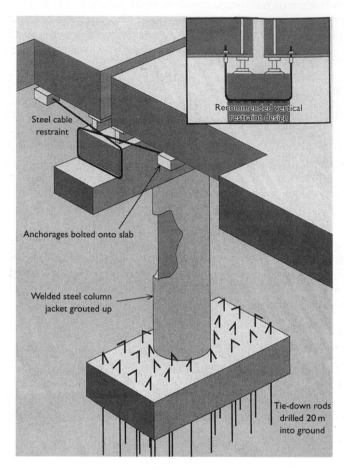

Steel cable restraint

Anchorages bolted onto slab

Welded steel column jacket grouted up

Recommended vertical restraint design

Tie-down rods drilled 20 m into ground

Figure 9.31 Retrofit measures to stop simply-supported spans lifting off their supports and also bursting failures of reinforced concrete columns.

warned about by Newmark and Rosenblueth (1971). The expressway had been built 30 years previously and the columns had not been strengthened since then. However, newer elevated highways, built to more stringent codes of practice, did survive at Kobe, and retrofitted bridges also survived the 1994 Northridge earthquake.

Planning aspects

Kobe is claimed to have been the most expensive earthquake to date. Apart from the structural problems mentioned above, buildings were damaged by

Figure 9.32 A toppled section of the Hanshin Expressway, Kobe, 1995. Notice the disintegrated reinforced concrete column base in the bottom right of the picture.

pounding (knocking against each other). Debris filled roads causing danger to people escaping from buildings and preventing access for emergency services to put out fires and conduct rescue work. The port of Kobe, which is Japan's largest, was severely disrupted because of damage associated with soil liquefaction.

In earthquake situations public buildings often have a key role. For instance hospitals are particularly important. Yet in some earthquakes it has been those very buildings which have collapsed. Mitigation of the worst effects of earthquakes is therefore an area where structural engineers need to work in conjunction with others involved in town planning and medical and emergency services to assess risks and to consider priorities with regard to construction and retrofit.

9.5 World Trade Center aftermath (McAllister, 2002)

Antisocial behaviour

The world's largest structural collapse occurred when two Boeing 767-200ER planes, which were each large enough to carry 200 passengers, were deliberately flown on a suicide mission at high speed into the twin towers of the World Trade Center (WTC) on 11 September 2001. It is a tragedy of civilisation that people sometimes behave antisocially and destroy things that are most precious, not only property but also the lives of others and even themselves. With advancing technology comes changes in the techniques of warfare and also possibilities for terrorism. To make buildings as safe as possible against smart bombs, crashing planes or whatever technological artefacts may be available in the future, it would be necessary to build them like tanks or put them underground. Leslie Robertson, the designer of the twin towers, after much heartsearching, thinks that the answer is not to build substantially stronger towers, but to find ways of reducing tension and conflict. However, it would be wrong not to take the opportunity of investigating what happened at the WTC in order to learn how to improve future design and construction techniques in case similar events occur in the future.

The World Trade Center

The WTC site, developed by the Port Authority of New York and New Jersey, contained six office buildings and a hotel surrounding a 5 acre plaza (Fig. 9.33). Below plaza level there were up to seven storeys which included two metro stations. Above plaza level twin towers (WTC1 and WTC2) rose to 417 m, each containing 110 floors of dimension 63 m square and providing almost one-third of the office space of the WTC. For a brief period when it was completed in 1970, WTC1 was the world's tallest building. WTC3 was a 22-storey hotel and WTC4–7 had 9, 9, 8 and 47 storeys, respectively, above plaza level.

Tower design

One of the problems of tall building design is the way space is interrupted by the heavy columns and bracing or shear walls required to carry the weight of building above and any side forces induced. The unique design of the WTC towers provided large open plan floor areas, despite the height, by adopting the following special features:

- The self-weight was kept low by using steel as much as possible in preference to concrete.

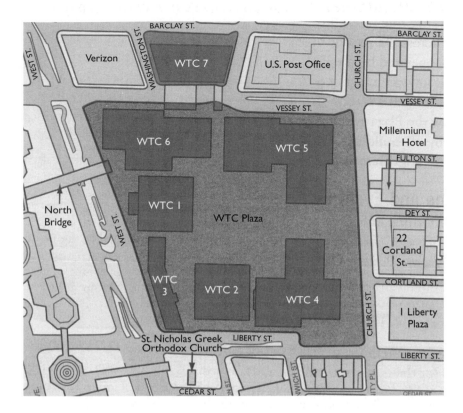

Figure 9.33 WTC plan.

- Columns were concentrated in the walls and the central core, so enabling the space in-between to be open plan (Fig. 9.34).
- Each wall comprised 59 closely spaced built up box columns approximately $0.35 \, m^2$ at 1.016 m centres, connected by spandrel plates at the floor levels with the intervening space glazed between floors.
- The core, which contained circulation space, services, elevators and staircases, was supported on box and wide flanged columns.
- The fire partitions around the core were made from plasterboard.
- Floor construction was typically 100–125 mm of lightweight concrete on a profiled sheet deck.
- Outside the core, floors were supported by steel trusses (called 'open-web joists' in the US) spanning between the central core and the exterior walls as shown in Fig. 9.34.
- These trusses were in pairs at 2.03 m spacing with spans of 18.3 m or 10.7 m depending on whether they connected to the sides or ends of

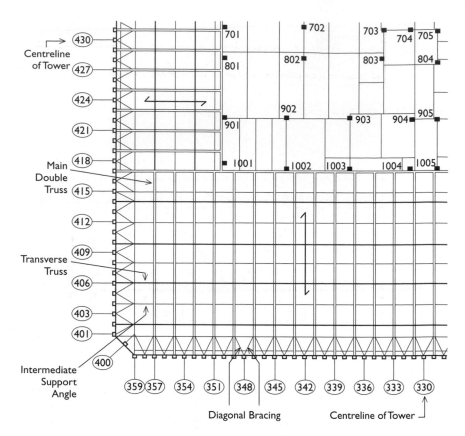

Figure 9.34 Typical plan of one-quarter floor area of a tower showing positions of wall columns, floor trusses as well as beams and columns within the core.

the core. Their diagonals extended above the top chord to act as shear connectors, thus enabling composite action to take place between the top chords and the concrete of the deck (Fig. 9.35).

- These trusses supported transverse trusses (Fig. 9.36) which in turn supported the steel deck whose profiling was aligned parallel to the main trusses.

The effectiveness of this structural design was dependent on the same wall and core dimensions being maintained throughout the towers so that there was continuity in the vertical members. A system of express and local lifts was adopted in such a way that the number of lifts within the core did not vary much from top to bottom, thus allowing the core to be of constant

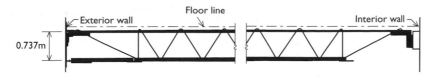

Figure 9.35 A main floor truss.

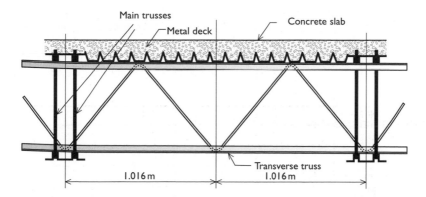

Figure 9.36 Part of a transverse floor truss (to a larger scale than Figure 9.35).

dimensions. The uniform dimensions of the exterior walls gave the towers their unique slender appearance which was functional rather than elegant.

With an aspect ratio of 6.6, the towers behaved as vertical cantilevers. When a tower was subjected to lateral force in the plane of one of the two centre-lines, two exterior walls acted as flanges and the other two as webs. Thus the exterior walls acted as a pierced tube providing a high degree of stability as well as carrying about 40% of the gravity forces.

In order to provide wider access openings in the exterior wall at ground level, each set of three columns was merged into one by means of fork-like junctions. The resulting massive columns carried on downwards to spread footings socketed into the bedrock. Diagonal bracing was also included to carry lateral forces in the lower sections. Although the appearance of the towers suggests a uniform structure, the wall thicknesses of columns varied with height from as little as 6 mm at the top to as much as 100 mm at the base.

Impact damage

The WTC towers were the first structures outside military installations and the nuclear industry where design studies were conducted to consider the impact of a jet liner. However the situation envisaged was that a plane, due to land at a nearby airport, would have been off course, being lost in fog. It would therefore have been flying at landing speed and low on fuel. The planes that hit the twin towers were somewhat larger and heavier than the one envisaged in the design study (and very much larger than the one that hit the Empire State Building on July 28th, 1945). They were also flying at 3 or 4 times the anticipated speed and with a much larger fuel load.

Figure 9.37 shows the damage to the north face of the North Tower (WTC1) between floors 94 and 98 where the first plane hit. Of the 59 columns on that face, 33 of the central ones were severed and there must have been severe local damage to the structures of the floors and core (landing gear from this plane passed right through the building). Debris from the plane and broken elements of the building must have been scattered around putting higher loads on undamaged parts of the structure. The other plane hit the south face of the South Tower (WTC2) between floors 78 and 84 and to the east side of the centreline of the building. More parts of this plane travelled through the building. Again there was much damage to the exterior wall and internal structure.

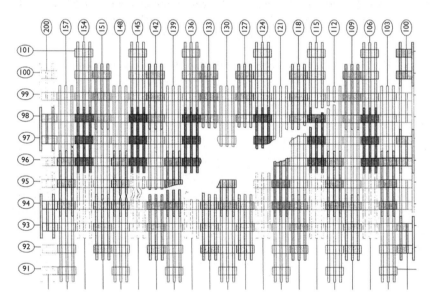

Figure 9.37 Impact damage to exterior columns on the north face of WTC1.

Alternative load paths

The fact that both towers withstood the force of the impacts despite considerable damage is a tribute to the strength and robustness of their design. There were an estimated 58,000 people in the WTC at the time, of whom almost one-third are likely to have been in each of the towers. If the towers had toppled on impact, the loss of life would have been very extensive indeed, not only in the towers but also in neighbouring buildings.

Floor loads transferring to external columns just above where they had been severed would have put them locally into tension. The gravity forces acting on the upper parts of these severed columns must have transferred, through Vierendeel truss action between the columns and the spandrel plates, to neighbouring continuous columns in the exterior wall. As a result these, neighbouring columns must have been very heavily loaded. Although the main and transverse floor trusses, on their own, had a statically deter-minate configuration, because they were interlaced in the form of a grid, alternative load paths were available in the event of local damage occurring. Also the horizontal bracing next to the exterior walls (shown in Fig. 9.34) would have helped to prevent distortion of the walls at other than the imme-diate impact zone. Damage within the core is likely to have incapacitated some interior columns. However, the columns in the core were linked to the outer shell of the building at roof level by ten outrigger trusses situated between floors 106 and 110 as shown in Fig. 9.38. Thus load carried by any damaged column in the core would have transferred through this truss system to undamaged columns.

The events on September 11, 2001 thus highlighted the importance of providing alternative load paths through redundancy in buildings.

Evacuation

With lifts dangerous to use in an emergency, evacuation from tall buildings requires adequate staircases. In each WTC tower there were three staircases of 1.12 m width or more. Thanks to the installation of emergency lighting and the calling of fire drills following a bomb attack in the basement area of the WTC in 1993, virtually all people in the towers below the points of impact exited safely.

Unfortunately, because of damage to the cores and also the ensuing fires, people above the impact zones were not able to descend and were killed in the subsequent collapses. Staircases were all situated within the cores but were not in the same positions throughout each tower. Figure 9.39 shows their position in the impact zone for WTC1 which were relatively close together and, as it happened, directly in the path of the aircraft debris. Even if the roofs of the towers had been clear enough for helicopters to land on, evacuation by helicopter would have been prevented by the palls of thick smoke. The death toll was 2830 of whom 157 were in the planes, 403 were firemen and other emergency staff. The other 2270 were mainly those trapped in the top portions of the towers.

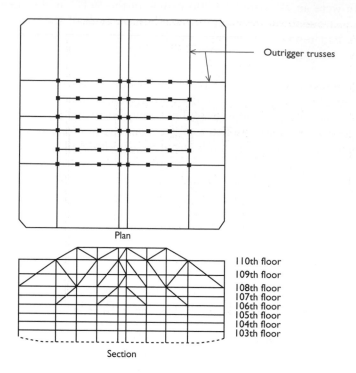

Figure 9.38 The outrigger truss system.

Fire damage

Fireballs at impact were estimated to have burnt off up to one-third of the aircraft fuel in the first two-seconds. The rest is likely to have burnt off within about five minutes. However, fire continued through, burning office furniture, papers, books, etc. over several floors with ventilation coming from damaged walls and blown out windows. Ceiling temperatures rose to about 1000°C at peak with 1.15 GW of heat generated for each tower (similar to that of a commercial power station). These fires continued for 102 min (WTC1) and 56 min (WTC2) until the towers collapsed.

Steel loses strength with temperature. Figure 9.40 shows the effect of temperature rise on one of the grades of steel used in the WTC. In order to delay loss of steel strength in a fire, a spray of mineral fibre was applied to all steel surfaces which would otherwise be exposed. Originally this had an average thickness of 19 mm for the steel floor trusses, but in 1990 a decision was made to upgrade this to 38 mm whenever floors became vacant. By September 11, 2001 all the floors in the impact zone of WTC1 had been treated, but only one in the impact zone of WTC2. Fireproofing was to

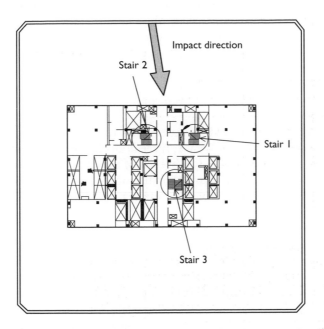

Figure 9.39 Position of staircases in the impact zone of WTC1.

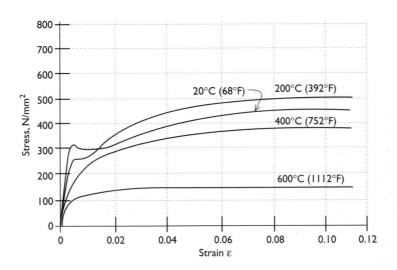

Figure 9.40 Effect of temperature on the characteristics of one of the grades of steel.

a two- or three-hour rating. However, the rating was more of a comparative measure with actual behaviour depending on local circumstances.

Compartmentation to restrict the spread of fires was achieved by sealing any holes through floor slabs due to cabling and plumbing and shielding staircases and elevator shafts by means of gypsum board walls. There were also standpipes, fire pumps, emergency water tanks, sprinkler systems and a ventilation system designed to keep smoke away from the core area in the event of a fire.

Accumulated debris including wall panels was reported by people in the core area of the 91st floor of WTC1, three floors below the impact level. This indicates that there was not only a breach of the compartmentation but also structural damage within the core. Furthermore, a pressure wave must have travelled down the elevator shafts at impact because windows were blown out at concourse level.

Collapse

There have been several reasons postulated why columns gave way causing total collapse after the fires had been raging for a while:

- Expansion of floors due to heat could have pushed columns outwards so introducing bending stresses in them.
- Sagging of floors due to heat or partial failure of supporting trusses could have rotated the columns or pulled them inwards. This would also have introduced bending stresses into the columns.
- Failure of floor/column joints could have allowed a floor to fall onto the one below. This not only would have put an extra load on this floor, but also would have removed lateral support from the columns, doubling their effective lengths (see Fig. 9.41).

The columns of the exterior walls were prefabricated in units of three columns and three spandrel plates which were bolted together on site. By staggering these units it was arranged that only one in three of the columns had bolted joints at any particular floor level (Fig. 9.42). Even so, the existence of bolted joints in the columns would have reduced their stiffness and may have contributed to their final collapse. On the other hand, the interior columns had less lateral support and could have failed in advance of the columns in the exterior wall.

There are a number of possible reasons why WTC2 collapsed after a shorter time than WTC1:

- The columns in the damaged region of WTC2 were under higher compressive load than those of WTC1.

- Any lean of the tower tops caused by the impacts would have had more of an effect on the structure of WTC2 than of WTC1 (NB   WTC2 is recorded as falling to the east and then the south).
- There was a lower level of fire protection for exposed steel in the impact area for WTC2 than there was for WTC1.

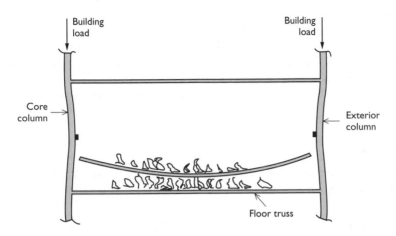

Figure 9.41 A possible failure in which floor/column joints give way and columns buckle.

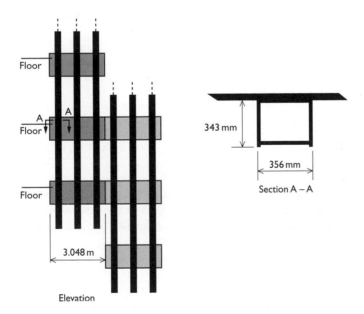

Figure 9.42 Two prefabricated exterior wall units (see also Figure 9.35).

Collapse of the towers down to ground zero would probably have been caused by debris falling on lower floors and shearing their joints. Collapses of this sort, where local damage to one part of a building causes consequential damage to other parts by means of a domino effect is called progressive collapse. Engineers in the UK became aware of this danger when a gas explosion on the 18th storey of a 22-storey block of flats at Ronan Point near London in 1968 took out a load bearing wall. The unsupported structure above the point of the explosion fell down and the debris destroyed parts of all the lower floors (Griffith *et al.*, 1968). Since then there have been other examples round the world of progressive collapse including at the Alfred P. Murrah Federal Building in Oklahoma City in 1995 where a truck bomb brought down a large swathe of its nine stories (Corley, 2001).

Collateral damage

Falling debris from the towers not only blocked roads making access difficult but also destroyed, partially or completely, neighbouring buildings with fires being started in some of these. A decision was taken not to attend a fire in the 47-storey WTC7 because of priorities elsewhere. After burning for approximately seven hours, it collapsed. Information about the real performance of steel structures in fire has been obtained from these buildings which were of similar construction to many others in the USA and elsewhere.

Other collateral damage occurred due to loss of support for the retaining wall keeping out the Hudson River from the substructure. This did not give way, but it did move and has had to be tied back until such time as it could be held in place by the floors of the reconstructed substructure.

Q.9.10
Discuss possible methods of providing bolted joints in the columns of the WTC towers between each prefabricated wall unit, taking account of their square box shape.

Q.9.11
In view of those events on September 11, 2001 discussed here, suggest what engineers should ensure or consider in the design of tall buildings required for human occupation. Organise your answer under the following headings:

a   structural design,
b   fire protection,
c   provision for escape.

Answers to questions

Q.1.1
315 kN at 7.111 m behind the leading axle.

Q.1.2
2700 N at 5.333 m from C.

Q.1.3
Force $= \int_0^4 10(11 - 1.5y)y\,dy = 560\,kN$
Moment about top of barrier $= \int_0^4 10(11 - 1.5y)y^2 dy = 1387\,kNm$
Hence centre of pressure is at $y = 2.476$ m.

Q.1.4
300.4 kN at 119.7° clockwise from the x axis.

Q.1.5
2150 kN.

Q.1.6
105 N, 100.6 N inclined upwards at 16.7° (i.e. perpendicular to the inclined part of the cord).

Q.1.7
90.05 N at 60.02° clockwise from the x axis together with a clockwise moment of 54 Nm.

Q.1.8
$P_1 = 39.14\,kN$, $P_2 = 39.45\,kN$, $\alpha = 55.3°$.

Q.1.9
The hydrostatic pressure block is triangular with a maximum pressure of 4.355 kN/m^2 at B and tapering to zero at 4.286 m from B. The depth of the pontoon at B is 444 mm giving an angle of tilt of 5.9°.

Q.1.10
The coordinates of F relative to A are (5.537 m, 2.519 m) and the slope of FE is 11.12°. Hence the perpendicular distance of A from EF is 3.539 m. G lies 3.695 m to the right of A. Hence by taking moments about A, $P = 271$ kN. Resolving forces gives $H_A = 266$ kN and $V_A = 208$ kN.

Q.1.11
a   53.3 kN.
b $W > 153$ kN.

Q.1.12
a   173.9 N.
b P_{AE} changes from pull to push if the intersection of AB and EF moves to the left of the centre of gravity of the signboard.

Q.1.13
a   The resultant of the 200 N and 600 N forces acts at a horizontal distance of 0.9375 m from A. Because the ground reaction must be directed towards the intersection of this resultant and the wall reaction, H_B, its inclination to the vertical must be $\tan^{-1}(0.9375/5.196) = 10.2°$.
b   21.7°.

Q.1.14
Their point of intersection is at infinity. For the other force to be concurrent it must therefore also be parallel.

Q.1.15
The two forces must be equal in magnitude, opposite in direction and have the same line of action (conforming with Newton's third law of motion).

Q.1.16
a   Rotation about the left-hand support is not prevented.
b   All body freedoms are restrained and the reactions may be obtained from equilibrium.
c   All body freedoms are restrained but there are too many reactions to obtain these from equilibrium alone.
d   All body freedoms are restrained and the reactions may be obtained from equilibrium (it needs to be assumed that the left-hand support will be able to develop a downwards reaction and therefore prevent lift off).

Q.1.17
Some possibilities are as follows:

- an aircraft or a submarine in steady motion;
- a hot air balloon in steady motion;
- a person or object, lying on a trampoline.

Q.1.18

a   Side loading on the swing changes the inclination of the chains and thus modifies the geometry. Confirmation that the principle of superposition does not apply in this case comes from the realisation that it is impossible to obtain an equilibrium solution for the swing subject only to the 40 N horizontal load.

b   The geometry of the pontoon does change when the 6 kN load is applied. However unlike (a), the rotations can be considered as small.

With only the self-weight acting, the hydrostatic pressure is uniform at 1.067 kN/m² and with only the 6 kN eccentric load acting the extreme hydrostatic pressures are $\bar{P}_1 = 0.267\,\text{kN/m}^2$ and $\bar{P}_2 = -0.267\,\text{kN/m}^2$. When added to 1.067 kN/m², the correct answer is given. Hence the theory of superposition holds, provided that the impossibility of a negative hydrostatic pressure occurring, when one of the two loads is applied on its own, is ignored.

Q.1.19

The gondolas will tilt slightly with their base further out than their roof when they are ascending and descending.

Q.1.20

The angular velocity, ω, is $\pi/16 = 0.196\,\text{rad/s}$. The horizontal centripetal acceleration is $\omega v = 18.85\,\text{m/s}^2$ giving a horizontal 'g' of $18.85/9.81 = 1.92$ and a lateral force requirement of 15.76 kN. The aerodynamic lift is therefore 17.75 kN with an angle of bank of 62.5°. Figure A.1 shows the pseudo-equilibrium configuration.

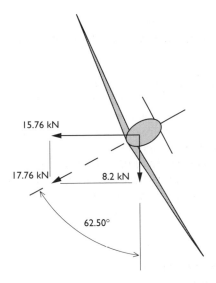

Figure A.1

Q.1.21

For a slope of 1 in 10, $\beta = \tan^{-1} 0.1 = 5.7°$. Forces are

a $\mu W \cos \beta + W \sin \beta = 120\,\text{kN}$
b $W \sin \beta = 40\,\text{kN}$
c $\frac{1}{2} W \sin \beta = 20\,\text{kN}$

The last result can be obtained by taking moments about the ground contact point.
(NB   Parry (2000) argues that (c) is the most likely way that large megalithic blocks were transported.)

Q.1.22

One possibility is shown in Fig. A.2.

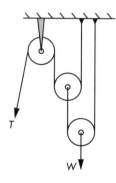

Figure A.2

Q.1.23

Because only two forces act on the wheel, these must be equal in magnitude, opposite in direction and have the same line of action (see Q.1.15). If A and B signify where the line of action intercepts the axle and the ground (Fig. A.3), the inclination of AB to the vertical is increased from zero as follows:

a   Ground softness produces a large contact area allowing the point B to move forwards.
b   A similar effect can be produced by soft wheels (such as pneumatic tyres).
c   If the axle/wheel contact is friction free the line AB must pass through the centre point of the wheel. However friction tends to move the point A backwards as the wheel tries to rotate (assuming that the driving force is applied to the wheel). The movement of A will depend on the diameter of the axle.

Hence factors which increase the slope of AB, thus reducing the efficiency of the wheel are ground or wheel rim softness, friction in the axle and a large ratio of axle diameter to wheel diameter. Note that ground friction is not generally important.

Direction of force trying to
move wheel

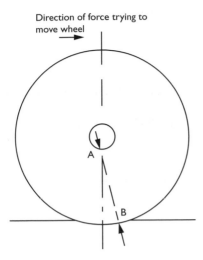

Figure A.3

Q.1.24

a   The free-body diagram is shown in Fig. A.4.
b   If $x = 6$ m, equilibrium of moments about A cannot be maintained.

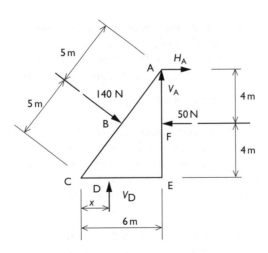

Figure A.4

Q.1.25

a $P = 1143\,N$ (see Fig. A.5(a)).

b $P = 1093\,N$, $H = 80\,N$ giving a resultant force on the nail of $1096\,N$ at $4.2°$ to the vertical (see Fig. A.5(b)).

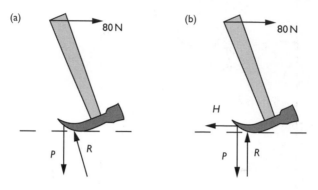

Figure A.5

Q.1.26

If the lower pipes are about to move away from each other, there will be no contact force between them. Hence the three forces acting on each lower pipe must be concurrent at the ground contact point (A for the LH pipe in the exploded free-body diagrams shown in Fig. A.6). From the geometry, R_B must be at $15°$ to the normal between the pipes. Hence the coefficient of friction between the pipes must be at least $\tan 15° = 0.268$. With R_A having components of $1.5\,W$ vertically and $0.5\,W \tan 15°$ horizontally, slippage will not occur between the pipes and the ground if the coefficient of friction is greater than 0.089. (Hence the lower pipes are much more likely to roll outwards than to slip at the ground contact points.)

Q.1.27

a   With forces defined in the free-body diagrams of Fig. A.7, $H_A = H_B = H_C = Wx/2r$, $V_B = V_C = Wx/s$ and $V_A = W - Wx/s$.

b   All but V_A are maximum when $x = 25\,m$ at values of $H_A = H_B = H_C = 333\,kN$, $V_B = V_C = 100\,kN$. V_A is maximum when $x = 0$ with a value of $200\,kN$.

Q.1.28

Taking moments about BD gives $V_A = 0$. The inclination of the chain has no effect on this result. V_A will change if loading is applied away from BD (e.g. at A or C) or if the positions of supports at B or D are moved.

Q.1.29

The equilibrium solutions are shown in Fig. A.8.

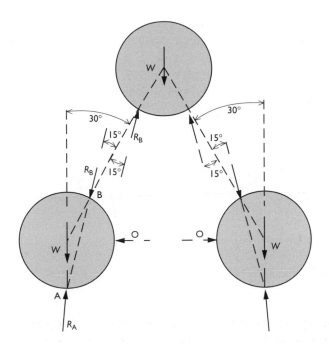

Figure A.6

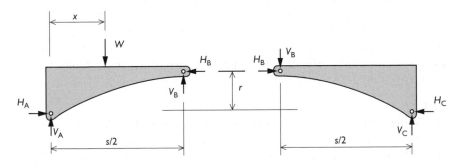

Figure A.7

Q.1.30
$R_B = -573.3\,\text{N}$, $R_C = 573.3\,\text{N}$, $R_F = 860\,\text{N}$. The hinges and catch are assumed to allow rotation.

Q.1.31
a 889 N,   b 1778 N placed on rim of table opposite a leg.

(a)

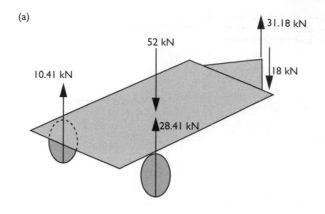

(b)

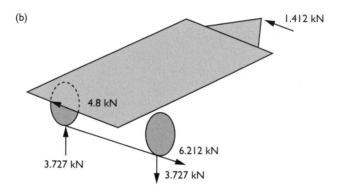

Figure A.8

Q.1.32

The rod lengths AB and BC carry the same loading in both configurations (namely W and $2W$), as do the connections at A and C. The difference is that, whereas connection B only carries a load of W, the connections B_1 and B_2 carry W and $2W$ respectively. It appears that the higher loading on connection B_2 in comparison with B had not been appreciated. This coupled with other weaknesses in the connection led to a failure of the form shown in Fig. A.9.

Q.2.1

- Reduces use of material.
- Reduces weight (easier to transport).

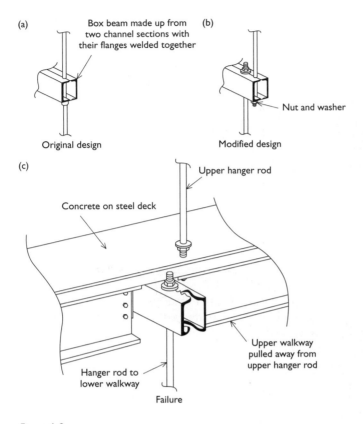

(a)   Box beam made up from   (b)
       two channel sections with
       their flanges welded together

Nut and washer

Original design   Modified design

(c)   Upper hanger rod

Concrete on steel deck

Upper walkway
pulled away from
upper hanger rod

Hanger rod to
lower walkway

Failure

Figure A.9

- Helps to avoid slippage at the joints through some mortar entering the voids.
- Helps to avoid distortion and cracking when cooled during manufacture.

Q.2.2
- Distributes local loads (load on an individual brick spreads on to two, three etc. on lower layers).
- Makes it difficult for vertical cracks to form in a wall (e.g. due to ground subsistence).

Q.2.3
The wall will be one brick thick at least. Bricks laid lengthwise are called 'stretchers'. Here every fourth row is of 'headers' which are bricks laid across to bind the sets of stretchers together. (This configuration is called 'garden wall bond'.)

Q.2.4

a $98.1\,\text{N/m}^2$, b $216.3\,\text{N/m}^2$. Thus the cavity wall requires only 45% of the load, compared with the other, to topple it.

Q.2.5

a   If P is the total lateral force and H the horizontal reaction at the top, equilibrium of both blocks is satisfied with the following force components or resultants (specified as horizontal, positive to the right; vertical, positive upwards).

On ABFE At A: $(H, -W)$, At the centre: $(-\frac{1}{2}P, -\frac{1}{2}W)$, At F: $(\frac{1}{2}P - H, 3W/2)$

On EFCD At F: $(H - P/2, -3W/2)$, At the centre: $(-P/2, W/2)$, At D: $(P - H, 2W)$

Solution of moment equilibrium equations for both blocks yield

$$H = 11\,Wt/2h \quad \text{and} \quad P = 12\,Wt/h$$

b   Two reasons why the above answers may not be true theoretical ones are as follows:

   (i) A different bedding joint other than EF could open up before the lateral load reaches its calculated value. This may be investigated by performing further equilibrium analyses with the blocks being of unequal height.

   (ii) A bedding joint may slip. Slippage of the top joint is the most likely to occur. In such a case, the whole wall below the top joint will rotate about D. Its equilibrium could be analysed by taking moments about D.

Q.2.6

For uniformly distributed vertical loading, the thrust line must be a parabola. If s is the span, a parabola touching the intrados at the springings and the crown has the formula $y = 4hx(s - x)/s^2$. But for $x < s/2$, the intrados has the formula $\bar{y} = 2hx/s$. The (vertical) thickness required of the arch is the maximum value of $y - \bar{y}$ which is $h/4$ occurring when $x = s/4$.

NB   This arch thickness does not allow for any safety factor against collapse.

Q.2.7

a H may be changed. V_1 (the vertical component of end reaction) may be changed. The thrust line may be raised (or lowered) bodily.

b   When subject to downwards loading, a flat arch comprising rigid blocks can only form a mechanism by pushing an abutment sideways. Thus hinges will form on the intrados at both springings and at the extrados close to the centre (for the uniform loading) which together with one abutment sliding gives a mechanism.

Q.2.8

The thrust line will comprise two straight lines meeting at an apex under the applied load. With the load applied between the centre and A on Fig. A.10, two straight lines can be drawn between the intrados and the extrados, so it is possible to support such a load. However, with the load to the left of A, the longer of the two lines will not fit, and so a large load there cannot be supported. From the triangle OBC, angle BOC = 28.07°. Hence angle AOX = 11.14° making A a distance 0.657 m from the centre-line. The situation regarding loading being applied to the right-hand side will be a mirror image of this result.

NB   Masonry arch bridges are usually most vulnerable when loaded near the quarter (or three-quarter) span positions.

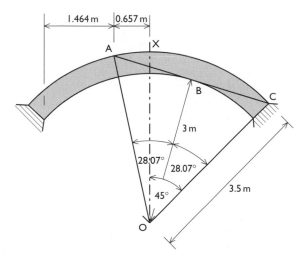

Figure A.10

Q.2.9

a

- The thrust line for the self-weight of the bridge should be as close as possible to the centre-line of the arch ring.
- The bridge should have a large self-weight.
- The arch ring thickness/rise ratio should not be small (if the rise/span ratio becomes very small, however, elastic effects become very important and abutment reactions become excessively high).

b

- Concentrated loads (over the haunches in particular) will disperse somewhat through the infill before they reach the arch.

- The spandrel walls and infill may carry compression load, allowing the thrust line to migrate above the extrados without collapse occurring.
- The formation of a collapse mechanism is resisted by material in the spandrel. This is particularly true above a hinge starting to form in the extrados where the material will be crushed.

Q.3.1

With a direction arrow clockwise, the internal forces are at A: $S_A = 0$, $P_A = 80\,\text{kN}$, $M_A = 2.4\,\text{kNm}$, and at B: $S_B = P_B = 56.57\,\text{kN}$, $M_B = 1.697\,\text{kNm}$ (see Fig. A.11).

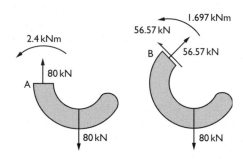

Figure A.11

Q.3.2

From overall equilibrium: $H_F = 60\,\text{kN}$ (to the right), $V_A = 65.14\,\text{kN}$ (downwards) and $V_F = 7.86\,\text{kN}$ (downwards). Hence from the method of sections: $P_B = 60.13\,\text{kN}$, $S_B = -7.45\,\text{kN}$, $M_B = -57.23\,\text{kNm}$, $P_C = 60.13\,\text{kN}$, $S_C = -39.95\,\text{kN}$, $M_C = 96.8\,\text{kNm}$, $P_D = 60\,\text{kN}$, $S_D = 40.14\,\text{kN}$, $M_D = 96.8\,\text{kNm}$, $P_E = 60\,\text{kN}$, $S_E = 16.14\,\text{kN}$, $M_E = 12.42\,\text{kNm}$.

Q.3.3

From overall equilibrium: $H_F = 18\,\text{kN}$ (to the left), $V_F = 100\,\text{kN}$ (upwards) and $M_F = 122\,\text{kNm}$ (anticlockwise). Hence from the method of sections: $P_B = -18\,\text{kN}$, $S_B = 40\,\text{kN}$, $M_B = -100\,\text{kNm}$, $P_C = 0$, $S_C = -60\,\text{kN}$, $M_C = -150\,\text{kNm}$, $P_E = -100\,\text{kN}$, $S_E = -18\,\text{kN}$, $M_E = -50\,\text{kNm}$ where member directions are to the right for AB and CD and upwards for FE. See Fig. A.12 for the equilibrium of joint B/C/E.

Q.3.4

a   Because there are four external restraints, one internal hinge is required to produce statical determinacy. This needs to be in member BC or CD (not AB or DE) and not close to B or D.

b and c   These are statically determinate.

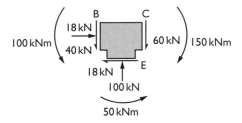

Figure A.12

d   This has the correct number of hinges for statical determinacy, but the hinge on AB will produce a mechanism allowing the outboard portion to rotate about the hinge. Instead this hinge should be placed within the hoop BCFE (e.g. near to B).

Q.3.5

With units of N/mm^2, the Mohr's circle is centred at (80, 0) with a radius of 92.195. Thus $\sigma_1 = 172.195$, $\sigma_2 = -12.195$, $\tau_{max} = 92.195$. Also $\tan 2\alpha = -4.5$ giving $\alpha = -38.74°$ with the Mohr's circle as shown in Fig. A.13(a). Since point $(\sigma_1, 0)$ is 77.47° anticlockwise from point (σ_x, τ_{xy}) on the Mohr's circle, the plane on which σ_1 acts will be $\frac{1}{2} \times 77.47°$ clockwise from the plane on which σ_x acts. Other planes may be determined in a similar way to identify the equivalent stress systems shown in Fig. A.13(b) in units of N/mm^2.

Q.3.6

The eigenvalue equations are

$$\begin{bmatrix} 110 & 50 \\ 50 & -45 \end{bmatrix} \begin{bmatrix} 0.9592 \\ 0.2826 \end{bmatrix} = \begin{bmatrix} 119.64 \\ 35.24 \end{bmatrix} = 124.73 \begin{bmatrix} 0.9592 \\ 0.2826 \end{bmatrix}$$

$$\begin{bmatrix} 110 & 50 \\ 50 & -45 \end{bmatrix} \begin{bmatrix} -0.2826 \\ 0.9592 \end{bmatrix} = \begin{bmatrix} 16.87 \\ -57.29 \end{bmatrix} = -59.73 \begin{bmatrix} -0.2826 \\ 0.9592 \end{bmatrix}$$

Thus $\sigma_1 = 124.73$ and $\sigma_2 = -59.73$.

Q.3.7

With p and q in the Lamé–Maxwell equations aligned with the hoop and radial directions, $\sigma_p = \bar{p}r/t$, $\partial\sigma_p/\partial p = 0$. Also $\partial\alpha/\partial p \simeq -1/r$, $\partial\alpha/\partial q = 0$. Hence the first L-M equation is automatically satisfied. The second equation gives $\partial\sigma_q/\partial q = -(\sigma_q - \sigma_p)/r$. Neglecting σ_q on the RHS gives the radial stress gradient as $\partial\sigma_q/\partial q = \sigma_p/r = \bar{p}/t$.

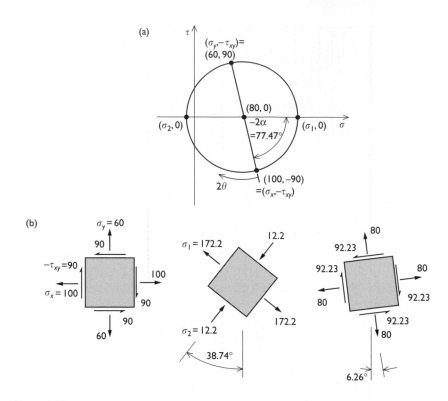

Figure A.13

Q.3.8
The original stress distribution (ADEF on Fig. A.14) was uniform. However, since the notch area is not taking load, the tensile stresses across the centre-line will need to be high next to the notch and lower on the LHS so that the resultant force still passes through the centre point B. A stress concentration is expected at C with crowding of primary stress lines there. The stress distribution curve has been estimated and the position at which the primary stress lines intersect the centre-line have been predicted (by making the area of the curve divide into ten equal parts). The way in which the primary stress lines bunch together has then been predicted together with the way in which secondary stresses act (shown in the figure). Note that the nature of the secondary stresses (whether tension or compression) depends on which way the primary stress lines are curving.

Q.3.9
In kNm units : $M_B = -240$, $M_C = -480$, $M_D = -180$, $M_E = 0$, $M_F = 60$, $M_H = 0$. See Fig. A.15 for the bending moment diagram and thrust line.

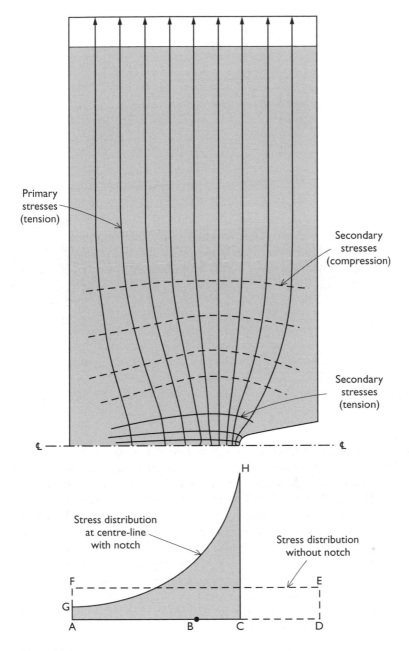

Primary stresses (tension)

Secondary stresses (compression)

Secondary stresses (tension)

Stress distribution at centre-line with notch

Stress distribution without notch

Figure A.14

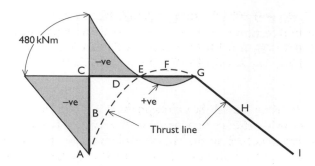

Figure A.15

Q.3.10
Figure A.16 shows thrust lines (dotted) and bending moment diagrams drawn on the tension side. The largest bending moments will be as follows: (a) and (b) at both eaves joints, (c) at base of columns, (d) under the concentrated load.

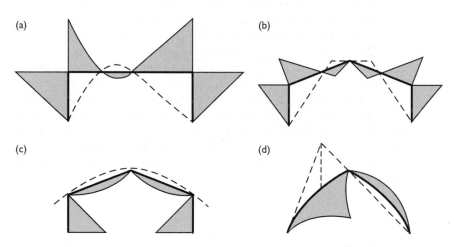

Figure A.16

Q.3.11
With vertical loading, the reaction under the pinned column must be vertical (so that there is no bending moment at the upper hinge). Hence a horizontal component of force is not present. A thrust line cannot be drawn and there is no reduction in bending moment in the beam due to H. An advantage of this arrangement is that side loads on the wheels and track are kept to a minimum.

Q.3.12
$R_A = 0.5\,W$, $R_C = 0.7071\,W$, $R_E = -0.2071\,W$
With x axis in direction A $\rightarrow$ B $\rightarrow$ C $\rightarrow$ D $\rightarrow$ E and y axis upwards,
$M_B = 0.3536\,Wr$, $T_B = 0.1464\,Wr$, $M_C = -0.2071\,Wr$,
$T_C = 0.2071\,Wr$, $M_D = -0.1464\,Wr$, $T_D = 0.0607\,Wr$.

Q.3.13
$\sigma_1 = \sigma_x = 226$. From the Mohr's circle in the y–z plane: centre is at $(-24, 0)$ and radius is $(58^2 + 100^2)^{1/2}$ giving $\sigma_2 = 91.6$ and $\sigma_3 = -139.6$. The maximum shear stress occurs on the largest Mohr's circle, that is on a plane at $45°$ to σ_1 and σ_3 with a value of $\frac{1}{2}(226 + 139.6) = 182.8$ (all units N/mm^2).

Q.3.14 (designating compressive forces as positive)
a   At 2000 m of water, the hydrostatic pressure is $2000 \times 9.81\,\text{kN/m}^2 = 19.62\,\text{N/mm}^2$. Hence the factored external pressure is $p = 39.24\,\text{N/mm}^2$. Consider the equilibrium of a half sphere. The hydrostatic force $\pi(r+t)^2 p$, which would act on a semi-circular cross-section of radius $r + t$ if the sphere were not present, must be carried by the hoop stress σ_h acting over an area $\pi[(r+t)^2 - r^2]$. Equating these gives a formula for hoop stress. Furthermore, the same hoop stress occurs in any direction in the sphere because of the complete symmetry. Hence σ_h is the maximum stress in the material. Solving for t with $r = 90\,\text{mm}$, $p = 39.24\,\text{N/mm}^2$ and $\sigma_h \leq 500\,\text{N/mm}^2$ gives $t \geq 3.92\,\text{mm}$.

b   With $t = 4\,\text{mm}$, $\sigma_h = 471\,\text{N/mm}^2$, ignoring the curvature of the sphere in the region of the ring, it will be subject to an inward pressure of $471 \times 4 = 1884\,\text{N/mm}$. Assuming that the outer diameter of the ring is 42 mm, the compressive force in the ring required to replace this is $1884 \times 21 = 39{,}560\,\text{N}$. For its stress to be no more than $471\,\text{N/mm}^2$, it is required to have a cross-sectional area of 84 mm^2 (as the depth has been assumed to be 6 mm, a thickness of 14 mm would give this result, see Fig. A.17).

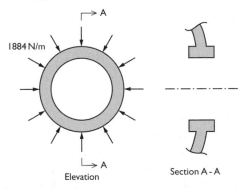

Elevation   Section A - A

Figure A.17

Q.3.16

(i) (a) or (b), (ii) (a) or (b), but for (a) the thinner vertical web will not contribute much, (iii) (c), but the cantilever ends of the deck, despite being thick, will not contribute much, (iv) (a), (v) (a) or (b), (vi) (c), with the shear flow likely to go mainly round the outer loop.

Q.3.17

a   This section is open with uniform thickness. The welds allow a shear loop to form round the whole section. The enclosed area $A \simeq 8 \times 332 = 2656 \,\text{mm}^2$. Hence, if the stresses are elastic, $q = 165.7 \,\text{N/mm}$, giving a maximum shear stress around the surface of $55.2 \,\text{N/mm}^2$.

b   Enclosed area (to centre-lines) is $114 \times 108 = 12{,}300 \,\text{mm}^2$. Hence $q = 35.7 \,\text{N/mm}$ and shear stress is $3.0 \,\text{N/mm}^2$ in the flanges and $6.0 \,\text{N/mm}^2$ in the webs. The maximum shear stress in the closed section is 11% of that for the open section.

Q.3.18

a   If x and y are measured from the centre of the opening, the equation of the ellipse is $x^2 + (1/2)y^2 = h^2$. Vertical equilibrium of the half ring shown in Fig. 3.67 gives $R = 2\sigma_\ell ht$. Consider equilibrium of a part to the right of a section at (x, y) as shown in Fig. A.18. Taking moments about the cut gives

$$M = \sigma_\ell t \times \tfrac{1}{2}y^2 + 2\sigma_\ell t \times \tfrac{1}{2}(h - x)^2 - R(h - x)$$

Hence

$$M = \sigma_\ell t \left(\tfrac{1}{2}y^2 - h^2 + x^2 \right) = 0.$$

b   Horizontal equilibrium of one quarter of the arch gives the axial force in the top centre of the ring to be $\sqrt{2}\sigma_\ell ht$ which is different from R.

Q.3.19

a   From the vertical equilibrium and symmetry $R_A = R_E = 2\sigma_\ell bt$. With $e = 0$, $M_B = -M_C = \tfrac{1}{2}\sigma_\ell b^2 t$.

b $M_B = \sigma_\ell bt(\tfrac{1}{2}b + 2e)$, $M_C = -\sigma_\ell bt(\tfrac{1}{2}b - 2e)$. Hence $|M_B| > |M_C|$ for $e > 0$.

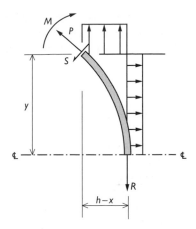

Figure A.18

Q.4.1
The SFD and BMD are shown in Fig. A.19. The maximum bending moment occurs at C where the shear stress changes sign. Also the area of the SFD is 36 to left of C and −36 to right of C totalling zero.

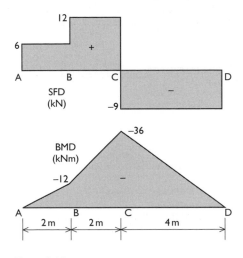

Figure A.19

Q.4.2
The reactions are $R_A = 26.4\,\text{kN}$, $R_C = 25.6\,\text{kN}$. On section AB, the shear stress varies linearly from $-26.4\,\text{kN}$ at A to $13.6\,\text{kN}$ at B and is constant at $25.6\,\text{kN}$ for BC. Hence the only zero shear stress is at distance $5.28\,\text{m}$ from A at which point the maximum bending moment is $69.70\,\text{kNm}$.

Q.4.3

The bridge can be considered to be a single continuous beam with two internal hinges where the central span is supported at C and D. The SFD and BMD for each of the two loading cases are shown in Fig. A.20.

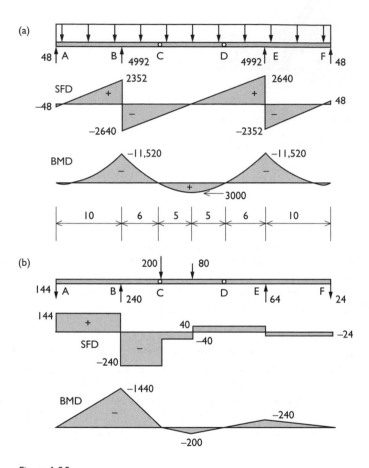

Figure A.20

Q.4.4

The horizontal reactions are $H_A = w\ell/6$ and $H_B = w\ell/3$. At distance x from A:

$$S = -\frac{w\ell}{6}\left(1 - \frac{3x^2}{\ell^2}\right), \quad M = \frac{w\ell^2}{6}\left(\frac{x}{\ell} - \frac{x^3}{\ell^3}\right)$$

Figure A.21 gives the shape of the internal force diagrams. For $S = 0$,

$$x = 0.57735\ell$$

Hence $M_{\text{max}} = 0.0642w\ell^2$.

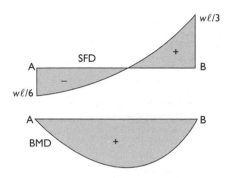

Figure A.21

Q.4.5
a   The internal forces are shown in Fig. A.22(a).
b   The modified internal forces are shown in Fig. A.22(b). The SFD crosses
    the axis at $x = 3.9$ and 4.1 m at which points the minimum and max-
    imum bending moments occur. Also the positive and negative areas of
    the SFD are 9.8×2 and -98×0.2 which sum to zero.

Q.4.6
The configuration is similar to that of Fig. 4.14 except that the loading
may or may not be present. The worst sagging (positive) bending moment
arises when BD only is loaded giving $(M_c)_{\text{max}} = w(\ell - 2\beta)^2/8$. The worst
hogging (negative) bending moment occurs when AB, DE or both are loaded
giving $(-M_B)_{\text{max}} = (-M_D)_{\text{max}} = w\beta^2/2$. Equating gives $\beta = \ell/4$. Hence
$|M|_{\text{max}} = w\ell^2/32$, one quarter of the value for an end-supported beam.

Q.4.7
$R_D = w(\ell - \beta)/2$, $M_B = w\ell\beta/2$ which is the maximum hogging bending
moment. The maximum sagging bending moment is at the centre of CD with
a magnitude of $w(\ell-\beta)^2/8$. The optimum occurs when $\beta(3-\sqrt{8})\ell = 0.172\ell$
with $|M|_{\text{max}} = 0.086w\ell$.

Q.4.8
a   (i) 172.8×10^6 mm^4, (ii) 187.3×10^6 mm^4, (iii) 159.5×10^6 mm^4, (iv) and
    (v) 109.4×10^6 mm^4.
b   (i) $Z = 0.96\times10^6$ mm^3, $\sigma_{\text{max}} = 104$ N/mm^2, (ii) $Z = 0.936\times10^6$ mm^3,
    $\sigma_{\text{max}} = 107$ N/mm^3, (iii) $Z = 1.14 \times 10^6$ mm^3, $\sigma_{\text{max}} = 88$ N/mm^2,

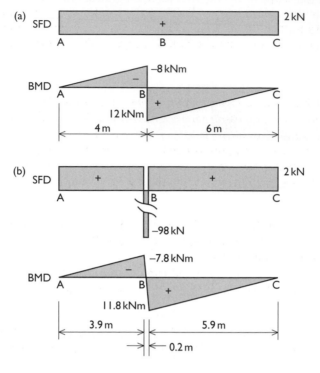

Figure A.22

(iv) $Z = 0.657 \times 10^6 \text{ mm}^3$, $\sigma_{max} = 152.3 \text{ N/mm}^2$ acting at the bottom fibres, (v) This is a type IV cross-section with no plane of symmetry.

Q.4.9

Ignoring the t^3 term, $I = 1.782t \times 10^6 \text{ mm}^4$. With $y_{max} = 90 \text{ mm}$, Eq. (4.5) gives $360 > 56 \times 10^6 \times 90/1.782 \times 10^6 t$. Hence $t > 7.86$ making $t = 8 \text{ mm}$ satisfactory.

Q.4.10

Advantages: Good two-way bending characteristics, good torsional characteristics, no projecting edges, low external surface area for corrosion protection, the internal volume could be utilised (e.g. for water to act as a heat sink in the event of fire).

Disadvantages: Difficult to make bolted connections, all surfaces are not accessible, not as efficient as universal beams and joists of same depth for one-way bending.

Q.4.11

Some possibilities are: Floor joists, lintels over windows or doorways, curtain rails, hand rails, horizontal bars of goalposts, horizontal members of gantries over motorways to support direction signs, railway rails, railway carriages.

Q.4.12

The distance to which a tape extends when inverted is approximately one half as much as when it is the correct way round (provided it is undamaged). Because the root bending moment is $w\ell^2/2$ for a cantilever, the ratio of bending moment capacities is approximately $1:4$. Failure in both cases is due to buckling which happens earlier when compressive stresses occur on the outsides of the cross-section than when they occur on the inside.

Q.4.13

Posts: AF: 25 N/mm^2 tensile at A $\rightarrow$ 0 at F; BE: 75 N/mm^2 compressive at
    B $\rightarrow$ 0 at E; CD: 50 N/mm^2 tensile at C $\rightarrow$ 0 at D.
Webs: ABEF : -1.5 N/mm^2; BCDE: 3 N/mm^2.
Flanges: 30 N/mm^2 compressive at B reducing linearly to 0 at A and C,
    48 N/mm^2 tensile at E reducing linearly to 0 at F and D.

Q.4.14

(i) Flange areas are 5027 mm^2 with their centres separated by 400 mm. Hence the flange forces are 250 kN and $\sigma = 49.73$ N/mm^2 (cf. 54.0 N/mm^2).

(ii) Flange areas are 3600 mm^2 separated by 420 mm. Hence $\sigma = 66.14$ N/mm^2 (cf. 52.8 N/mm^2).

There are two influences on accuracy: (a) the bending contribution of the web is neglected making the bending stresses larger in the elementary beam; (b) variation in stress across each flange is neglected making the maximum bending stresses less in the elementary beam. In (i), because of the flange depths of 80 mm each, effect (b) is larger than effect (a). However, in (ii) with a flange depth of only 20 mm, the position is reversed.

Q.4.15

This may be checked either by forming the integral $b \int_{-d/2}^{d/2} \tau \, d\tilde{y}$. Alternatively the graph 4.37(b) can be used knowing that the area of the parabola is $\frac{2}{3} \times$ base $\times$ height.

Q.4.16

The maximum bending moment for a single concentrated load is when the load is at the centre. Thus $M_{max} = P\ell/4 = 1250P$ mm. However, $I = 104.8 \times 10^6$ mm^4. Hence the maximum bending stress is $0.001909P$ N/mm^2. For this not to exceed 150 N/mm^2, $P \leq 78.6$ kN.

The maximum shear force is when the load is next to a support. Thus $S_{max} = P$. The maximum shear is at the neutral axis. $\int y \, d\tilde{A}$ for the area above the neutral axis is $378,400 \text{ mm}^3$. Hence $\tau_{max} = 0.0004515P \text{ N/mm}^2$. For this not to exceed 30 N/mm^2, $P \leq 66.4 \text{ kN}$. Hence the maximum allowable concentrated load is 66 kN.

Q.4.17
The centroid is at 179.5 mm from the bottom, $I = 69.8 \times 10^6 \text{ mm}^4$, $\tau_{max} = 12.0 \text{ N/mm}^2$.

Q.4.18
Adding the two plates to the flanges increases I from $397.8 \times 10^6 \text{ mm}^4$ to $778.9 \times 10^6 \text{ mm}^4$. For one of the plates $\int y \, d\tilde{A} = 3600 \times 230 \text{ mm}^3$. Hence shear flow across the lines of bolts on one flange is 241.9 N/mm. With two lines of bolts, a spacing of no greater than 0.248 m is required.

Q.4.19
a   Table A.1 shows the values obtained in the computation of flange forces and web shear flows for the two beams. The values for P_U and P_L at the support have been obtained from the triangle of forces there. The values of shear flow over the supports have been predicted by examining their trends when cross-sections are chosen close to the supports. These results are shown in Fig. A.23.
b   To be within the required stress bounds for the uniform beam $A_{flange} \geq 796 \text{ mm}^2$ and $t_{web} \geq 7.2 \text{ mm}$ and for the partly tapered beam $A_{flange} \geq 749 \text{ mm}^2$ and $t_{web} \geq 1.6 \text{ mm}$.
c   The maximum flange forces could be reduced by expanding the central section. Thus if only the outer 2 m lengths are tapered, the maximum depth of section would need to be 0.9 m and the maximum flange force would be 247.5 kN at centre span (P_L at the support would be 241.3 kN). This is a reduction of 22.4% from the value for the uniform beam.

Table A.1

Distance from support (m)	Internal forces		Uniform beam		Partly tapered beam			
	S (kN)	M (kNm)	P_U, P_L (kN)	q (N/mm)	P_U (kN)	P_L (kN)	q_d (kN)	q (N/mm)
0	−99	0	0	141.4	282.9	299.7	0	(31)
1.5	−66	123.8	176.8	94.3	235.7	249.7	16.5	31.4
2.99	−33	198	282.9	47.1	188.6	199.8	33.0	31.4
3.01	−33	198	282.9	47.1	188.6	188.6	−33.0	−31.4
4.5	0	222.8	318.2	0	212.1	212.1	0	0

Q.4.20
The weight is expected to be reacted by elastic bending theory shear in the cylinder which will be mainly at the sides as shown in Fig. A.24. For case

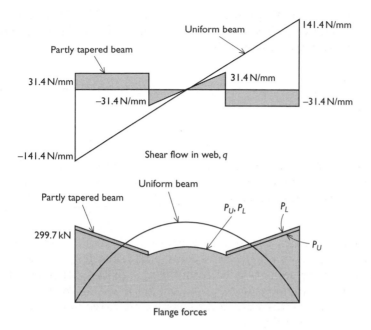

Figure A.23

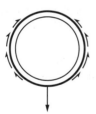

Figure A.24

(a), a significant bending moment will be produced at the bottom of the ring stiffener due to misalignment of the vertical force components. In case (b), some of the load will be transferred to the top of the ring stiffener, thus reducing the maximum bending moment by appropriately one half. In case (c), however, the loading is applied to the cylinder close to the shear reactions, thus creating much less bending moment in the ring stiffener. Also horizontal movement of the loading ring is not prevented in cases (a) and (b). This could lead to flexing of the hanger and subsequent damage to its top connection. In case (c), flexing of the hanger would not occur unless large horizontal forces acting on the loading ring caused buckling of a hanger.

Q.4.21

The general formulae are $I = a^3t/12 + a^2bt/2$, $S_F = 3Sb^2/a(a+6b)$, the distance of the shear centre outside the web is $e = 3b^2/(a+6b)$.

Q.4.23

If I_{yz} is the product second moment of the area about its own centroid which is situated at coordinates (s,t), the contribution to the product second moment of area of the whole cross-section is $\bar{I}_{yz} = I_{yz} + Ast$.

Q.4.24

a   The maximum bending moment is $w\ell^2/8 = 10\,\text{kNm}$. The centroid is at 16.67 mm to the right of B and 66.67 mm above it. $I_z = 1.333t \times 10^6$, $I_y = 0.250t \times 10^6$, $I_{yz} = 0.333t \times 10^6$ in mm units. Hence $\sigma t = -11.25y + 15z$. At A, $(y,z) = (133.3, 16.7)$. Therefore, $\sigma_A = -1250/t\,\text{N/mm}^2$. Similarly, $\sigma_B = 1000/t\,\text{N/mm}^2$ and $\sigma_C = -500/t\,\text{N/mm}^2$. Hence $|\sigma|_{max} = 1250/t\,\text{N/mm}^2$.

b   Because the stress in the lower leg varies linearly from $1000/t$ to $-500/t$, the zero position must be at 2/3 span. This could have been inferred from equilibrium of the stress distribution when seen in plan.

Q.4.25

The centroid is at 26.5 mm to the right of B and 76.5 mm above it. $I_z = 26.867 \times 10^6$, $I_y = 5.167 \times 10^6$, $I_{yz} = 6.584 \times 10^6$ in mm^4 units. Hence $\sigma = -0.541y + 0.690z$ (this compares with $-0.563y + 0.750z$ if $t = 20$ is substituted in the corresponding formula from the answer to Q.4.24). The most critical stress is $\sigma_E = -67.77\,\text{N/mm}^2$.

Q.4.27

The weight of water is $6.6 \times 1.85 \times 9.81 = 120\,\text{kN/m}$ span. Hence the maximum bending moment (at centre span) is $120 \times 21.45^2/8 = 6902\,\text{kNm}$. Treating the cross-section as an elementary beam with a distance between flanges of 1.855 m gives an estimate for flange forces of $6902/1.855 = 3720\,\text{kN}$. With flange areas of $2 \times 20 \times 1038 = 41{,}520\,\text{mm}^2$ (top) and $10 \times 7740 = 77{,}400\,\text{mm}^2$ (bottom), the estimates for flange stresses are $89.6\,\text{N/mm}^2$ (compression in top flange) and $48.1\,\text{N/mm}^2$ (tension in bottom flange).

NB   These stresses will add to the stresses due to self-weight.

Q.4.28

Lateral loading on a gondola will be reacted by lateral bending in which the base plate acts as the main web. If only lateral bending were to take place the shear reaction would be as shown in Fig. A.25. However being a channel section, the shear centre will be below the base plate. Hence some torsional restraint must also occur and Bredt–Batho shear flow of the form seen in Fig. 4.84(c) must add to the ones shown in Fig. A.25 in order to maintain overall equilibrium.

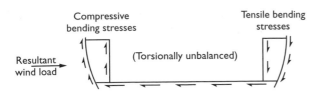

Figure A.25

Q.4.29
Because the boat displaces an equal weight of water, the forces acting on the gondola will be unaffected by its presence. Hence there is no increase in bending moment.

Q.5.1
a   Possibilities include: roof trusses in domestic houses, public halls, offices, warehouses, shops, railway stations, air terminals, sports stadia, etc.; electricity pylons; cranes including tower cranes; gantries (e.g. at docks or electric railway tracks); scaffolding; off-shore oil rigs; gates (particularly farm gates); bridges.

Q.5.2
b   No (there are two more members and one more joint).

Q.5.3
a Statically determinate truss;   b truss with one redundant member; c frame with one sway mechanism;   d frame with one sway mechanism; e frame with one sway mechanism and one redundant member;   f frame with one sway mechanism (despite the fact that this configuration is often called a queen-post truss). Figure A.26 shows sway mechanisms present in (c)–(f).

Q.5.4
The answer is given in Section 5.8.

Q.5.5
a $P_{CD} = 50.48\,\text{kN}$ tension, $P_{CB} = 70.0\,\text{kN}$ compression, $P_{AC} = 59.40\,\text{kN}$ tension, $P_{AB} = P_{BD} = 42.0\,\text{kN}$ compression.
b $P_{AB} = P_{CB} = 15.56\,\text{kN}$ compression, $P_{AC} = 11\,\text{kN}$ tension, $P_{BD} = P_{DF} = 22\,\text{kN}$ compression, $P_{CD} = P_{DE} = P_{CE} = P_{EF} = 0$.

Q.5.6
The vertical load applied to D has been applied upwards instead of downwards (the overall equilibrium equations should give a clue to this error).

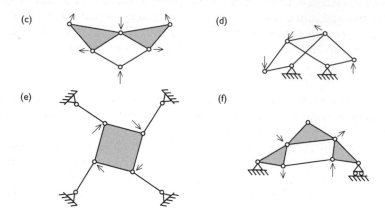

(c)   (d)

(e)   (f)

Figure A.26

Q.5.7

a   A sway mechanism is present which can be illiminated by restraining horizontal movement of the right-hand support.

b   A sway mechanism is present which can be illiminated by restraining the lower support point from horizontal rather than vertical movement.

c   A body movement is possible which can be illiminated by supplying one support with horizontal restraint.

d   A sway mechanism is present which can be illiminated by adding a horizontal or vertical member to the central diamond shaped panel.

e   No mechanism is present, hence, as a next step, the data input should be checked.

Q.5.8
8.5 kN tension.

Q.5.9

a

- A dividing line needs to be found which cuts the individual member and not more than two others.
- If the dividing line cuts three members, they should not be collinear.
- The external reactions on one side or other of the dividing line should all be computable (e.g. from overall equilibrium).

b

- Draw a free-body diagram for the part of the truss for which the external forces are all computable.
- Determine any required external forces.

• Take moments about the intersection point of the line of action of the other two unknown forces to obtain the required equation. However, if the other two forces are parallel to each other, resolve forces in a direction perpendicular to their lines of action (see Section 1.6).

Q.5.10

a Make a section cutting members JM, JK, KL and LO (there is no need for sections to be straight). Take moments about J for the top part of the truss to give $P_{LO} = -86\,kN$.

b Make a section cutting members JM, KM, KO and LO. Resolve horizontally to give $0.3846\,(P_{KO} - P_{KM}) + 10\,kN = 0$. However, vertical equilibrium of joint K gives $P_{KM} = -P_{KO}$. Hence, $P_{KM} = -P_{KO} = 13\,kN$. Resolving horizontally at joints M and O give: $P_{NO} = -P_{MN} = 5\,kN$.

Q.5.11

a This has a sway mechanism and so should not really be classified as a truss. The sway mechanism has a symmetric shape with the supports spreading as shown in Fig. A.27(a). (If the joints are not stiff, but the four members are continuous, sway produced by vertical loading will need to be resisted by bending in the members as shown in Fig. A.27(b)). Hence, the members need to be strong in bending even to support dead load (this truss being only suitable for small spans).

b This truss is statically determinate (the vertical post eliminates the sway mode). To have a large span (as at the Old Faithful Inn), the members must be strong and stiff in bending because of the long lengths between joints of the compression members. One advantage of this truss is that spans somewhat greater than the lengths of timber can be produced without the need for a splice.

c The queen-post truss has a mechanism, but this has an antisymmetric movement (see Fig. A.26(f)). Bending strength is required through the joints, but only to resist antisymmetric loads such as wind. This is not very suitable for large spans because of the long lengths required of rafter members. Its advantage is that it provides a clear rectangular internal roof space.

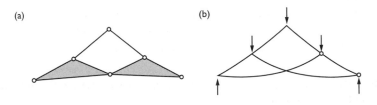

(a)

(b)

Figure A.27

Q.5.12

In Fig. A.28, the compression members are shown as a full line and the tension members as a dotted line.

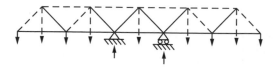

Figure A.28

Q.5.13

a   Quick inspection of joint equilibrium can be used to identify the sign of all but two members, as shown in Fig. A.29 (unloaded members are not shown).

b   By examining the direction of horizontal components of forces acting on the lower chord, it is clear that segment IJ must carry the largest tension load. For the upper chord, the largest compression load must be in either segment AB or BC depending on the sign of the force in the upper part of member BI. (Using the method of sections, it is found that $P_{BC} > P_{AB}$. Hence member BI must be in compression along its whole length and BJ must be in tension.)

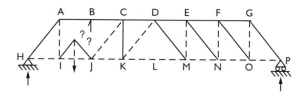

Figure A.29

Q.5.14

a   The column EA carries compression load and the other members are unloaded.

b and c   The loaded members are shown in Fig. A.30.

Q.5.15

There are four self-equilibrating stress systems in Fig. 5.11 (two for the last truss) and two in Fig. 5.16 (in frames (b) and (e)). The force systems involved are shown in Fig. A.31. Other self-equilibrating stress systems are possible for the third frame of Fig. 5.11; but only as linear combinations of the two given.

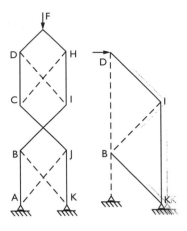

Figure A.30

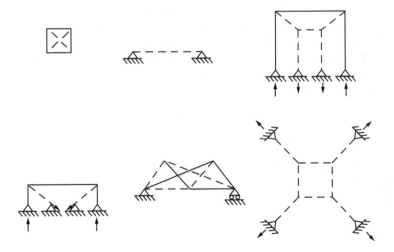

Figure A.31

Q.5.16

The answer to this question is left as an interesting puzzle.

Q.5.18

- Check overall equilibrium.
- Identify members loaded in tension and compression to see if any joint appears to have an unbalanced set of forces.
- Spot check the equilibrium of specific joints.

- See if there are suitable cross-sections to do spot checks.
- Reanalyse the truss using a completely independent program.

Q.5.19

a   This has one sway mechanism and one self-equilibrating force system as shown in Fig. A.32.

b   This is a statically determinate truss.

c   If m is even, there are n sway mechanisms and n self-equilibrating force systems. However, if m is odd, it is a statically determinate truss.

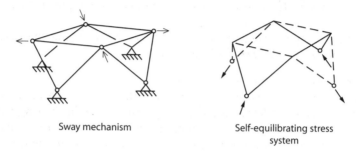

Sway mechanism

Self-equilibrating stress system

Figure A.32

Q.5.20

- As jibs of cranes including the horizontal jibs of tower cranes;
- As pipe bridges;
- As masts for radio communication, etc. (sometimes guyed);
- As beams to support flat roofs.

Q.5.22

Factors favouring trusses above the deck:

- There may not be sufficient headroom for a truss below deck level.
- Inspection of the structure is simpler.
- Truss members may be used to support parapets and crash barriers.

Factors favouring trusses below the deck:

- There could be difficulty bracing trusses above the deck whilst maintaining clearance for vehicles.
- There is risk of damage to the primary trusses through collision with vehicles on the deck.
- Truss members above the deck can increase the gustiness of cross-winds.
- The view of passengers is impeded if trusses are placed above the deck.
- A truss below the deck provides protected space for pipes and cables.

Q.5.23
Footballs are often made from 12 five-sided and 20 six-sided panels in the form of a truncated icosahedron (an icosahedron with its vertices chopped off).

Q.6.1
- Generally people don't like a structure to feel shaky or vibrate.
- Deflections can cause cracking of plaster, brickwork, etc.
- Bolts and fastenings may work loose and beams may fall off their bearings.
- Equipment may malfunction, particularly measuring equipment such as telescopes.
- Deflection can sometimes have a feedback effect enhancing the loads (discussed in Section 6.12 and Chapter 7).

Q.6.2 (Compression forces designated positive)
From the answer to Q.3.14 with a factored load, the stress in the sphere of thickness 4 mm was 471 N/mm^2 and the compressive force in the ring was 39.56 kN. For the sphere, the circumferential stress, σ_c is the same in two orthogonal directions and a Poisson's ratio effect will be present. Hence the circumferential strain ε_c is given by

$$\varepsilon_c = \sigma_c(1 - v)E = 471 \times 0.7/200 \times 10^3 = 0.00165$$

However, the ring is subject to hoop compression without any significant Poisson's ratio effect. For this strain, the hoop stress needs to be $0.00165 \times 200 \times 10^3 = 330$ N/mm^2. Hence its cross-sectional area would have to be $39560/330 = 120$ mm^2.

Q.6.3
a   The cross-sectional area of the bar is $\pi \times 10^2 = 314.2$ mm^2. Hence the (unidirectional) stress is $8800/314.2 = 28.0$ N/mm^2 giving $E = 28.0/0.00103$ N/mm$^2 = 27.2$ kN/mm^2. Poisson's ratio $v = 0.00026/0.00103 = 0.252$.
b   For the 2D stress system at the surface, assuming no shear stress is present,

$$E\varepsilon_1 = \sigma_1 - v\sigma_2, \quad E\varepsilon_2 = \sigma_2 - v\sigma_1$$

$$\therefore \sigma_1 - 0.252\sigma_2 = 6.53 \text{ N/mm}^2, \quad -0.252\sigma_1 + \sigma_2 = 18.22 \text{ N/mm}^2$$

Solving gives $\sigma_1 = 11.88$ N/mm^2 and $\sigma_2 = 21.21$ N/mm^2.

Q.6.4
a   Tensile strain increases the length of filaments. Also, through the Poisson's ratio effect, the cross-sectional area of the filament is decreased. Both of these increase the resistance of the circuit.

b   The broad hairpin bends reduce the resistance round the bends effectively eliminating any possible sensitivity to strain in the lateral direction (before printed circuits were used, wire gauges tended to have a small amount of cross-sensitivity caused by the bends in the wire).

Q.6.5

With x measured from one end, the differential equation for bending is

$$EI\frac{d^2v}{dx^2} = M = \frac{1}{2}w(\ell x - x^2)$$

Integrating and using the end conditions $v = 0$ at $x = 0$ and ℓ gives

$$v = \frac{w}{24EI}(2\ell x^3 - x^4 - \ell^3 x)$$

with the maximum, $v_{max} = -5w\ell^4/384EI$, occurring when $x = \frac{1}{2}\ell$.

Q.6.6

a   The BMD and deflection diagram for the beam are shown in Fig. A.33. Using similar triangles, $(-v_B + \delta_A^B)/a = \delta_A^C/\ell$.
Evaluating δ_A^B from triangle A′B′E gives: $EI\delta_A^B = (M_Ba/2) \times (a/3)$ and evaluating δ_A^C from triangles AB′E and B′C′E gives

$$EI\delta_A^C = \frac{M_Ba}{2}\left(\frac{a}{3} + b\right) + \frac{M_Bb}{2} \times \frac{2b}{3}$$

Hence $-EIv_B = M_Bab/3$, giving $v_B = Pa^2b^2/3EI\ell$.
b   When $a = b = \frac{1}{2}\ell$, $-v_B = P\ell^3/48EI$.

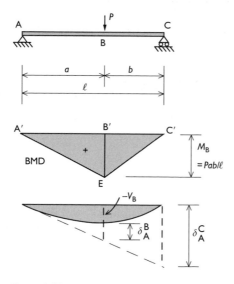

Figure A.33

Q.6.7
For the equivalent timber section $I = 21.40 \times 10^6 \text{ mm}^4$. Results should agree with the last row of Table 6.2.

Q.6.8
$I = 250 \times 129^3/3 + 15 \times 603.2 \times 231^2 = 661.7 \times 10^6 \text{ mm}^4$
Deflection $= 5w\ell^4/384E_cI = 13.0 \text{ mm}$

Q.6.9
The end conditions are

$$v = 0 \quad \text{and} \quad dv/dx = 0 \qquad \text{at } x = 0$$

and

$$v = 0 \quad \text{and} \quad M = 0 \qquad \text{at } x = \ell$$

Integration of the loading curve gives Eq. (6.14), but with constants $c_1 = -5w\ell/8$, $c_2 = w\ell^2/8$. Hence $-EIv = wx^2 \ (\ell - x)(3\ell - 2x)/48$ and $M = -w(\ell - x) \ (\ell - 4x)/8$. See Fig. A.34.

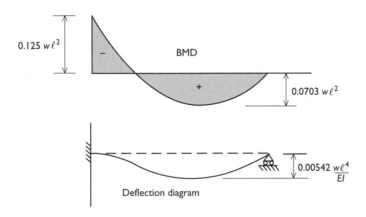

Figure A.34

Q.6.10
With $\overline{M}$, the moment at the built-in end as the redundancy, the moment area method (see Fig. A.35) gives

$$EI\delta_A^B = \frac{P\ell^3}{16} - \frac{\overline{M}\ell^2}{3} = 0$$

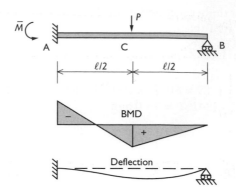

Figure A.35

Hence $\overline{M} = 3P\ell/16$. The maximum hogging bending moment equals $\overline{M} = -0.188P\ell$. The maximum sagging bending moment $= M_c = 0.156P\ell$. The central deflection $-v_c = \delta_A^C = 0.0091P\ell^3/EI$.

Q.6.11
- Railway rails spanning over sleepers.
- Floor beams in some buildings.
- Floor boards spanning across joists.
- Bookshelves or curtain rails with more than two supports.
- Chords of a truss when continuous and loaded in the plane of the truss.

Q.6.12
The moment–area equations are the same as Eq. (6.15) except that the last three right-hand sides are zero. The solution is

$$\{\overline{M}_B \overline{M}_C \overline{M}_D \overline{M}_E\} = \frac{w\ell^2}{836}\{56 \quad -15 \quad 4 \quad -1\}$$

with the BMD and approximate deflected form shown in Fig. A.36.

Q.6.13
a   The largest BM will be hogging when spans AB, BC and DE are loaded.
b   The largest BM will be sagging when spans BC and DE are loaded.
c   The largest reaction will be when spans BC, CD and EF are loaded.
(Any load acting on the other spans has a reverse effect.)

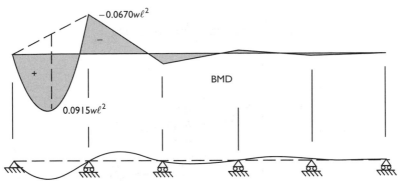

-0.0670$w\ell^2$

−

BMD

0.0915$w\ell^2$

+

Approximate deflections

Figure A.36

Q.6.14

$$\overline{M}_B\ell_{BC} + 2\overline{M}_C\left(\ell_{BC} + \ell_{BD}\right) + \overline{M}_D\ell_{CD} = 6\left(\frac{A_{BC}\vec{x}^B}{\ell_{BC}} + \frac{A_{CD}\vec{x}^D}{\ell_{CD}}\right)$$

Q.6.15

a   From the Williot diagram, Fig. A.37, A displaces 7.4 mm horizontally and C displaces 5 mm horizontally.

b
$$\begin{bmatrix} -1 & 0 & 0 & 0 & 1 & 0 \\ 0 & 1 & 0 & 0 & 0 & 0 \\ & & 1 & 0 & -1 & 0 \\ & & 0 & 1 & 0 & 0 \\ & & & & 0.6 & 0.8 \\ & & & & -0.6 & 0.8 \end{bmatrix} \begin{bmatrix} u_A \\ v_A \\ u_C \\ v_C \\ u_B \\ v_B \end{bmatrix} = \begin{bmatrix} e_{AB} \\ e_{AD} \\ e_{BC} \\ e_{CE} \\ e_{BD} \\ e_{BE} \end{bmatrix} = \begin{bmatrix} -2.4 \\ 0 \\ 0 \\ 0 \\ 3.0 \\ -3.0 \end{bmatrix}$$

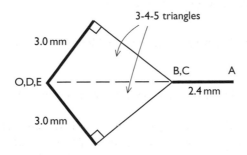

3-4-5 triangles

3.0 mm

B,C   A

O,D,E

2.4 mm

3.0 mm

Figure A.37

Q.6.16

a

$$\begin{bmatrix} e_{AB} \\ e_{BC} \\ e_{AC} \end{bmatrix} = \begin{bmatrix} 1 & 0 & 0 \\ 0 & 1 & -1 \\ 0 & 0 & -0.8 \end{bmatrix} \begin{bmatrix} u_B \\ v_B \\ v_C \end{bmatrix}$$

b

$$\begin{bmatrix} H_B \\ V_B \\ V_C \end{bmatrix} = \begin{bmatrix} 1 & 0 & 0 \\ 0 & 1 & 0 \\ 0 & -1 & -0.8 \end{bmatrix} \begin{bmatrix} p_{AB} \\ p_{BC} \\ p_{AC} \end{bmatrix}$$

Q.6.17

a 14×14

b   See Fig. A.38.

c   Diagonal submatrices for the members are

$$\text{CD:} \begin{bmatrix} 39.7 & 0 \\ 0 & 0 \end{bmatrix} \quad \text{DF:} \begin{bmatrix} 0 & 0 \\ 0 & 39.7 \end{bmatrix} \quad \text{DE:} \begin{bmatrix} 14.0 & -14.0 \\ -14.0 & 14.0 \end{bmatrix}$$

which add together on the diagonal. The four non-zero submatrices on rows 7 and 8 are

$$\begin{array}{cccc} C & D & E & F \end{array}$$
$$\begin{bmatrix} -39.7 & 0 & \bigm| & 53.7 & -14.0 & \bigm| & -14.0 & 14.0 & \bigm| & 0 & 0 \\ 0 & 0 & \bigm| & -14.0 & 53.7 & \bigm| & 14.0 & -14.0 & \bigm| & 0 & -39.7 \end{bmatrix}$$

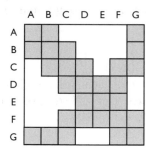

Figure A.38

Q.6.18

b   The answer should represent the case where joint B is given a 1 mm displacement to the right.

Q.6.19
A bending moment is really a pair of equal and opposite moments and hence requires a different sign convention to single moments. See 'Sign conventions for 2D force systems' in Section 3.1.

Q.6.22
See Fig. A.36.

Q.6.23
The one point of inflection must occur at the apex. Hence the member BMDs will be as shown in Fig. A.39. From overall equilibrium, the vertical components of base reaction will be $\pm\frac{1}{2}P$. Taking moments about the apex for one half of the frame gives the horizontal components of base reaction also as $\pm\frac{1}{2}P$. Hence the maximum bending moment is at the eaves joints with a magnitude $\frac{1}{2}Ph$.

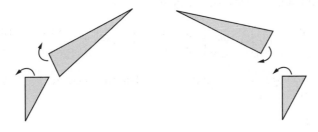

Figure A.39

Q.6.25
This statement is invalid. Shear force and slope diagrams are both antisymmetric. (Integration and differentiation change a symmetric diagram into an antisymmetric one or vice versa.)

Q.6.26
a   See Fig. A.40(a).
b   If an analysis is carried out also of the antisymmetric loading with half-frame modelling as shown in Fig. A.40(b), the required solution is the sum of this result and the symmetric one (see Fig. A.40(c)).

Q.6.28
There will be a constant difference between the horizontal positions of corresponding joints. This is because the position of the horizontal restraint has been transferred to centre span in the half-frame modelling.

Q.6.29
The eaves and ridge joints could be assembled first with the frame lying flat. It could then be lifted into place with a crane. Tolerances in the bolt

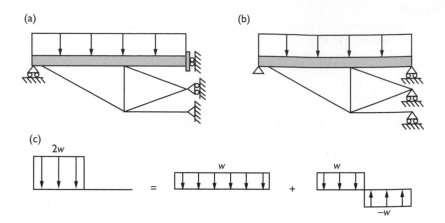

Figure A.40

holes of the baseplates (using slotted holes where appropriate) could be used to cater for any possible misalignments in the horizontal plane without recourse to forcing the joints into place. If the baseplates end up not being perfectly horizontal, they could be packed or grouted to obtain a flush fit.

Q.6.30

a   Expansion required $= 80 \times 10^3 \times 3.4 \times 10^{-6} \times 70 = 19\,\text{mm}$.

b   The required compressive strain is $40 \times 3.4 \times 10^{-6} = 0.136 \times 10^{-3}$.
Hence the stress is $0.136 \times 10^{-3} \times 23 = 3.13 \times 10^{-3}\,\text{kN/mm}^2$ and the force is $3.13 \times 10^{-3} \times 2.8 \times 10^6 = 8760\,\text{kN}$.

Q.6.31

If there is a support at one end (say A) restraining horizontal movement, roller supports could be used at B and C with an expansion joint at the right-hand end of the deck. An alternative to a roller support at B would be to hinge the column at B and D. However, neither roller nor hinges may be necessary for the column, if the column is flexible enough and bending moments resulting from temperature movements of the deck are acceptably low. It would not normally be very suitable to put the expansion joint at B because of loss of continuity in the deck structure.

Q.6.32

The compatibility equations are

$$e_s + e_t = Au$$

but $p = \overline{K}e_s$ and $f = A^T p$. With no externally applied forces, $f = 0$.

Hence

$$A^T\overline{K}(Au - e_t) = 0$$

With $K = A^T\overline{K}A$ the equations which can be used to obtain u are

$$Ku = A^T\overline{K}e_t$$

The member forces can then be obtained from

$$p = \overline{K}(Au - e_t)$$

Q.6.33

(i) Absolutely wrong. The beam will need to adopt a specific curvature to meet the deflection criterion. In view of the relationship $\sigma/y = E/R$ increasing y will increase the maximum stress, not reduce it.

(ii) This does not increase the stress, but by the same criterion as before, it does not reduce it either. Therefore it is not a viable solution.

(iii) This will reduce the maximum stress due to subsidence. However, it will increase the maximum stress due to the other loading cases. The I value will need to be increased (by increasing flange areas for instance) to meet these criteria.

(iv) Moving to a statically determinate structure means that movement of supports do not affect the stresses. However, a redesign would be necessary to cater for the other loading cases.

Q.6.34

Some possible advantages:

- The inside space is kept clear.
- There may be less wind resistance.
- The primary structure does not clash with the space required for secondary and tertiary structure.
- The structural support is visible rather than hidden.

Some possible disadvantages:

- The main structure is more subject to corrosion.
- The result could be aesthetically clumsy unless the visible structure is simple.
- Roof loads need to be taken by tension rather than compression fittings and thus are subject to possible fatigue.

Q.7.1

a   End conditions are fixed at A and pinned at B. Hence $\ell_e = 0.7\ell$.

b   If the cable forces always acted vertically downwards, the end conditions would be fixed–free with $\ell_e = 2\ell$. However, the cable forces always point towards the centre-line of the deck as shown in Fig. A.41. Hence, with no bending moment at the bottom or top, this can be likened to Euler buckling with $\ell_e = \ell$.

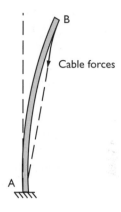

B

Cable forces

A

Figure A.41

Q.7.2
a   A pin-jointed frame analysis gives

$$P_{AB} = P_{BD} = P_{DF} = P_{FG} = 1.155W \text{ (in compression)}$$

and all the other members loaded in tension or unloaded. For the four compression members $I = 12.18 \times 10^6 \text{ mm}^4$, giving $P_E = 962\,\text{kN}$. Since there are no sway modes, the truss would buckle when $1.155W = 962\,\text{kN}$, that is $W_{cr} = 833\,\text{kN}$ if it were pin-jointed. Because it is rigidly jointed, the buckling load should be marginally above this value.

b   Member forces are modified to $P_{AB} = P_{BD} = 0.77W$, $P_{DE} = P_{DF} = P_{FG} = 0.385W$ (all in compression) with the rest in tension. Since AB, BD, DF and FG carry less load, the buckling load is likely to be increased. However, member DE, previously unloaded, now carries a compression load. If this has a lower I value than the top chord and end diagonals, the buckling load could be reduced.

Q.7.4
From overall equilibrium, the horizontal reactions at A and C will be zero. From equilibrium of the segment AB, the bending moment at B will be $M_B = Pv_B$. However, from the stiffness of the joint, $M_B = 580\theta_B$ where θ_B, the rotation at B is such that $\theta_B = 2v_B/12$. Thus $P_{cr} = 96.7\,\text{kN}$.

Q.7.5

With v_B and v_C as the deflection of joints B and C and θ_B and θ_C their changes in angle:

$$\theta_B = \frac{(2v_B - v_C)}{12}, \quad \theta_C = \frac{(-v_B + 2v_C)}{12}$$

However $Pv_B = 580\theta_B$ and $Pv_C = 580\theta_C$. Hence for equilibrium

$$\begin{bmatrix} 2 & -1 \\ -1 & 2 \end{bmatrix} \begin{bmatrix} v_B \\ v_C \end{bmatrix} = \frac{P}{48.3} \begin{bmatrix} v_B \\ v_C \end{bmatrix}$$

The modes are

$$\{v_B \ v_C\} = \{1 \ 1\} \text{ with } P = 48.3 \text{ kN} \quad \text{and}$$
$$\{v_B \ v_C\} = \{1 \ -1\} \text{ with } P = 145 \text{ kN}$$

The first, being the lowest, is the buckling load.
(NB Unlike for Ex.7.4, it is the symmetric mode which is critical.)

Q.7.6

The ground displacement due to tilt at distance y from the axis of rotation is θy and hence the ground pressure is $k\theta y$. The building will tilt about its long centre-line as shown in the plan of Fig. A.42, hence a strip of width dy will have a moment about the axis of rotation of $28k\theta y\,dy$. The total stabilising

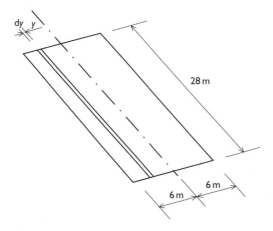

Figure A.42

moment due to the change in ground reaction is in metre units:

$$28k\theta \int_{-6}^{6} y^2 \, dy = 4032k\theta$$

(assuming that no part of the building lifts off its foundations).

Equating this to the destabilising moment of $24{,}000 \times 32\theta$ kNm gives $k_{cr} = 190 \, \text{kN/m}^3$.

Q.7.7
In the original design, there are four horizontal members crossing the main vertical post of each truss. If the lower two and the upper one are removed, the double cantilever truss like that of the replacement bridge have just one redundancy caused through the cross bracing in the central panel. Likewise the trusses of both suspended spans have just one redundancy occurring in the central panels. In the original double cantilever design, the extra horizontal members are there to stabilise the main vertical posts. The lowest of these, being connected to two lower chord joints creates an additional redundancy. The other two, not being connected to any main joint of the truss, are likely to be less effective at stabilising the post and would lead to a mechanism being present in any pin-jointed analysis. In the second design, the horizontal top chord members linking the suspended span to the cantilevers are also redundant.

Q.7.9
a   With B at the same level as A (see Fig. A.43), $V_A = 0$.
    Taking moments about E for the whole cable gives $H = 40$ kN.
    Taking moments about C for AC gives $y_C = 1$ m.
    Taking moments about D for AD gives $y_D = 4$ m.
    The lengths of the cable segments are AB: 8 m, BC: 8.062 m, CD: 8.544 m and DE: 8.944 m. The maximum force in the cable is on segment DE $= (40^2 + 20^2)^{1/2} = 44.72$ kN.

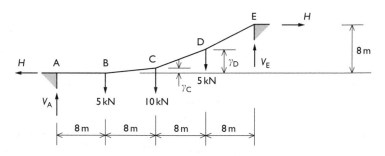

Figure A.43

b   The BMD of the equivalent simply-supported beam has $M_B = M_D = 80\,\text{kNm}$ and $M_C = 120\,\text{kNm}$. These are $40\,\text{kN}$ times the vertical distances of B, C and D below the line AE.

Q.7.10
Consider the support reactions to be a force F acting along the line AB plus vertical reactions V_A and V_B. Moment equilibrium about the supports shows that V_A and V_B are the same as the end reactions of the simply-supported beam. By sectioning the cable at any point (x, y), where y is measured vertically downwards from the line AB, and taking moments about this point, the moment of the vertical forces must equal $F\bar{y}$ where $\bar{y}$ is measured perpendicular to AB as shown in Fig. A.44. However the moment of the vertical forces is equal to M, the bending moment at spanwise position x in the simply-supported beam. If α is the inclination of AB, $\bar{y} = y\cos\alpha$. Hence $M = Fy\cos\alpha = Hy$ where H is the horizontal component of F.

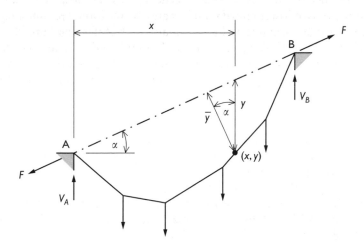

Figure A.44

Q.8.1
a   The deflection form with $\Delta = 1\,\text{mm}$ is shown in Fig. 6.36. Virtual work gives $P_{BD} + 0.707P_{CD} - 0.5547P_{DE} = 0$ (i.e. the fourth of Eq. (6.20)).
b   If joints C, D and E all move Δ to the right, the only member to change in length is BC. As there is no external work done, $P_{BC} = 0$.
c   Rotating the whole truss about B such that A drops by Δ requires E to drop by 3Δ (because its horizontal distance from B is three times that of A). Hence, from virtual work, $100 \times 3\Delta - R_A\Delta = 0$ giving $R_A = 300\,\text{kN}$.

Q.8.2

If H_A and H_I are horizontal components of reactions at A and I (acting to the left), from overall equilibrium $H_A + H_I = 55.55$ kN. Taking moments about C and G for each column gives $M_C = 4H_A - 60$ and $M_G = -4H_I + 40$ in kNm units. Eliminating H_A and H_I gives the required equation.

Q.8.3

If C drops a distance Δ, because ABC pivots about B, A rises a distance $4\Delta/5$. The rotation introduced at C is $2\Delta/5$. Hence $M_C \times 0.4\Delta = 50\Delta - 20 \times 0.8\Delta$ giving $M_C = 85$ kNm.

Q.8.4

When a unit load is placed at D, the only non-zero member forces are $\bar{P}_{AD} = 1.25$ and $\bar{P}_{ED} = -0.75$. Replacing the column for P_i in Table 8.1 by their new values gives the deflection of D to be 4.1 mm.

Q.8.5

Due to a unit force acting to the right at D, the non-zero member forces are $P_{BC} = -P_{CD} = 1.414$, $P_{BD} = 1$ and $P_{AC} = -2$. Hence from virtual work

$$u_D = (0 + 1.414 \times 169.7 \times 5.657 + 1 \times 200 \times 8 + 2 \times 300 \times 4)/EA$$

giving $u_D = 8.93$ mm (this agrees with the solution for Eq (6.19) given earlier in the text).

Q.8.6

The maximum downward deflection will occur at A. Hence applying a unit load there, the graphs for m and $\bar{m}$ are as shown in Fig. A.45. To evaluate the virtual work Table 8.2 may be used, if the curve for m between B and C is split into two parts (e.g. a rectangle plus a triangle). The required deflection is found to be $333/EI$ in m units.

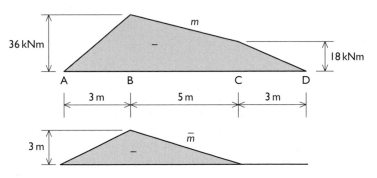

Figure A.45

Q.8.7

From overall equilibrium the reactions, as depicted in Fig. A.46 are $V_A = 15\,\text{kN}$, $V_D = 5\,\text{kN}$, $H_A = 0$. Hence $M_B = M_C = 20\,\text{kNm}$ (sagging).

For a unit load applied horizontally to the right at D, $H_A = -1$, $V_A = 0.25$, $V_D = -0.25$. Hence the bending moment at C is -1 in m units.

Figure A.46 shows the bending moment curves for m and $\bar{m}$. Using Table 8.2, the integral is evaluated in Table A.2, thus giving the horizontal deflection at D as $69.8/EI$ to the right in m units (where EI is in kNm^2).

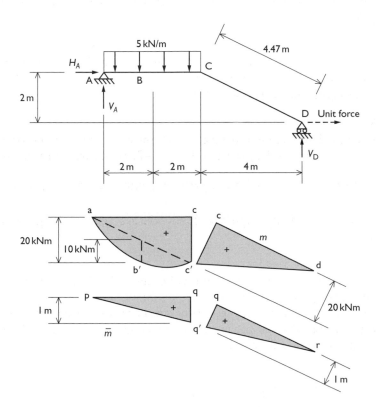

Figure A.46

Table A.2

m'	m	ℓ' (m)	s (m)	t (kNm)	integral (kNm³)
$\triangle pqq'$	$\triangle acc'$	4	1	20	26.7
$\triangle pqq'$	ab'c'	4	1	10	13.3
$\triangle qq'r$	$\triangle cc'd$	4.47	1	20	29.8
Σ					69.8

Q.8.8

a For member CD, $P = 50.48\,\text{kN}$, $\ell = 10.82\,\text{m}$, $EA = 176{,}000\,\text{kN}$. Hence the strain energy stored in it is $\frac{1}{2}P^2\ell/EA = 78.3\,\text{kNmm}$. Strain energies in kNmm for other members are CB:83.5, AC:85.1, BC:75.2. Summing gives $U = 322\,\text{kNmm}$.

b Equating U to $\frac{1}{2} \times 28v_\text{D}$ gives $v_\text{D} = 23\,\text{mm}$.

Q.8.9

a The bending moment at distance x from A is $\frac{1}{2}Px$. Hence the strain energy for length AB is

$$\frac{1}{2} \int_0^{\ell/2} \frac{P^2 x^2}{4EI}\, dx = \frac{P^2 \ell^3}{192EI}$$

Since length BC is symmetric with AB, $U = P^2\ell^3/96EI$.

b Equating U to $\frac{1}{2}Pv_\text{B}$ gives $v_\text{B} = P\ell^3/48EI$.

Q.8.10

a With t constant round the section $\oint ds/t = 2\pi r/t$. Also $A = \pi r^2$. Hence from Eq. (8.5) the torsional stiffness $T\ell/\phi = 2\pi G r^3 t$.

b Consider a cylinder of radius $\bar{r}$ and thickness $d\bar{r}$ as shown in Fig. A.47. Its torque contribution is

$$dT = \frac{2\pi G\phi}{\ell}\bar{r}^3\, d\bar{r}$$

If ϕ/ℓ is the same for all parts of the solid cylinder integrating to obtain the total torque gives

$$T = \frac{2\pi G\phi}{\ell} \int_0^r \bar{r}^3\, d\bar{r} = \frac{\pi G r^4 \phi}{2\ell}$$

and the torsional stiffness is $\frac{1}{2}\pi G r^4$.

Q.8.11

The diagonals of the two wall segments are inclined at $x = 2t/h$ to the vertical. Hence if F moves a distance v to the left, it will rise vertically a distance $2vt/h$ and A will rise twice this amount. From the position of the three vertical loads, their potential energy will increase by

$$\frac{W}{2} \times \frac{vt}{h} + \frac{W}{2} \times \frac{3vt}{h} + W \times \frac{4vt}{h} = \frac{6Wvt}{h}$$

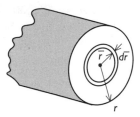

Figure A.47

But the potential energy of the lateral pressure will reduce by $\frac{1}{2}Pv$. Because this is an inelastic problem involving no strain energy, for stability

$$\tfrac{1}{2}Pv < 6Wvt/h$$

Hence collapse is predicted at $P = 12Wt/h$.

Q.8.12
From the theorem of least work, the central portion of the beam, being stiffer, will attract more bending moment. To minimise the strain energy, the point of inflection will be close to the 1/8th span positions. The anticipated BMD conforms with a deflection diagram in which the central part remains almost rigid (Fig. A.48).

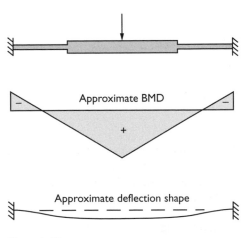

Figure A.48

Q.8.13

For the first strut, strain energy of the joint $U = 580(v_B/6)^2/2\,\text{kNm}$. The change in potential energy due to axial shortening is $V = -Pv_B^2/12$. Hence the condition $U + V \geq 0$ for stability gives $P < 69.7\,\text{kN}$.

For the second strut,

$$U = \frac{580}{2}\left[\left(\frac{2v_B - v_C}{12}\right)^2 + \left(\frac{2v_C - v_B}{12}\right)^2\right] \quad \text{and}$$

$$V = -\frac{P}{2 \times 12}\left[v_B^2 + (v_B - v_C)^2 + v_C^2\right]$$

Hence, with $\lambda = 12P/580$, the eigenvalue equations are

$$\begin{bmatrix} (5 - 2\lambda) & (-4 + \lambda) \\ (-4 + \lambda) & (5 - 2\lambda) \end{bmatrix}\begin{bmatrix} v_B \\ v_C \end{bmatrix} = \begin{bmatrix} 0 \\ 0 \end{bmatrix}$$

The lowest solution is $\lambda = 1$ corresponding to $P = 48.3\,\text{kN}$ with mode shape $v_C = v_B$.

Q.8.14

$$U = \frac{4EI_C\,(2a)^2}{2\ell^2}\int_0^\ell x\,(\ell - x)\,\mathrm{d}x = 4EI_C a^2\ell/3$$

and

$$V = -\frac{Pa^2}{2}\int_0^\ell (\ell - 2x)^2\,\mathrm{d}x = -Pa^2\ell^3/6$$

Hence for stability $P < 8EI_C/\ell^2$.

NB This is the correct theoretical mode shape and hence the correct buckling load.

Q.8.15

(ii) has a negative diagonal element
(iv) has a negative value for $(a_{11}a_{22} - a_{12}^2)$
(v) is unsymmetric.

Q.8.16

$$K = \frac{2EI}{\ell}\begin{bmatrix} 1/\ell & -1/\ell \\ 0 & 1 \end{bmatrix}\begin{bmatrix} 2 & -1 \\ -1 & 2 \end{bmatrix}\begin{bmatrix} 1/\ell & 0 \\ -1/\ell & 1 \end{bmatrix} = \frac{2EI}{\ell}\begin{bmatrix} 6/\ell^2 & -3/\ell \\ -3/\ell & 2 \end{bmatrix}$$

Hence $F = \dfrac{\ell}{6EI}\begin{bmatrix} 2\ell^2 & 3\ell \\ 3\ell & 6 \end{bmatrix}$

(This checks against the flexibility matrix (8.13) if the first row and column are omitted.)

Q.8.17
The ILs are given in Fig. A.49

a (i) 152 kN, (ii) 312 kN, (iii) 152 kN, (iv) −200 kN, (v) −160 kNm
b (i) −64 kN, (ii) 216 kN, (iii) 104 kN, (iv) −112 kN, (v) −320 kNm

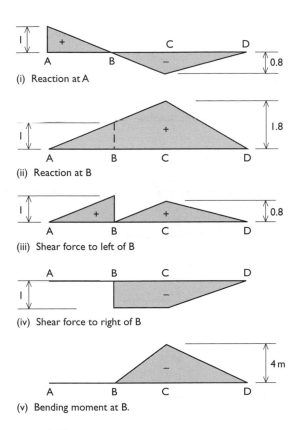

(i) Reaction at A

(ii) Reaction at B

(iii) Shear force to left of B

(iv) Shear force to right of B

(v) Bending moment at B.

Figure A.49

Q.8.18
For the reaction ILs, lift one support and let the hinge at C rotate. For the other ILs, place the distorted beam ABC from Fig. 8.27 onto the supports A and B and let the hinge at C rotate. The localised distortion needs to be in the negative direction to obtain the correct IL if plotted normally. Thus the unit shear displacement in Fig. 8.27 corresponding to a negative shear stress gives the correct shear IL. However the unit rotation in Fig. 8.27 corresponds to a sagging (positive bending moment) so that the IL looks correct when plotted upside down.

Q.8.19
The ILs are sketched in Fig. A.50.
Worst loading cases are (i) Loading on AC and EF to obtain maximum reaction or on CE only for maximum uplift, (ii) Loading on AE only, (iii) Loading on CE only for maximum hogging bending moment or on AC and EF for maximum sagging bending moment, (iv) Loading on CE only, (v) Loading on AE only, (vi) Loading on CE only for maximum sagging bending moment.

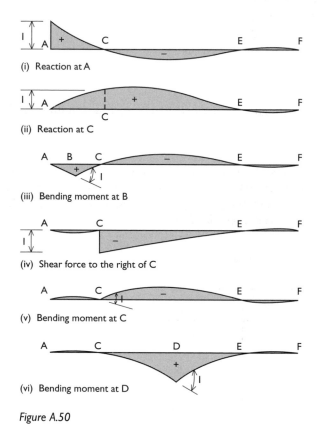

(i) Reaction at A

(ii) Reaction at C

(iii) Bending moment at B

(iv) Shear force to the right of C

(v) Bending moment at C

(vi) Bending moment at D

Figure A.50

Q.8.20
The ILs are sketched in Fig. A.51.

Q.8.21
From Maxwell's reciprocal theorem, the IL for deflection at point of a linear elastic structure is the deflection diagram of the deck due to a unit load

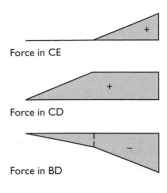

Force in CE

Force in CD

Force in BD

Figure A.51

applied at the point (the sign may need to be changed depending on sign conventions used).

Q.9.1
The two-bar truss remains elastic until $e = 12$ mm and $P = 143$ kN. For the three-bar truss, member BD yields when $e = 6$ mm and $P = 122$ kN. After that, loading continues until $e = 12$ mm and $P = 193$ kN when members AB and BC also yield, see Fig. A.52.

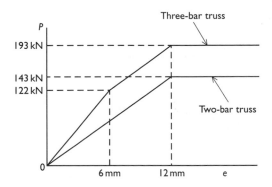

Figure A.52

Q.9.2
Because the section is doubly symmetric, the neutral axis will pass through the centre at all times. Treating the section as a full section with a cut-out gives $I = (240 \times 520^3 - 180 \times 320^3)/12 = 2.321 \times 10^9$ mm^4. Hence the elastic modulus $Z = 8.93 \times 10^6$ mm^3.

The plastic modulus is

$$Z_P = \left(240 \times 520^2 - 180 \times 320^2\right)/4 = 11.62 \times 10^6 \text{ mm}^3$$

Hence the shape factor $Z_P/Z = 11.62/8.93 = 1.30$ (close to case (c) of Table 9.1).

Q.9.3

In order to have the same area above and below, the neutral axis must be at 15 mm from the top of the section. Splitting the cross-section into three areas as shown in Fig. A.53, provides a means for calculating the plastic moment. The result, shown in Table A.3, is 3.33 kNm. In the second case there are two stress blocks, each of 54,000 N at 85 mm apart giving $M_P = 4.59$ kNm (a 38% increase).

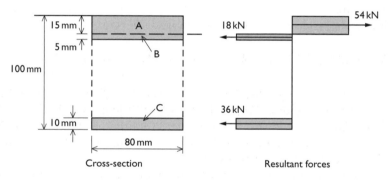

Figure A.53

Table A.3

Region	Area (mm^2)	Force (kN)	Arm (mm)	Moment (kNm)
A	1200	54	7.5	0.405
B	400	−18	−2.5	0.045
C	800	−36	−80	2.88
Total				3.33

Q.9.4

The steel force at yield $= 3 \times \pi \times 17^2 \times 440 = 1,198,000$ N.
The concrete force at ultimate strength is $300 \times 0.667 \times 32h$ where h is the depth of the concrete stress block. Equating these gives $h = 187$ mm.
(Note: Since $0.45 \times 460 = 207$ mm is greater than h, the beam is under-reinforced.)

The lever arm is $460-93 = 367$ mm and $M_{ult} = 119.8 \times 0.367 = 44.0$ kNm. But for a simply-supported beam of length ℓ, subject to a uniformly distributed load w, $M_{max} = w\ell^2/8$. Hence for a safe design $\ell^2 \leq 8 \times 44.0/5$ giving $\ell \leq 8.4$ mm.

Q.9.5
At the ultimate bending moment, the strain at the outer concrete fibres is 0.0035. With the neutral axis at half of the depth to the reinforcement, the strain in the steel will be 0.0035 also. But $\sigma_Y = 500$ N/mm^2. With $E = 200$ kN/mm^2, $\varepsilon_Y = 0.0025$. Hence the steel is beyond its yield strain.

Q.9.6
a   The BMD for the beam is shown in Fig. A.54(a). To satisfy the static theorem of plastic collapse, the plastic moment needs to be no less than 126 kNm.

b   If M_A is considered as a variable, from equilibrium

$$M_B = 132 + 0.75M_A, \quad M_C = 176 + 0.5M_A, \quad M_D = 132 + 0.25M_A$$

Previously the largest bending moment was M_C. To reduce this till it is the same magnitude as $|M_A|$, solve the second equation with $M_C = -M_A$ which gives $M_A = -117.3$ kNm, $M_B = 44.0$ kN, $M_C = 117.3$ kNm and $M_D = 102.7$ kNm. With $M_P = 117.3$ kNm, $|M| \leq M_P$ everywhere giving a more economical design.

c   The three possible mechanisms are shown in Fig. A.54(b). For the first mechanism, let B deflect Δ, the plastic hinge rotations will be $\theta_A = \Delta/4$ and $\theta_B = \Delta/4 + \Delta/12 = \Delta/3$. The forces at B, C and D drop distances Δ, $2\Delta/3$ and $\Delta/3$. Hence $(\Delta/4 + \Delta/3)M_P = 22(\Delta + 2\Delta/3 + \Delta/3)$ giving $M_P = 75.4$ kNm. The second and third mechanisms give $M_P = 117.3$ kN and 105.6 kNm, respectively. Since the beam needs to be strong enough to prevent any of the three mechanisms activating, it is necessary for $M_P \geq 117.3$ kNm.

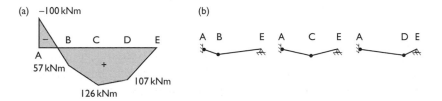

Figure A.54

Q.9.7

The only possible mechanism involves plastic hinges forming at A and B as shown in Fig. A.55. With B deflecting by Δ, the work equation is

$$M_P\left[2\Delta/x + \Delta/(\ell - x)\right] = P\Delta$$

Hence $M_P(2\ell - x) = P(\ell x - x^2)$. Either by differentiation or by trying different values of x/ℓ, the maximum for M_P is found to occur at $x/\ell = 2 - \sqrt{2} = 0.586$ with $M_P = 0.172P\ell$. This defines the required bending strength.

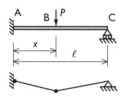

Figure A.55

Q.9.8

With three redundancies, four plastic hinges would be needed to form a mechanism (partial collapse not being feasible). Keeping one of the five joints rigid in turn, gives the only possible mechanisms as those in Fig. A.56. However, with the direction of loads shown, the fourth and fifth mechanisms will be less critical than the second and first. Hence only the first three would need to be analysed.

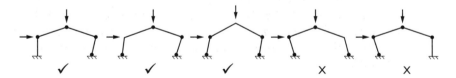

Figure A.56

Q.9.9

Because the axial lengths of beams and columns cannot change, joints C and F are fixed in position. It is not possible to have complete collapse which would require four plastic hinges to form. Critical mechanisms will involve collapse of the beams by plastic hinges forming at B, C, E and F. However, less energy will be required to form plastic hinges in the columns than the beams, so the mechanisms will be as shown in Fig. A.57. The critical loads are found to be $P_B = 48\,\text{kN}$ and $P_E = 56\,\text{kN}$.

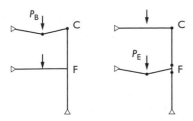

Figure A.57

Q.9.10
Some possibilities, shown in Fig. A.58, could have been as listed here.

(i) By welding flange plates onto both ends to be joined, bolting could have been external to each column. However, this would have been an inelegant solution.

(ii) Cover or 'splice' plates could have been bolted to the outsides of the column. However, this would have caused difficulties in fabrication because of the need to grind the ends of the columns and the possibility that the bolt tightening forces would have distorted the cross-sections of the columns (Fig. A.58 shows tubes of the correct lengths inserted round the shafts of the bolts to prevent this happening). Hence it is impractical.

(iii) The method actually adopted was to weld cover plates to the ends of each column length so that abutting cover plates could be bolted together. Holes needed to be drilled in the internal faces of the columns to provide access to the heads of the bolts. This technique provided the columns with clean external faces. However, it resulted in a relatively flexible joint if sufficient bending moment or tension were to be present to open it up.

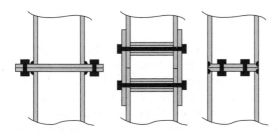

Figure A.58

Q.9.11

a *Structural design*

- It is very important to provide alternative load paths in buildings so that areas of local damage may be circumvented.
- Joints should be sufficiently strong to carry unusual loads arising from thermal strains, blast loading and falling debris.
- If extra strong floors were to be installed at regular intervals in tall buildings, these could be designed to withstand the progressive collapse of floors above.

b *Fire protection*

- Fire protection of structural members and joints should be sufficiently robust not to be easily disturbed by blast loading.
- Compartmentation should be as effective as possible in order to prevent the spread of fires.
- The amounts of combustible material should be kept as low as possible.

c *Provision for escape*

- Staircases need to be as far apart as possible.
- Escape routes need to have sufficient strength and fire protection to keep them intact as long as possible.
- Ventilation measures are required to avoid smoke infiltrating escape routes.
- Consideration should be given of whether there is any viable alternative means of escape from the top of tall buildings.

d *Other considerations*

There are many other factors which could be included (e.g. measures to make it difficult for terrorists to plant large bombs near to vulnerable parts of buildings). For a more complete discussion see Roberts, J.M. (2002).

Units and constants

Prefixes
k = kilo = 1000×
m = milli = 0.001×

SI units
kg = kilogram   N = Newton
m = metre   Pa = Pascal

$1 \, N$ = force required to accelerate 1 kg through 1 m/s^2
$1 \, Pa = 1 \, N/m^2$
$1 \text{ litre} = 10^{-3} \, m^3 = 10^6 \, mm^3$
1 hour = 3600 s
1 km/hour = 0.2778 m/s

Imperial and US units
in = inch   1 mile = 5280 ft
ft = foot   1 ton (UK) = 2240 lb
lb = pound   1 short ton (US) = 2000 lb
1 kip = 1000 lb   1 mile/hour = 1.476 ft/s
1 ft = 12 in

Constants
Acceleration due to gravity at earth's surface = 9.807 m/s^2
1 atmosphere = 101.3 kN/m^2, 1 bar = 100 kN/m^2
Water density = 1 kg/litre = 1000 kg/m^3

Conversion factors

Measurement	1 in = 25.4 mm	1 mm = 0.0394 in
	1 ft = 0.3048 m	1 m = 3.281 ft
	1 mile = 1.609 km	1 km = 0.6214 miles
Area	1 in^2 = 645 mm^2	1 mm^2 = 0.00155 in^2
	1 ft^2 = 0.0929 m^2	1 m^2 = 10.764 ft^2
Volume	1 ft^3 = 0.02832 m^3	1 m^3 = 35.31 ft^3

Second moment of area	$1 \, \text{in}^4 = 0.4162 \times 10^6 \, \text{mm}^4$	$1 \, \text{mm}^4 = 2.403 \times 10^{-6} \, \text{in}^4$
Mass	$1 \, \text{lb mass} = 0.4536 \, \text{kg}$	$1 \, \text{kg} = 2.205 \, \text{lb mass}$
Force	$1 \, \text{lb force} = 4.448 \, \text{N}$ $1 \, \text{ton} = 9.964 \, \text{kN}$ $1 \, \text{short ton} = 8.897 \, \text{kN}$	$1 \, \text{N} = 0.2248 \, \text{lb force}$ $1 \, \text{kN} = 0.1004 \, \text{ton}$ $1 \, \text{kN} = 0.1124 \, \text{short ton}$
Distributed force	$1 \, \text{lb/ft} = 14.59 \, \text{N/m}$ $1 \, \text{ton/ft} = 32.69 \, \text{kN/m}$ $1 \, \text{short ton/ft} = 29.19 \, \text{kN/m}$	$1 \, \text{N/m} = 0.0685 \, \text{lb/ft}$ $1 \, \text{kN/m} = 0.03059 \, \text{ton/ft}$ $1 \, \text{kN/m} = 0.03426 \, \text{short ton/ft}$
Pressure and stress	$1 \, \text{lb/ft}^2 = 47.88 \, \text{N/m}^2 \, (\text{Pa})$ $1 \, \text{ton/ft}^2 = 107.3 \, \text{kN/m}^2$ $1 \, \text{short ton/ft}^2 = 95.76 \, \text{kN/m}^2$ $1 \, \text{lb/in}^2 \, (\text{psi}) = 6.895 \, \text{kN/m}^2$	$1 \, \text{N/m}^2 = 0.02089 \, \text{lb/ft}^2$ $1 \, \text{kN/m}^2 = 0.00932 \, \text{ton/ft}^2$ $1 \, \text{kN/m}^2 = 0.01044 \, \text{short ton/ft}$ $1 \, \text{kN/m}^2 = 0.145 \, \text{psi}$

References

Alexander, D., 1993, *Natural Disasters*, UCL Press, London.

Allen, H.G. and Bulson, P.S., 1980, *Background to Buckling*, McGraw-Hill (UK), Maidenhead, Berkshire.

Argyris, J.H. and Kelsey, S., 1960, *Energy Theorems and Structural Analysis*, Butterworths, London (Originally published in Aircraft Engineering, Oct. 1954–May 1955).

Arnold, D., 2001, 'Ignored design codes blamed for devastation', *New Civil Engineer*, 15 Feb., pp. 12–15.

Ballinger, G., 2003, 'The Falkirk Wheel – from concept to reality', *The Structural Engineer*, Vol. 81, No. 4, pp. 24–28.

Barber, E.H.E., Bull, F.B. and Shirley-Smith, H., 1971, *Report of Royal Commission into the Failure of West Gate Bridge*, Government of Victoria, Melbourne.

Barbey, M.F., 1981, *Civil Engineering Heritage: Northern England*, Thomas Telford, London.

Bazant, Z.P. and Cedolin, L., 1991, *Stability of Structures*, Oxford University Press, New York.

Beckett, D., 1980, *Brunel's Britain*, David and Charles, Newton Abbot, Devon.

Bignell, V., Peters, G. and Pym, C., 1978, *Catastrophic Failures*, Open University Press, Milton Keynes.

Brebbia, C.A. (ed.), 1996, *The Kobe Earthquake: Geodynamical Aspects*, Computational Mechanics Publications, Southampton.

Brown, D.J., 1993, *Bridges*, Mitchell Beadey, London.

Castigliano, A., 1919, *Elastic Stresses in Structures*, Translation by E.S. Andrews, Scott Greenwood.

Chandler, A. and Pomonis, A. (eds), 1997, *The Hyogo-Ken Nanbu (Kobe) Earthquake of 17 January 1995*, Earthquake Engineering Field Investigation Team, IStructE, London.

Clark, G.M. and Eyre, J., 2001, 'The Gateshead Millennium Bridge', *The Structural Engineer*, Vol. 79, No. 3, pp. 30–35.

Coburn, A. and Spence, R., 1992, *Earthquake Protection*, John Wiley & Sons, Chichester, West Sussex.

Cohen, Lord, 1955, *Report of the Court of Inquiry into the Accident to Comet G-ALYP on 10th January 1954 and Comet G-ALYY on 8th April 1954*, CAP127, HMSO, London.

Corley, W.G., 2001, 'Lessons learned from the Oklahoma City bombing', in *Learning from Construction Failures: Applied Forensic Engineering*, Campbell, P. (ed.), Whittles Publishing, Latheronwheel, Caithness.

Coulton, J.J., 1977, *Greek Architects at Work*, Elek Books, London.

Dallard, P., Fitzpatrick, A.J., Flint, A., LeBourva, S., Low, A., Ridsdill Smith, R.M. and Willford, M., 2001, 'The London Millennium Footbridge', *The Structural Engineer*, Vol. 79, No. 22, pp. 17–33.

Das, B.M., Kessimali, A. and Sanni, S., 1994, *Engineering Mechanics: Statics*, Richard O. Irwin, Burr Ridge, IL.

Drewry, C.S., 1832, *A Memoir on Suspension Bridges*, Longman, Rees, Orme, Brown, Green and Longman, London.

Dwight, H.B., 1961, *Tables of Integrals and Other Mathematical Data*, 4th edn, Macmillan Pub. Co., New York.

Frampton, K., Webster, A.C. and Tishhauser, A., 1996, *Calatrava Bridges*, 2nd edn, Birkhauser Verlag, Basel.

Friedman, D., 1995, *Historic Building Construction*, W.W. Norton & Co., New York.

Frocht, M.M., 1941, *Photoelasticity*, Vol. 1, John Wiley, New York.

Gould, M.H., Jennings, A. and Montgomery, R., 1992, 'The Belfast roof truss', *The Structural Engineer*, Vol. 70, No. 7, pp. 127–129.

Griffiths, H., Pugsley, Sir A. and Saunders, Sir O., 1968, *Collapse of Flats at Ronan Point, Canning Town*, HMSO, London.

Handy, C., 1994, *The Empty Raincoat, Making Sense of the Future*, Hutchinson, London.

Harriss, J., 1975, *The Tallest Tower*, Houghton Mifflin Co, Boston.

Heinle, E. and Leonhardt, F., 1989, *Towers, A Historical Survey*, Butterworth Architecture, London (Translation of 1988 German publication).

Hendry, A.W., Sinha, B.P. and Davies, S.R., 1997, *Design of Masonry Structures*, Spon, London.

Home, G., 1931, *Old London Bridge*, John Lane, The Bodley Head, London.

Hopkins, H.J., 1970, *A Span of Bridges*, David and Charles, Newton Abbot, Devon.

Horne, M.R. and Morris, L.J., 1981, *Plastic Design of Low-Rise Frames*, William Collins, London.

Illston, J.M. and Domone, P.L.J. (eds), 2001, *Construction Materials*, Spon Press, London.

Janner, M.M., Lutz, R., Moerland, P. and Simmonds, T., 2001, 'The world's largest self-supporting enclosed space', *The Structural Engineer*, Vol. 79, No. 21, pp. 25–29.

Jennings, A., 1983, 'Gravity stiffness of classical suspension bridges', *ASCE, Journal of Structural Engineering*, Vol. 109, No. 1, pp. 16–36.

Jennings, A., 1986, 'Stability fundamentals in relation to masonry arches', *The Structural Engineer*, Vol. 64B, No. 1, pp. 10–12.

Jennings, A. and McKeown, J.J., 1993, *Matrix Computation*, 2nd edn, Wiley, New York.

Johnson, J. and Curran, P., 2003, 'Gateshead Millennium Bridge – an eye-opener for engineering', *Proc. ICE, Civil Engineering*, Vol. 156, Issue 1, pp. 16–24.

Jones, A.C., Hamilton, D., Purvis, M. and Jones, M., 2001, 'Eden Project, Cornwall: design, development and construction', *The Structural Engineer*, Vol. 79, No. 20, pp. 30–36.

Koerte, A., 1992, *Two Railway Bridges of an Era, Firth of Forth and Firth of Tay*, Birkhauser Verlag, Basel.

Lancaster, J., 2000, *Engineering Catastrophes*, 2nd edn, Abington Publishing, Cambridge.

Lee, D.J., 1994, *Bridge Bearings and Expansion Joints* (2nd edn), Spon Press, London.

Leliavsky, S., 1958, *Uplift in Gravity Dams*, Constable & Co. Ltd, London.

Levy, M.P. and Salvadori, M.G., 1992, *Why Buildings Fall Down*, W.W. Norton & Co., New York.

Livesley, J.K., 1953, 'Analysis of rigid frames by an electronic computer', *Engineering*, Vol. 176, p. 230.

Mairs, D. and Lomax, S., 2002, 'Footbridge design: respecting context and relating to users', *The Structural Engineer*, Vol. 80, No. 5, pp. 15–18.

Martin, T. and Macleod, I.A., 1995, 'The Tay rail bridge disaster – a reappraisal based on modern analysis methods', *Proc. ICE, Civil Engineering*, Vol. 108, pp. 77–83.

Matheson, J.A.L., 1971, *Hyperstatic Structures*, 2nd edn, Butterworths, London.

McAllister, T. (ed.), 2002, *World Trade Center Building Performance Study: Data Collection, Preliminary Observations, and Recommendations*, Federal Emergency Management Agency, Washington, DC.

McCullough, D.G., 1982, *The Great Bridge*, Simon and Schuster, New York, 1982.

Mendelssohn, K., 1974, *The Riddle of the Pyramids*, Praeger Publishers, New York.

Merrison, A.W., 1973, *Report of the Committee of Inquiry into the Basis of Design and Method of Erection of Steel Box Girders*, HMSO, London.

Naesheim, T., 1981, *The 'Alexander Kielland' Accident*, Norwegian Public Report Nou 1981: 11 (English translation by Translatørservice A/S, Stavanger).

Newmark, N.M. and Rosenblueth, E., 1971, *Fundamentals of Earthquake Engineering*, Prentice-Hall, Englewood Cliffs, New Jersey.

Page, J., 1993, *Masonry Arch Bridges*, HMSO, London.

Parker, D., 1993, 'Madness in their method', *New Civil Engineer*, Oct. 28, p. 8.

Parry, R.H.G., 2000, 'Megalith mechanics', *Proc. ICE, Civil Engineering*, Vol. 138, pp. 183–192.

Petroski, H., 1985, *To Engineer is Human*, Macmillan, London.

Roberts, A.P., 2003, *Statics and Dynamics with Background Mathematics*, Cambridge University Press, Cambridge, UK.

Roberts, G., 1968, 'Severn Bridge: design and contract arrangements', *Proc. ICE*, Vol. 41, pp. 1–48.

Roberts, J.M., 2002, *Safety in Tall Buildings and Other Buildings with Large Occupancy*, Institution of Structural Engineers, London.

Sandström, G.E., 1970, *Man the Builder*, McGraw-Hill, New York.

Schlager, N. (ed.), 1994, *When Technology Fails*, Gale Research International Ltd, Detroit, Michigan.

Schneider, C.C., 1908, *Royal Commission Quebec Bridge Inquiry Report*, Ottawa.

Shah, B.V., 1983, 'Is the environment becoming more hazardous? A global survey 1947–80', *Disasters*, Vol. 7, pp. 202–209.

Shirley-Smith, H., 1964, *The World's Great Bridges*, 2nd edn, Phoenix House, London.

Short, W.D., 1962, 'Accidents on construction work with special reference to failures during erection or demolition', *The Structural Engineer*, Vol. 40. No. 2, pp. 35–43.

Simiu, E. and Scanlan, R.H., 1996, *Wind Effects on Structures: Fundamentals and Applications to Design*, 3rd edn, Wiley, New York.

Skempton, A.W., 1981, *John Smeaton FRS*, Thomas Telford, London.

Sparkes, J.J., 1991, 'Quality in continuing education and training', in *Innovative Teaching in Engineering*, Smith, R.A. (ed.), Ellis Horwood, London, pp. 27–31.

Sprague de Camp, L., 1970, *Ancient Engineers*, MIT Press, Cambridge, MA.

Stansfield, K., 1999, 'Glass as a structural engineering material', *The Structural Engineer*, Vol. 77, No. 9, pp. 12–14.

Steinman, D.B., 1929, *A Practical Treatise on Suspension Bridges*, 2nd edn, J. Wiley, New York.

Sutherland, R.J.M., 1983, '1780–1850' in *Structural Engineering – Two Centuries of British Achievement*, Collins, A.R. (ed.), Tarot Print Ltd, Chislehurst, Kent.

Tarkov, J., 1986, 'A disaster in the making', *American Heritage of Invention and Technology*, Vol. 1, Pt 3, pp. 10–17.

Thomas, H.H., 1976, *The Engineering of Large Dams*, John Wiley & Sons, London.

Timoshenko, S.P., 1983, *History of Strength of Materials*, Dover Publications, New York.

Timoshenko, S.P. and Goodier, J.N., 1970, *Theory of Elasticity*, 3rd edn, McGraw-Hill, New York.

Upton, N., 1975, *An Illustrated History of Civil Engineering*, Heinemann, London.

Walther, R., Houriet, B., Isler, W. and Möia, P., 1988, *Cable Stayed Bridges*, Thomas Telford Ltd, London (Translation of a 1985 publication).

Wells, M., 2002, *30 Bridges*, Lawrence King Publishing, London.

Westhofen, W., 1890, *The Forth Bridge*, The Engineer, London.

Williams, M.S. (ed.), *The Erzincan Turkey Earthquake of 13 March 1992*. Earthquake Engineering Field Investigation Team, IStructE, London.

Index

eBooks – at www.eBookstore.tandf.co.uk

A library at your fingertips!

eBooks are electronic versions of printed books. You can store them on your PC/laptop or browse them online.

They have advantages for anyone needing rapid access to a wide variety of published, copyright information.

eBooks can help your research by enabling you to bookmark chapters, annotate text and use instant searches to find specific words or phrases. Several eBook files would fit on even a small laptop or PDA.

NEW: Save money by eSubscribing: cheap, online access to any eBook for as long as you need it.

Annual subscription packages

We now offer special low-cost bulk subscriptions to packages of eBooks in certain subject areas. These are available to libraries or to individuals.

For more information please contact webmaster.ebooks@tandf.co.uk

We're continually developing the eBook concept, so keep up to date by visiting the website.

www.eBookstore.tandf.co.uk

Guildford College
Learning Resource Centre

Please return on or before the last date shown.
No further issues or renewals if any items are overdue.
"7 Day" loans are **NOT** renewable.

- 6 JAN 2005

2 9 JUN 2005

1 5 DEC 2005

2 3 APR 2009

1 4 MAY 2009

- 4 FEB 2010

Class: 624.17 JEN

Title: STRUCTURES – FROM THEORY TO PRACTICE

Author: JENNINGS, Alan